Luminescence in Chemistry

Edited by

E. J. BOWEN F.R.S.

Physical Chemistry Laboratory
Oxford

D. VAN NOSTRAND COMPANY LTD.
LONDON
PRINCETON, NEW JERSEY TORONTO MELBOURNE

D. VAN NOSTRAND COMPANY LTD.
Windsor House, 46 Victoria Street, London

D. VAN NOSTRAND COMPANY INC.
Princeton, New Jersey

VAN NOSTRAND REGIONAL OFFICES:
New York, Chicago, San Francisco

D. VAN NOSTRAND COMPANY (CANADA) LTD.
Toronto

D. VAN NOSTRAND AUSTRALIA PTY. LTD.
Melbourne

Published simultaneously in Canada by
D. VAN NOSTRAND COMPANY (Canada) LTD.

Library of Congress Catalog Card No. 68–16129

MADE AND PRINTED IN GREAT BRITAIN BY THE CAMELOT PRESS LTD.
LONDON AND SOUTHAMPTON

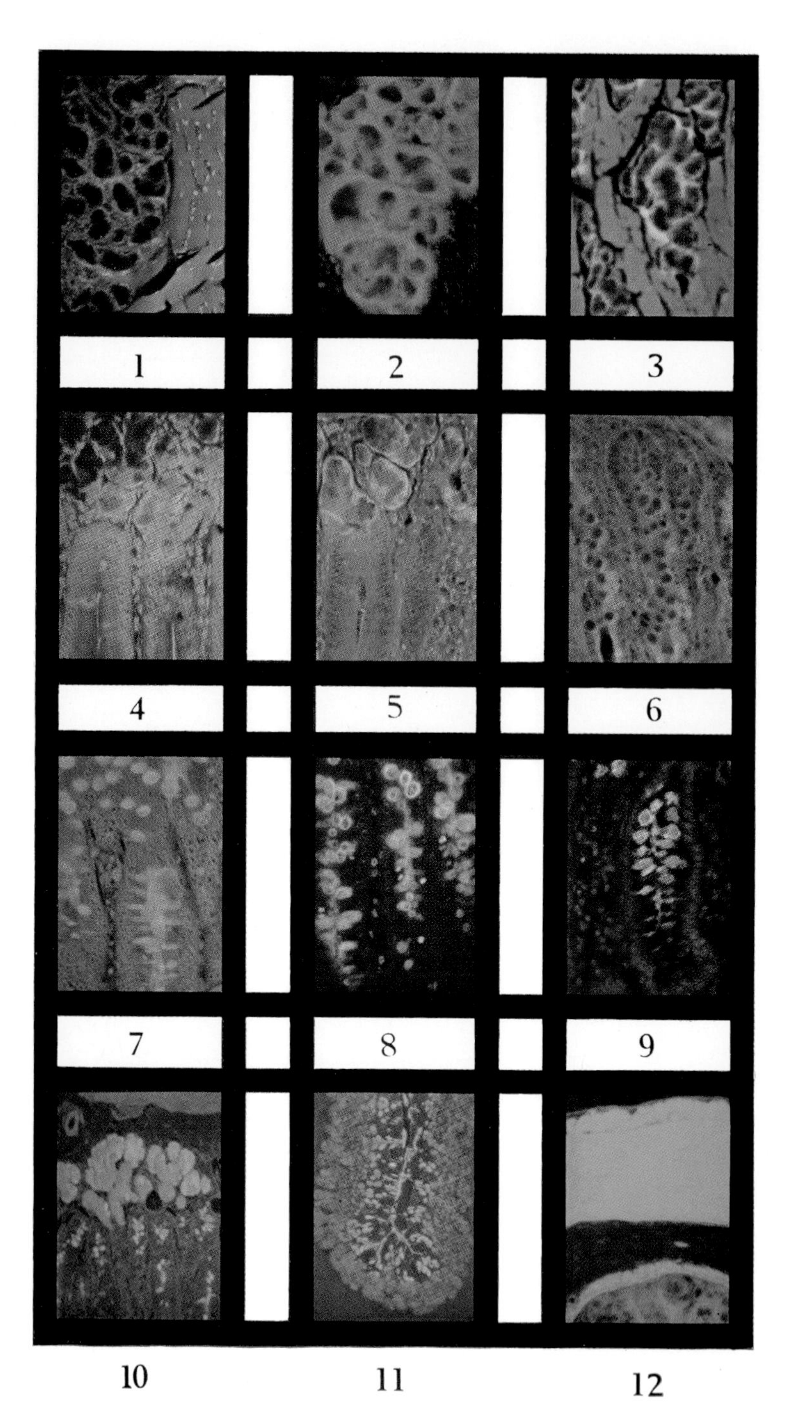

For key, see pages vii–ix.

LUMINESCENCE IN CHEMISTRY

THE VAN NOSTRAND
SERIES IN PHYSICAL CHEMISTRY

Edited by
T. M. SUGDEN, F.R.S.
Director, 'Shell' Research Ltd.,
Thornton Research Centre, Chester

This series aims to provide a group of fundamental books covering the whole field of physical chemistry. It will be the aim of each volume to summarize its topic in a modern fashion for senior honours undergraduates and postgraduates, and for all practising chemists whether engaged in industry, teaching or research.

R. M. GOLDING—*Applied Wave Mechanics*

N. E. HILL *et al.*—*Dielectric Properties and Molecular Behaviour*

M. LEFORT—*Nuclear Chemistry*

J. P. SUCHET—*Chemical Physics of Semiconductors*

T. M. SUGDEN and C. N. KENNEY—*Microwave Spectroscopy of Gases*

E. J. BOWEN *et al.*—*Luminescence in Chemistry*

Additional titles will be listed and announced as published

Ex luce lucellum

UNITS

$1\ \text{nm} = 1\ \text{m}\mu = 10\text{Å} = 10^{-9}$ metre

1 einstein $= 6{\cdot}02 \times 10^{23}$ quanta

$E(\text{kcal/einstein}) = 2{\cdot}86 \times 10^{4}/(\text{wavelength in nm})$

LEGENDS FOR FRONTISPIECE

Material: Figs. 1–6. Sections of tissues taken from a female Syrian hamster and fixed overnight in Carnoy's fluid at 5°.

Figs. 3–9. As above except tissues are from a male rat.

Figs. 10–12. As above except tissues are from a male guinea pig.

The treated tissues in all figures have been excited with u.v. illumination.

Figure 1. Part of the tongue, stained in 0·01% aqueous solution of impure (i.e., as supplied commercially) coriphosphine at pH 4·0.

The black cells on the left are glossal glands and contain highly sulphated mucosubstances. Although these mucosubstances have here taken up coriphosphine, the fluorescence of the adsorbed dye is quenched completely (cf. Fig. 2). Magnification: ×80.

Figure 2. Part of the tongue, stained in 0·01% aqueous solution of purified coriphosphine at pH 4·0.

The coriphosphine adsorbed on to the sulphated mucosubstances in the glossal glands here emits an intense brownish-orange fluorescence. The proteins surrounding the glossal glands actually emit a green fluorescence similar to that in Fig. 1, but here it appears black because the exposure time required for this photomicrograph was too short to record the protein fluorescence correctly. Note that small nuclei (bottom right corner) are apparent by their distinctive orange fluorescence. Magnification: ×80.

Figure 3. Part of the tongue, first esterified with a methanolic solution of thionyl chloride, second reduced with a solution of lithium aluminium hydride in dioxane for 48 hours at 60°, third saponified with an alcoholic solution of potassium hydroxide, and lastly stained in a 0·01% solution of purified coriphosphine at pH 4·0.

The sulphated mucins in the glossal glands emit a brown fluorescence whose intensity is less than that in the untreated section shown in Fig. 2, but more important here, is the fact that the affinity of nuclei for coriphosphine has been abolished as shown by the abolition of their fluorescence (cf. Figs. 1 and 2). The intensity of the green fluorescence emitted by the connective-tussue proteins here appears greater than it actually is. When stained at pH 2 instead of 4 the proteins do not fluoresce at all, thus showing that most of their carboxyls have been reduced. Magnification: ×80.

Figure 4. Duodenum (small intestine), stained in 0·01% aqueous solution of purified coriphosphine at pH 4·0. It is very difficult to distinguish the fluorescence or coriphosphine adsorbed on to sulphated mucosubstances from that of dye adsorbed on to the DNA of nuclei and the carboxyls of protoplasmic proteins (Cf. Fig. 5). Magnification: ×160.

Figure 5. Exactly the same tissue as in Fig. 4 except that after staining in coriphosphine it has been stained for 1 minute in a 0·001% aqueous solution of thiazol yellow at pH 2·05. The sulphomucins in the upper half of the Figure emit a distinctive orange or brownish-orange fluorescence which in no way can be confused with the yellowish-green fluorescence emitted by nuclei or the bluish-green fluorescence of proteins. Magnification: ×160.

Figure 6. Large intestine, first hydrolysed in 5N HCl at room temperature for 5 minutes, secondly immersed overnight in a solution of N,N-dimethyl-*m*-phenylene diamine in aqueous phosphate buffer at pH 6, and lastly stained in a 0·1% aqueous solution of purified coriphosphine at pH 5·2.

Without the intermediate condensation with meta-diamine, the entire sections would have emitted a uniform reddish-orange fluorescence. Here the coriphosphine adsorbed by the sulphomucins in the goblet cells emit an almost unchanged brownish-orange fluorescence, whereas the nuclei surrounding these cells are clearly *not* fluorescent. The fluorescence emitted by proteins is diminished in intensity and its colour is changed to dull green. Magnification: ×80.

Figure 7. Large intestine, first oxidized with periodic acid and second treated with a 0·1% solution of benzoflavine saturated with sulphur dioxide (final pH: 3·1). The goblet cells emit a light green fluorescence because the aldehydes engendered from the *vic*-glycols of their mucosaccharide content have reacted with sulphur dioxide and benzoflavine. Unfortunately the fluorescence emitted by the product thus formed is here almost indistinguishable in colour and intensity from the fluorescence emitted by dye taken up by nuclei and protoplasmic proteins. It is very difficult to assign a particular fluorescence colour to sites containing originally a high concentration of *vic*-glycols (especially in tissues other than this one) whereas in Fig. 8 all the fluorescence can be so assigned. Magnification: ×400.

Figure 8. Exactly as Fig. 7 except that before periodate-oxidation, the tissue section was methylated with a 2% solution of thionyl chloride in methanol for four hours at room temperature. For comments, see Fig. 7. Magnification: ×400.

Figure 9. Large intestine, first methylated with a 2% solution of thionyl chloride in methanol, second oxidized in periodic acid, third treated with sulphurous acid, fourth treated with sulphurous acid of pH 3·1, and lastly stained in an aqueous 0·01% solution of acridine orange at pH 3·0.

The cells secreting mucosubstances whose *vic*-glycols are oxidizable with periodic acid here emit a brownish-orange fluorescence. That such cells emit any fluorescence at all with this sequence of reactions substantiates the theory that Schiff's reagent reacts with aldehydes through alkylsulphonic acids as intermediates. The nuclei here emit a green fluorescence because their DNA content has undergone a partial Feulgen hydrolysis during the sulphurous acid treatment and the aldehydes thus exposed have combined with sulphurous acid and acridine orange in exactly the same way as the aldehydes engendered from the *vic*-glycols of the mucosaccharides. Magnification: ×400.

Figure 10. Duodenum (small intestine) oxidized in periodic acid, condensed with salicylhydrazide and lastly complexed with aluminium ions.

The *vic*-glycols of the sulphated mucosubstances in the upper half of the Brunner's glands (top half of Figure) and in goblet cells (the small cells in the lower half of Figure) can be correlated here with the intense light blue fluorescence and those of neutral mucosubstances in the bottom half of the Brunner's glands with the less intense and darker blue fluorescence. The connective-tissues give off only their weak, so-called primary fluorescence. Magnification: ×80.

Figure 11. Large intestine, oxidized in periodic acid and then treated with salicylhydrazide.

The small, light blue fluorescent cells towards the centre of the Figure contain a sulphomucin. The less intense, royal blue fluorescence emanates from deep mucous cells which contain another and unusual type of sulphomucin whose periodate-engendered 'dialdehydes' exist, it is thought, not in the free aldehyde form but in a hemiacetal or a hemialdal form. Possibly salicylhydrazide forms with such hemiacetals or hemialdals fluorescent cyclic hydrazano derivatives. Magnification: ×40.

Figure 12. Large intestine, treated first with a 50:50 mixture of acetic anhydride and pyridine at 60° for 1 hour, second with salicylhydrazide, and third with a solution of zinc acetate at pH 6.

In this procedure, the C-terminal carboxyl groups of muscle proteins (upper half of Figure) have been converted into methyl ketone derivatives whose salicylhydrazones do *not* fluoresce until they have, as here, been complexed with zinc ions. Magnification: ×80.

PREFACE

THIS book is intended for students interested in luminescence and for research workers, particularly in the biological and biochemical field, who need a background knowledge of the various forms of luminescence. Not many years ago the subject seemed confined to a few curious observers. It now receives concentrated study both by academic scientists and by industrial organizations, and has grown to such an extent that sub-specializations within it become necessary. Major applications are in the lamp industry, for television tubes, and for optical brighteners. The commercial uses have reacted back upon and stimulated academic studies, and these have proved exceedingly fruitful in developing knowledge of the properties and behaviour of the electronic levels of molecules and crystals. Luminescence work presents challenging difficulties which have lately been greatly eased by the use of modern apparatus; its characteristic feature is its sensitivity to trace effects, which has opened up new fields in analysis and in detectional applications, and has made sensitive observations possible over a wide range of scientific fields. These matters are developed by the several authors of the chapters of this volume.

E. J. BOWEN

Physical Chemistry Laboratory,
Oxford.

CONTENTS

Chapter 1

GENERAL PRINCIPLES OF LIGHT EMISSION

E. J. BOWEN

ELECTROMAGNETIC radiations, varying over enormous ranges of wavelength from radio to radioactivity waves, carry energy through space at a fixed speed of about 3×10^{10} centimetres per second *in vacuo*. The narrow range perceived by the eye, together with a small extension into the ultraviolet region, is commonly called 'light', and its emission is the subject of this volume. Wavelengths are measured in ångströms, Å, mμ, or in modern usage nm ($=1$ m$\mu = 10$ Å $= 10^{-9}$ metre), varying from about 800 nm to 250 nm. Frequencies ν s^{-1} are usually expressed as wavenumbers ν cm^{-1}, ranging from 125,000 to 40,000. As molecules are numbered in *moles*, so quanta are measured in *einsteins*, $\boldsymbol{Nh}\,\nu$, of magnitude $2{\cdot}86 \times 10^4/\lambda$, kcal, where λ is the wavelength in nm.

Conceptual Ideas of Light Absorption and Emission

Atoms and molecules have outer shells of electrons, and the energy contents of these shells do not assume values of any magnitude, but are confined to sets of values called *energy levels*. Basically the reason for this is that electrons bound to atoms, because of the operation of the Uncertainty Principle, have to be treated as possessing a *wave* rather than the *particle* character they show when free. Since the wave-motion has to be of 'stationary' nature, like vibrations in an organ pipe (fixed nodes), a limited set of wave-forms only is allowed. The physical picture of these wave-forms is that the square of the wave amplitude at any point represents (negative) charge density; antinodes are positions of high density, and nodes those of zero density. As the wave-forms or patterns are three-dimensional, their nodes are usually surfaces separating regions of high charge density.

For the simple case of the hydrogen atom, with a single electron, the differential equations put forward by Schrödinger can be solved, and the energies and shapes of the electron patterns (called *orbitals*) can be precisely deduced by using only the fundamental constant $\boldsymbol{h}$ of Planck and the electron charge and mass. Jumps of the electron from one orbital to another, with emission or absorption of radiation according to the quantum law: (energy difference) $= \boldsymbol{h} \times$ (frequency), then provide a complete interpretation of the line spectrum of the hydrogen atom. These matters are fully discussed in numerous textbooks, and will not be elaborated here. It is necessary, however, to note that the lowest energy level of the hydrogen atom, of symmetry type denoted *s*, has a spherical shape with high charge density round the positive nucleus. This is related to a higher energy level of symmetry type *p* by plane-polarized light absorption or emission as diagrammatically represented in Fig. 1.1. The vertical arrows indicate the

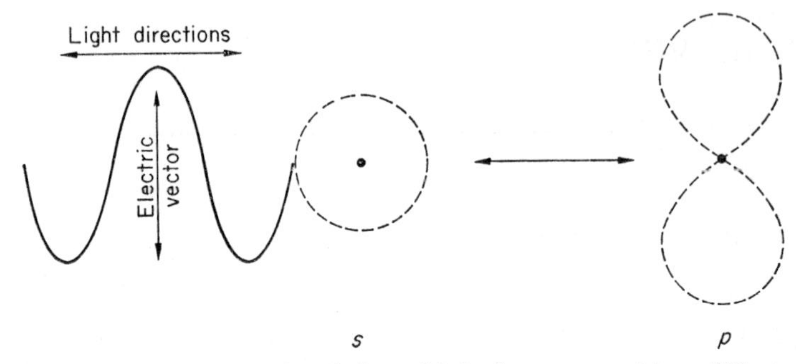

FIG. 1.1 Absorption and emission of light for *s–p* transition of H-atom.

electric vector direction of the incoming or outgoing light, which by the absorption process (left to right), converts the spherical *s* form to the hour-glass shaped *p* orbital with axis along the vector direction, whose wave-form vibrates up and down. Such an oscillating separation of charge, with development of a new node in the wave-form, is an essential part of light absorption or emission. The oscillating charge separation can be expressed by a mathematical quantity called the 'dipole moment of transition' (given by the distance separating the 'centres of gravity' of the two halves of a *p* orbital). An *s–s* transition is 'forbidden' because there would be negligible transition moment. Diagrams such as Fig. 1.1 omit an important property of the electron, a magnetic character known as 'spin', which can combine 'with' or 'against' the orbital motions. Energy levels of atoms with single outer electrons are therefore slightly 'split', and spectral lines appear as 'doublets' (e.g., the yellow lines of sodium).

In atoms and molecules with more electrons the wave-patterns become exceedingly complex and true calculation of levels impossible. However, as an approximation the problem can be broken up by the assignment of electrons to individual orbitals with qualitative shapes based on those of hydrogen atoms. In molecules, as the simple case of H_2, the electrons group round a two-centre positive interior, and *s* and *p* type orbitals (each able to take two electrons of opposite spin—Pauli Principle) become stretched along the molecular axis to give σ and π symmetries. Figure 1.2 represents absorption and emission for H_2. The π-pattern has two regions of negative charge separated by a nodal plane containing the molecular axis. Note the geometrical relationship of the electric vector and the nodal plane, which indicates that absorption and emission are *polarized.* In a molecule like benzene pairs of electrons are imagined to form localized C–H and C–C bonds and the remainder (six for benzene) are *delocalized* in six-centre orbitals of π-type, since they all have a nodal plane in the plane of the ring, as well as nodes at right-angles as the Pauli Principle dictates. Light absorption or emission from transitions between the lower electronic levels is, then, as a simplification, regarded as single electron jumps between assumed individual orbitals: ground orbital $\leftrightarrow$ low energy excited

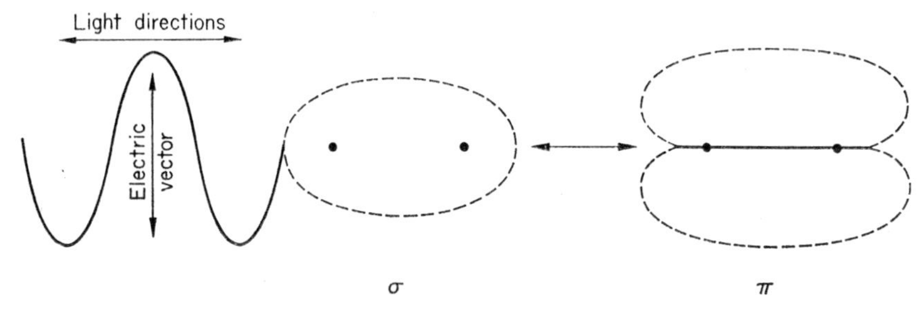

FIG. 1.2 Absorption and emission of light for σ–π transition of H_2 molecule.

orbital; excitation involving the appearance of a new node as described above. Such a node, for lower electronic states of benzene, will be at right-angles to the ring since one already occupies the ring plane. Transitions of this type are called π–π*. Another type, referred to as n–π* is found in ketones, quinones, and many heterocyclics. Here an electron, of a 'lone pair' on an oxygen or nitrogen atom and occupying a p-type orbital, transfers to an 'empty' orbital, the transfer involving a 90° twist of axis. This is a weak form of transition because of the ill-matched geometry of the orbitals. Still another type is a transfer of an electron from one part of a complex molecule to another, as in metal chelates or aromatic hydrocarbon-nitrocompound complexes. This is in effect a virtual photochemical redox reaction which often leads to permanent change, and its reverse process is not commonly accompanied by luminescence.

When the effect of electron spin is taken into account for molecules with even numbers of electrons, it is seen that each orbital picture as shown in Fig. 1.2 represents several energy levels. The electrons are normally spin-paired in ground state orbitals (except for O_2). It is imagined that, by absorption of a light quantum, one electron is raised, without change of spin, to a higher orbital. Such a transition is called singlet-singlet. If the transition to the higher orbital is accompanied by a reversal of spin, theory indicates that the energy change is *less* (Hund's rule) and that *three* closely spaced levels arise from the different combinations of spin and orbital motion. These groups are called *triplets,* with symbols ^3T. The nomenclature is necessary, but in luminescence studies triplet separation is unimportant, and a singlet-triplet transition appears as a single level change. Its significant characteristic is that some intense magnetic influence is required to reverse a spin, and therefore singlet-triplet transitions are associated with light absorption or emission in a very feeble manner, save where heavy atoms are involved (e.g., the strong $^3P \leftrightarrow {}^1S$ line of mercury at 254 nm). This 'forbidden' character can be visualized in another way. The wave-function of two electrons (see Fig. 1.2) of paired spin represents both swinging across the central nodal plane *together*, allowing of a large transition moment; if the spins are unpaired, however, the electrons swing in opposite phases and give zero moment.

Characteristics of Spectra

The emission or absorption spectra of gaseous atoms are composed of narrow lines because of the sharpness of their energy levels. At low pressures the line width is set by the Döppler effect of movement in the line of sight. At higher pressures collisional interactions cause some 'detuning', and the lines are broadened. The electronic spectra of gaseous molecules, however, show wide bands due to the simultaneous excitation of motions of the atomic nuclei. Electronically-excited molecular states normally differ from ground states in having somewhat larger dimensions, different angles between the chemical bonds, and dipole moments different in magnitude and direction. Excitation is a very rapid process ($\sim 10^{-15}$ s), and a newly-formed excited molecule finds itself in a compressed state and so begins to execute vibrations of the atomic skeleton as well as rotations. A more or less complex pattern of sub-bands associated with the electronic transition thus appears. The quantized vibrational levels have energies 10^{-1} to 10^{-2} that of the electronic energy change, and the rotational levels 10^{-4} or so. The full resolution of such spectra into their sharp line components for molecules such as the halogens, SO_2, etc., requires instruments of very high dispersion, and the analysis of spectra is often difficult and involved. These features become blurred for larger molecules because of multi-band overlap, and with the additional 'detuning' caused by variable solvent environments the spectra of dissolved molecules often show merely smooth wide bands. Other reasons, such as short life of a state, or the fact that the excited state differs little in size and shape from the ground state, as in conjugated condensed ring molecules with many delocalized electrons, explain lack of fine-structure of molecular spectra. Spectra from crystals, where lattice as well as molecular vibrations may be excited, are usually smooth and structureless except in special instances, as in the atom-like lines from lanthanide ions or the vibrational structure found for aromatic hydrocarbon molecules. Low temperatures, by reducing vibratory motions, have the effect of narrowing band-widths, unless these are due to variable molecular environments; liquid helium is frequently used to secure significant sharpening. In many examples of luminescent solids a simple chemical picture of the light absorption and emission is complicated by both theoretical and experimental obscurities. Here a different nomenclature of 'centres', 'traps', etc., is used, as described in Chapter 5.

Modes of Excitation

By Light Absorption

The reciprocal relationships of absorption and emission of light quanta, and their connection with electronic energy levels, have been discussed above. For quantitative experimental work, or for clear thinking on the subject, one deals with the absorption of *monochromatic* light (and for crystals it should be *plane-polarized* light too). The governing law is then that of Beer–Lambert:

$$\text{Fraction of light absorbed} = (1 - 10^{-\varepsilon cd}) \qquad (1.1)$$

where ε is the molar extinction coefficient for the wavelength used, c is the concentration in moles per litre, and d is the thickness traversed by a parallel beam. Variations in the formulation of the law may be expressed using the exponential function and different definitions of concentration. The practical effect of this law is that for reasonably uniform absorption throughout a volume of substance the product εcd must be small; high values mean that almost total absorption occurs in the very thin first layers after light entry. If light is re-emitted after absorption, observation at right-angles to the incoming beam best avoids errors due to reflections, etc., but an angular view from the front is necessary where the beam's penetration is confined to the surface. For gaseous systems with sharp-structured line spectra attention must be paid to the degree of monochromatic nature of the exciting light; unless this has a narrower line-width than that of the absorbing substance the application of the Beer–Lambert law will lead to serious error. Low-pressure mercury or other gas lamps are here required, but where broad absorption spectra obtain, the lines of hot, higher pressure gas lamps are sufficient and have a much higher intensity.

The emission processes following light absorption are discussed in a later section.

By Electron Impact

This is the means by which the luminescence from mercury- and sodium-vapour lamps, cathode-ray screens, etc., is excited. The electrons (generated usually from a heated filament and accelerated by an applied voltage) have energies very much larger than the quantized levels of the outer electrons of atoms and molecules, and by their passage they detach bound electrons which enter the accelerated stream. In the plasma (ion-electron gas) so formed, recombination occurs, and by downward jumps between energy levels light quanta are emitted. Molecules exposed to electron impact, unless very stable, become chemically broken up, and any luminescence observed is due to those products, usually radicals, which have good radiative powers. Identification of emitters and of detailed processes is then usually difficult. The action of electrons impinging on crystal phosphors is somewhat complex, and is dealt with in Chapter 5. Impacts of heavy gaseous ions, X-rays, etc., act indirectly, again by detaching electrons in which localized geometrical details may become important; on the other hand, the effects of γ-rays in generating secondary electrons are very uniformly distributed through considerable thicknesses of material, as when they produce luminescence in 'scintillator' crystals, plastics, or solutions.

The term 'electroluminescence' has been used to describe the effect of electron impact, but has also been applied to several other phenomena of quite different nature. One is the glow at electrodes observed in some electrolyses, which is due to ion-radical recombination. Another is the luminescence excited by application of voltages to thin layers of specially prepared zinc sulphide, where the light emission is probably due to electronic effects at points of local high-potential gradient. Electroluminescence also describes the light produced when a current is passed across a p–n junction of a semiconductor such as gallium arsenide or phosphide; the effect is here due to 'recombination of the charge carriers', or, in chemical terms, the restoration of electrons to crystal ions whose valency has been changed.

By Heat

Energy quanta in the luminescence range, expressed in chemical terms per mole of substance, are quite large, comparable to bonding energies in molecules: $Nh\nu = 40$–100 or more kcal/mole. Only at very high temperatures can an appreciable fraction of temperature-equilibrated molecules possess such energy. Temperature radiation from hot, totally absorbing substances is calculable by Planck's theory and is expressed by the well-known 'black body' curves of intensity against wavelength. Imperfectly absorbing materials are equally imperfectly emitting, and their temperature radiation (called *incandescence* in the luminous region) never exceeds in intensity the contours of the corresponding 'black body' curve. True *luminescence* is the emission of greater quantities of light than thermal radiation theory allows, and therefore always involves some momentary disturbance of the thermal equilibrium; in fact, it interferes with the concept of a true temperature. It is found that the Na-line emission from a bunsen flame is close to temperature radiation, but emission from certain newly-formed radicals such as C–H and N–H in combustion zones may be (chemi)luminescence. Very high temperatures are produced in shock-tubes, and bright emission from atoms and radicals observable; strict separation of incandescence from luminescence may here be impracticable.

The ambiguous term 'thermoluminescence' does not refer to normal temperature emission, but to a special effect found in some crystal phosphors. After these have been excited by light, radioactive exposure, etc., electrons liberated within the crystal become 'trapped' (especially at low temperatures) in some way, as at lattice imperfections or foreign ion inclusions. Energy so stored may then sometimes be re-emitted as luminescence when heating drives electrons out of their trapped positions. Illumination by infra-red radiation may produce a similar result; this is not due to heating but to the specific absorption of infra-red by the traps. The phenomenon is complex because of the several imperfectly understood stages involved, some of which lead to energy degradation instead of emission.

By Chemical Reaction

Radiative recombination of ground state atoms or small radicals is rare and weak. The collision time for a bimolecular encounter, $\sim 10^{-13}$ s, is very short compared with radiative times of 10^{-8} s or more. Even if collisions are stabilized by a third body, interaction in pairs of, say, H-atoms or CH_3-radicals, give singlet ground state and triplet excited state molecules, between which radiation is 'forbidden'. In cases where the potential energy curves 'cross' (come close together with interaction), 'inverse predissociation' may lead to formation of an excited molecule, but the probability of its radiating becomes large only when a number of conditions are fulfilled. Higher energy states of atoms or small radicals, capable of forming excited molecules which radiationally 'combine' with the ground state, are needed for strong emission effects. For complex oxidation reactions or organic molecules in solution, potential energy surface considerations become less useful. Chemiluminescent reactions in this field are uncommon either because the reaction energy is insufficient or because the molecules present are incapable of radiating, i.e., non-fluorescent. This is dealt with more fully in Chapters 4, 9, and 10.

Triboluminescence

When certain crystals are crushed, luminescent sparkles appear. These are due to electric discharges following charge separation at the intense stress points of growing cracks. Ultra-violet light from the discharges then may cause fluorescence within the crystals, as with uranyl nitrate. So-called *crystalloluminescence*, observed when certain salts (e.g., strontium bromate) crystallize from hot solutions, is similarly due to the relief of stresses in newly growing crystals. Frequently the emission comes from some electronically-excited radical fragment produced in crystal fracture, like the green luminescence of friction between cane sugar crystals. Electric discharges also occur at the point of stripping of plastic tapes from a roll, and in related ways. 'Air-lines' from the sparks then form the emission.

There is a totally different, and rare, form of crystalloluminescence which has a co-operative origin. During the rapid crystallization of sodium chloride, for example, large numbers of ions orient themselves near the crystal surface in the supersaturated solution, and so sudden deposits of molecularly thick layers occur, with large energy release. If a small concentration of fluorescent ions, like Tl^+ is also present, this energy may concentrate there in sufficient amount to give electronic excitation followed by fluorescence. The phenomenon illustrates the need to regard the energy levels of ionic crystals as whole crystal levels rather than individual molecular features, except at points where a foreign ion is present.

Modes of Emission

The terms *fluorescence* and *phosphorescence* have no strictly agreed meaning, and have been used variously at different times. Emission from atoms excited by light absorption is usually called atomic fluorescence; if the atoms start from and return to the ground state, the emission is termed *resonance radiation,* easily demonstrable with the yellow sodium lines or the 254 nm line of mercury. The atomic electron-waves can be said to be in resonance with the absorbed or emitted light waves, of an identical sharp frequency. If light of a different, lower frequency is shone on the gas the electrons are merely subjected to 'forced vibrations' which slow up the light without real energy transfer (refractive index effect). With molecules effects of intramolecular atomic vibrations intrude into the picture. Under conditions of forced vibrations (use of longer-wave non-absorbed light) the refractive index effect operates for the greater part of the light beam, but a few molecules behave differently. They abstract from the light a small quantity of energy to set their nuclei into vibration, leaving a few light quanta of diminished energy and therefore frequency, whose waves cannot combine with the incident beam, and so emerge in all directions. The spectrum of such light therefore shows a relatively strong line of the incident (monochromatic) light, scattered sideways by local density fluctuations, and having a weak line or lines on the longer-wave side; the frequency differences give molecular vibration frequencies. This is the Raman effect, and it is a common phenomenon for all molecules because the 'lifetime' of the radiation-matter interaction is less than 10^{-15} s—too short for any degradatory processes to occur.

Matters are very different when incident light of a frequency within an electronic absorption band of a molecule is used. The light is absorbed and the molecule raised to an excited state. The re-radiation from electronically excited states is governed by theory which indicates that molecules in a gas or liquid, under usual conditions, perform their downward energy transitions, emitting a light quantum, in an individual manner according to the law of a unimolecular reaction, characterized by an exponential time decay and a *mean life* τ. Further, in the luminescence region of the spectrum this 'life' cannot be shorter than about 10^{-9} s, and may be much longer. The time between collisions in a liquid is about 10^{-12} s. Consequently degradatory processes have a fairly long opportunity, on the molecular time scale, to destroy the excitation before radiation occurs. Collisions may play a part in systems more condensed than gases at low pressure (<10 mm), but even where the influence of collisions is negligible, internal energy conversions (electronic$\rightarrow$vibrational) in multi-atom molecules rob the system of radiating power. Quantum theory shows that electronic and atomic vibrational wave-functions are not strictly separable, but are combined so that energy redistributions within a molecule can take place. Re-emission after absorption by molecules is an uncommon phenomenon. Usually degradation of absorbed energy to final heat is complete, as with all ordinary coloured substances. At very low pressures, resonance radiation (resembling the Raman effect but more intense) may be observed in the case of diatomic molecules (as I_2), but re-emission is usually of a different character. As explained in an earlier section, electronic excitation produces molecules with extra vibrational as well as electronic energy. This vibrational energy is, in condensed systems, very rapidly removed by collisions, so that the excited molecule becomes thermally equilibrated ($\sim 10^{-12}$ s). If no further energy degradation occurs the excited molecule radiates in its own good time (10^{-8} s or longer), dropping, not to the unvibrating ground state, but to a stretched one which begins to vibrate. The emitted quantum therefore has less energy than the absorbed one. Typical emission bands of this type, lying on the long-wave side of absorption bands, are shown in Fig. 1.3. When properly plotted on a frequency scale, the emission band, tailing towards the red, often appears as a fairly close mirror-image of the longest-wave absorption band, which tails towards the violet. This simply results from the fact that excitation, being spread over a conjugated system, as in a dye molecule, does not greatly alter molecular angles and distances. Luminescence of this kind is called *fluorescence*; being related to the longest-wave absorption band the transition is not of a strongly forbidden type. In contrast with the Raman effect, the position of a fluorescence band in the spectrum is independent of the exciting wavelength. Raman emission is very feeble; that from a *solvent* has roughly the same intensity as fluorescence from a *solute* at 10^{-8} mole/litre or less.

Fluorescent molecules in the dissolved state show a larger separation of absorption and fluorescence bands than the vibrational energy losses mentioned above would allow; a gap appears, which depends on temperature and the polarity of the solvent. (See Fig. 1.3.) This is caused by external energy loss to the solvent environment. The excited molecule is ill-adjusted, after sudden formation, to fit its original 'hole' in the solution, both in respect to size and to dipole moment, but it has time to relax to a better orientation before radiating. As it drops back into another unequilibrated situation a double loss of energy is suffered. From a study of the effect it is possible to obtain values for the dipole moment of the excited state (not to be confused with the oscillating 'dipole moment of transition').

An important feature of molecular fluorescence from solutions or solids is that only one fluorescence band is observed for each substance, and it forms the approximate mirror-image of the longest-wave absorption band. If shorter-wave light is used for excitation, so that a higher electronic level is reached, the same fluorescence band appears. It is a general rule that higher excited states, particularly where plenty of collisions are available to remove energy, tend to tumble back to the ground state. The mechanism is a changeover from a higher electronic state with little vibrational energy, iso-energetically to a lower state with larger vibrational energy; the latter is then quickly lost by thermal equilibration with the solvent. For most substances the entire 'cascade' return process takes less than 10^{-12} s, making emission negligible, but with fluorescent substances the return is

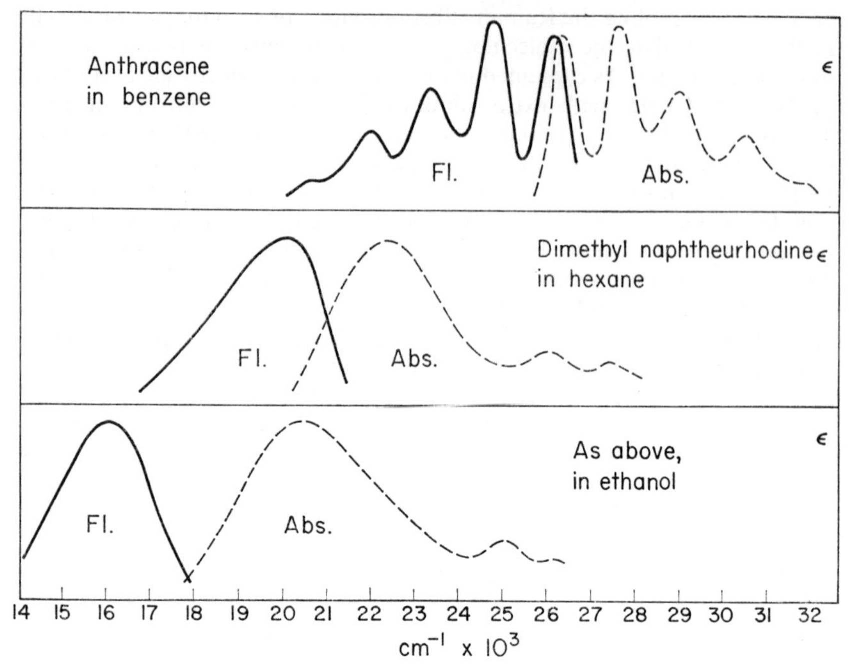

FIG. 1.3 Examples of absorption and fluorescence bands for substances in solution, showing various peak separations due to relaxation effects.

checked at the lowest excited state long enough for light to be emitted. The cascade return through upper states is not hindered by the considerable electronic adjustments involved. For example, phenol (in viscous solvents where molecular rotation is impeded) gives a fluorescence band which is related to the absorption band at 290 nm, and the emitted light is positively polarized when plane-polarized light at 290 nm is used for excitation, i.e., the electric vector is unchanged in direction. If light at 220 nm is used, however, taking the phenol molecule to a higher electronic level, the observed fluorescence is negatively polarized, showing that the transition moment directions of the two levels are at somewhere near 90° to each other; nevertheless the higher level drops very rapidly to the lower with electron re-orientation.

For the special case of organic molecules the term *phosphorescence* is used to describe luminescence from electron jumps between levels of different multiplicity, usually from the lowest triplet level to the singlet ground level. This is a 'forbidden' transition of long radiational life and is therefore greatly exposed to deactivating collisional effects, as by traces of oxygen, and, for lowest triplet n–π^* transitions, by photochemical hydrogen-atom abstraction from the solvent to give monoradicals. Phosphorescence is therefore greatly enhanced when collisions are minimized by the use of low temperatures and rigid glass-like solvent media. The triplet states are produced by extension of the 'cascade' process, after absorption giving an upper singlet state. Details of this, and of other phenomena such as various forms of 'delayed' and 'sensitized' fluorescence and phosphorescence are given in Chapters 7 and 8.

Quantum Yields

Because of the non-radiational ways in which excited molecules can return to the ground state, fewer quanta are usually emitted as luminescence than are absorbed; the ratio is called the *quantum yield* (usually denoted by Φ). It is peculiarly difficult to measure directly because emission is in all directions (not usually equally), and for molecules is a wide band which has to be measured with detectors of unequal wavelength sensitivity. Once agreed values for certain standard substances are available, however, matters are simplified, and direct comparison of spectral intensities under equal conditions of illumination and light absorption gives the magnitudes sought.

Figure 1.4 is a typical energy-level scheme of electronic levels for an organic molecule having n–π^* as well as π–π^* transitions. These levels have associated vibrational energy, stretching them upwards into bands overlapping those of higher levels. Transitions between levels may occur non-radiationally by horizontal, isoenergetic passage from band to band, or radiationally by emission of a photon. The replacement of electronic energy by vibrational energy is rapidly followed by thermal equilibration (energy degradation). Quantum yields of emission depend on the relative rates of the processes; rates of emission depend on the degree of 'allowedness' of the transition, and rates of degradation depend on more complex factors which include the proximity of the levels. If n–π^* transitions have lower energies than the lowest π–π^* transition, degradation is favoured because of the longer time the former are exposed to deactivation. Since these two types of transition are shifted in opposite directions with change of solvent polarity, substances for which the levels are nearly equal may vary greatly in fluorescence yield in different solutions. Theoretical treatment of degradation is very difficult. We can qualitatively explain facts about yields from a knowledge of the nature and relative positions of energy levels, which is obtained from absorption and emission spectral data, but quantitative prediction has yet to be developed.

Since fluorescence is often reduced by rise of temperature, and radiational processes are not expected to be temperature dependent, there must be an 'energy barrier' effect in degradation processes. The easiest passage to a triplet state must therefore be from a singlet excited state with excess vibrational energy. This may be either that a triplet state lying somewhat higher than the excited singlet is first attained, or that there are some special vibrational energy and structural shape conditions favouring isoenergetic changeovers. Activation energies of degradation can be obtained, and usually lie between 2 and 8 kcal/mole. A non-radiational cascade descent down a series of levels of the same multiplicity is commonly

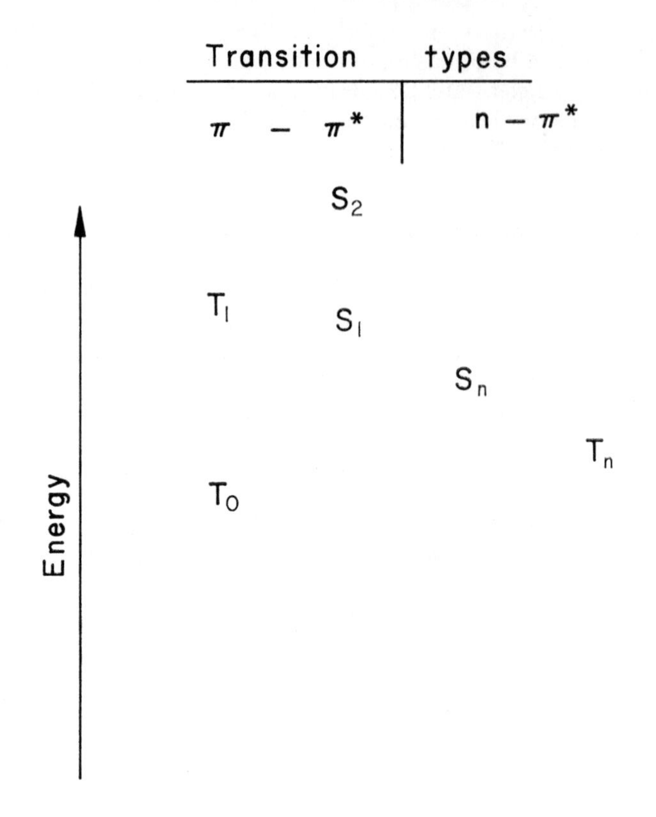

FIG. 1.4

Light absorption (strong) = $S_0 \to S_1$ or S_2; (weak) = $S_0 \to S_n$.
Fluorescence = $S_1 \to S_0$.
Phosphorescence = $T \to S_0$.
Internal conversion = $S_2 \to S_1$ or $S_1 \to S_n$.
Intersystem crossing = $S_1 \to T_0$ or $S_n = T_n$.
The relative orders of position of S_1 and S_n and of T_0 and T_n vary in different molecules; aromatic ketones often have the T_n level lowest, a state with strong H-abstracting properties.

referred to as *inner conversion*, and a switch to a level of different multiplicity, as singlet→triplet, as *intersystem crossing*. The two processes are not basically different, since the concept of 'pure' singlets and triplets is not strictly applicable to molecules and atoms above helium. Singlets, mathematically speaking, contain some triplet character and *mutatis mutandis*.

Mean Lives of Excited States

Theoretically, mean radiational lives τ are related to the degree of allowedness of the electron transition, and this is experimentally given by the absorption band area. For atomic line emission from transitions between two states a formula can be derived which can be used with fair approximation for molecular band emission:

$$1/\tau = 3 \times 10^{-9} n^2 \nu_0^2 \int \varepsilon \, d\nu \tag{1.2}$$

Here n = refractive index, ν_0 = wave-number of band centre, and ε = molar decadic extinction coefficient. (This expression omits statistical weight factors where there is change of multiplicity; where there is such a change, however, the 'forbidden' nature of the transition makes the band area too small to measure.) The band whose area is measured is that part of the full absorption band curve (on its long-wave side) which is due to the ground state—lowest excited state transition—and which may have to be separated off from shorter-wave bands by means of the approximate 'mirror-image' rule. For bands usually encountered in solutions of organic substances τ is of the order of 10^{-8} s, but if for some reason the band is weak, as it is for coronene because of its somewhat symmetry-forbidden character, values are longer. Variations of the above formula have been proposed, to make better allowance for the considerable widths of absorption and fluorescence bands.

Mean lives can be directly measured by devices known as fluorometers (Chapter 2). Such measured lives only equal the theoretical radiational lifetime if the quantum yield is unity; otherwise the measured life is shorter by the factor of the yield. Radiational lifetime is therefore distinguished as τ_0 from the measured lifetime τ; $\tau/\tau_0 = \Phi$.

While fluorescence lifetimes for solutions of organic molecules are of the order of 10^{-8} s phosphorescence lifetimes are 10^6–10^9 times longer because of the 'forbidden' nature of the transitions involved. Their values are easily measured, but values of quantum yields necessary to calculate radiational lifetimes are difficult to obtain accurately, because of practical errors associated with work at low temperatures and with congealed systems not entirely optically clear.

Quenching

In addition to the above mentioned 'internal' factors reducing luminescence yields below unity the presence of certain substances in a system may 'quench' luminescence. Effects due to permanent photochemical reaction, complex formation, or absorption of incident light by the added substance (inner-filter effect) should be excluded from the term 'quenching'. True quenching is a reversible effect in which removal of the quenching agent restores the original luminescence intensity, and requires collisional or close interactions between the two molecules involved. Many observations indicate that, commonly, a quencher molecule acts by either extracting an electron from, or supplying one to, an excited molecule, and subsequently, after energy loss, reversing the process. During the interaction interval the electron spin becomes reversed, either by becoming subjected to magnetic influences near the nucleus of a heavy atom, or because of intermediate 'doublet' or 'radical' states. In solids, where luminescence often depends on the return of electrons detached by light, 'electron traps' such as variable valency ions may be effective quenchers.

The kinetics of quenching of fluorescence in solutions (ideally) follow the Stern–Volmer law:

$$R - 1 = K[Q]$$

where R is the ratio of fluorescence intensity, without and with the quencher whose concentration is $[Q]$.

This equation is easily deduced as the result of two competing reactions, a unimolecular light emission of mean life τ and a bimolecular quenching encounter of constant k; $K = k\tau$. From the constant K and a value of τ the absolute value of k in units of litre $mole^{-1}$ s^{-1} is calculable. For certain 'strong' quenchers, as dissolved oxygen often is for solutions of fluorescent organic molecules, k for solvents such as hexane or benzene is about 10^{10}. This value is close to that predicted by theory for the fastest possible rate for a bimolecular reaction, where the molecular encounters are limited by diffusion. Quenching constants are here viscosity dependent. Weaker quenchers have constants independent of viscosity because the rate is limited by the inefficiency of the quenching process; they behave, in fact, like an ordinary thermal bimolecular reaction, and are subject to the usual salt effects if in ionic solution.

Fluorescence yields of concentrated solutions are often less than those of dilute ones. Sometimes, as with certain aqueous dye solutions, changes in the absorption spectra are observed, which are due to ion-counterion interactions or dimer formation, with consequent inner-filter action. Some solutions, however, show true *concentration quenching*, there being no change in the absorption spectrum. Facts observed for the fluorescence of pyrene solutions have pointed to an interpretation. Here the fluorescence band characteristic of dilute solutions is diminished for stronger solutions and becomes replaced by a new band at longer wavelengths. A fluorescent 'excimer', or short-lived combination between an electronically excited and a ground state molecule, is responsible for the effect. Sometimes, as with anthracene solutions, the van der Waals type interactions of the excimer pass over further into chemical linking to give permanent (ground state) dimer formation. If an excimer does not fluoresce, its presence is a matter for speculation, although favoured by theory.

The principle of competing reactions has been used to estimate rates of fast protonation reactions by their effect on fluorescence emission from suitably chosen organic acid or base solutions.

If two luminescent substances of appropriate characteristics and sufficient concentration are both present in a system, *energy transfer* may occur from one to the other, so that the emission from one is quenched and the other enhanced. Chapter 8 describes this type of effect.

Illumination under laboratory conditions with ordinary lamps may deliver about 10^{17} quanta per second on to about 10^{21} molecules. For excited state lifetimes of a tiny fraction of a second, it is obvious that the stationary concentration of excited molecules must be very small. Under these conditions re-radiation from excited molecules is a spontaneous process, each molecule behaving independently, the whole radiation from the system showing an exponential decay with time. A different situation arises when the illuminating intensity is much larger, as from powerful flash sources. So-called 'stimulated' emission may then occur. Wave-mechanical theory deals with transitions along the following lines. The differential equation of Schrödinger is the governing one; its solutions define the energy levels of any system. For simplicity, consider a system with only two levels. Since linear combinations of solutions of the differential equation are also solutions, a possible state of the system is to be 'in both levels at once'. This corresponds to the molecules emitting and absorbing electromagnetic energy in the form of light waves and being equilibrated with them (as in a closed reflecting container). The 'light bath' surrounding the molecules introduces an additional energy term into the governing equation, which has the effect of making the coefficients of the linear combination time dependent. For the emission process an expression with the sum of two terms is derived; one gives the mean life of 'spontaneous' emission τ_0, and the other relates to the 'stimulated' or 'radiation-induced' emission. It is thus realized that, besides the two processes, of the lower state absorbing a photon and the upper state spontaneously emitting a photon, there is a third, namely the effect of a photon interacting with the *upper* state. Such an interaction induces the upper state to *emit* a photon, with the same probability as an interaction with the lower state would give *absorption*. The interacting and the liberated photons have identical phases in their combined waves. For total emission to exceed total absorption in a system of molecules interacting with light waves, there must be more molecules in the upper state than in the lower; this state of affairs, of course, can never be produced by simple heating, but may be by some form of intense excitation by an external source. A further condition is required in practice to make 'stimulated' emission prevail over the normal 'spontaneous' process; a resonant 'cavity' must be created so that the electromagnetic field of the light waves can build up to a sufficient intensity to overcome energy losses. These principles are the basis of laser construction; excitation is by a flash or electrical discharge, and resonance is achieved by having reflecting ends of the laser. (See Chapter 3.)

USE OF LUMINESCENCE IN ANALYSIS

Because of the great sensitivity of modern photomultipliers it is possible, in favourable instances, to estimate concentrations of fluorescent materials down to 10^{-10} M, or one molecule in 10^{12}. Inorganic ions are examined as fluorescent organic complexes. In the organic and biochemical field, applications are many and rapidly increasing; for example, the 1966 *Annual Review of Analytical Chemistry* lists nearly 500 references to developments of a single year. (See Chapter 6.)

Chapter 2

INSTRUMENTATION

I. H. MUNRO

Introduction

THE observation and measurement of luminescence from organic and inorganic materials may provide the chemist with one of the most sensitive techniques for the detection and determination of compounds [1], and offer a technique for the analysis of chemical kinetics in the nanosecond (10^{-9} s) time region.

Fluorescence measurements fall conveniently into two categories: the measurement of emission spectra, called fluorimetry, and measurement of the time dependence of the emission, called fluorometry. The two categories sometimes overlap in studies of the long-lived excited states in phosphorescence or delayed fluorescence, when simultaneous observations are made of the emission spectrum and of its time decay.

Fluorimetry

GENERAL PRINCIPLES

The essential components of an instrument required to excite and measure fluorescence are indicated in Fig. 2.1. Excitation of the sample is achieved using an appropriate emission band from the source, selected either by a filter or by a monochromator. The light emitted from the sample is viewed through a second filter or monochromator by a detector whose output signal is proportional either to the total intensity or to the total energy of the signal falling upon it. The signal may be displayed on a meter dial, a chart or X–Y recorder or on an oscilloscope.

It is convenient to distinguish between instruments which use filters for selection of the excitation and emission wavelengths and are called filter fluorimeters, and those which use two monochromators for wavelength selection and are called spectrofluorimeters or fluorescence spectrometers.

Despite the limited wavelength selectivity inherent in filter fluorimeters, they have the compensation that greater light intensities are transmitted, giving them an effective sensitivity which can be much greater than that of an equivalent fluorescence spectrometer. Filter fluorimeters are therefore ideally suitable for trace analysis and as comparators for comparing the fluorescence intensities of extract samples with those of standards.

Fluorimeters of this type are relatively cheap and simple to set up, and once suitable filters are obtained for a given assay, no wavelength calibration is necessary.

A fluorescence spectrometer, on the other hand, may be an instrument of considerable mechanical and electronic sophistication and can be extremely expensive. However, when corrected for fluctuations in source

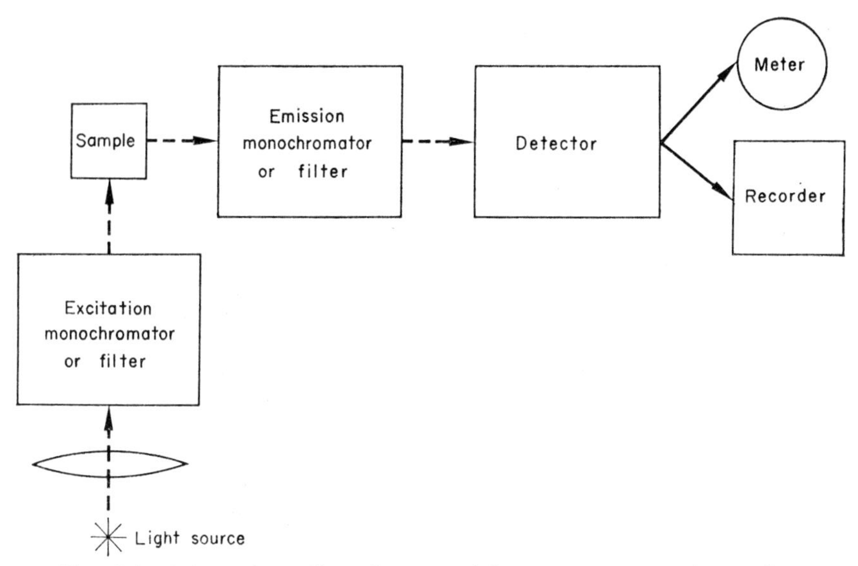

FIG. 2.1 Schematic outline of a general fluorescence measuring device.

intensity and for the wavelength response of its various optical components, an instrument of this type is of the maximum utility in the identification and study of luminescent materials. For example, both the fluorescence excitation spectrum, which is identical in profile to the absorption spectrum, and the fluorescence emission spectrum of many dissolved substances can be measured easily in solutions of concentration less than 1 μg/ml.

EXPERIMENTAL DETAILS

1. Light Sources

Since the measured fluorescence signal will be proportional to the intensity of the exciting light (see 5, p. 23), it is clearly important to obtain a light source which provides maximum intensity in the region of greatest absorption by the solute (usually in the ultra-violet), and which can be used conveniently with either a filter or monochromator. The type of light source chosen will depend primarily on whether a fluorescence excitation spectrum or fluorescence emission spectrum of the sample is to be measured.

Measurement of fluorescence excitation spectra requires a source which emits a high-intensity continuum throughout the ultra-violet and visible regions and, ideally, has a level quantum output at all frequencies, to minimize correction factors.

Incandescent sources of radiation (such as high-power tungsten filament lamps) used to produce broad continua are both weak and inefficient ultra-violet generators. Although they can be used for excitation in the visible and infra-red regions, they are of little practical value for excitation at wavelengths less than 450 nm.

The brightest and most powerful sources of ultra-violet radiation are high-pressure xenon or mercury arcs. The emission from such sources is approximately continuous because of the pressure and temperature broadening of the spectral lines which occurs under the lamp operating conditions, when the lamp pressure may be as much as 100 atmospheres. Many details such as spectral emission characteristics, operating conditions and relative costs of currently commercially available arc sources are obtainable [2]–[4]. Although a high-pressure mercury lamp is more efficient and powerful in the ultra-violet (particularly below 300 nm) than an equivalent xenon lamp, the emission spectrum contains many prominent lines superimposed upon a weak continuum (see Fig. 2.2) and is not very suitable for fluorescence excitation spectroscopy.

A high-pressure xenon lamp, however, emits a rather smooth continuum between about 250 nm and 450 nm (Fig. 2.2) and is ideal for excitation spectroscopy in this region. In addition, the ultra-violet output is sufficiently high for use in conjunction with a monochromator as a source in fluorescence emission spectroscopy. Consequently the majority of commercially available spectrofluorimeters use a single xenon arc of 100–500 W power both for excitation and emission measurements. Increased intensities can be gained by using larger lamp powers of up to 6 kW; these require more expensive associated control equipment, such as a large transformer to produce the low-voltage, high-current supply (of the order of 100 amps) necessary for stable lamp operation.

An ideal source for fluorescence emission spectroscopy would possess high intrinsic brightness and emit monochromatic light. The development of d.c. operated lasers may provide precisely such illumination, although at present these are restricted to mean powers of less than 1 W and to emission at wavelengths usually longer than 300 nm [5]. High intensities of nearly monochromatic radiation are produced by 'resonance' sources. A 30 W hot cathode, low-pressure, mercury lamp of this type will radiate about 24% of the energy input to the lamp in the 254 mn line [3]. Such lamps, unfortunately, are necessarily of low intrinsic brightness because of their large surface area, and are not suitable for high-power operation. For many routine applications, such as in filter fluorimetry, effective, reliable and cheap sources are found in 80 W or 125 W medium-pressure mercury vapour lamps operated from mains a.c. through a choke.

The very highest intensities in the ultra-violet are probably emitted by the high-power, high-pressure mercury arc sources, and dose rates from such sources have been measured [23]. The use of such lamps requires an efficient filter as monochromator to eliminate the many lines emitted in the visible region. A typical light energy output from a 1 kW water-cooled mercury lamp in a quartz envelope is 169 W radiated at wavelengths shorter than 380 nm [6].

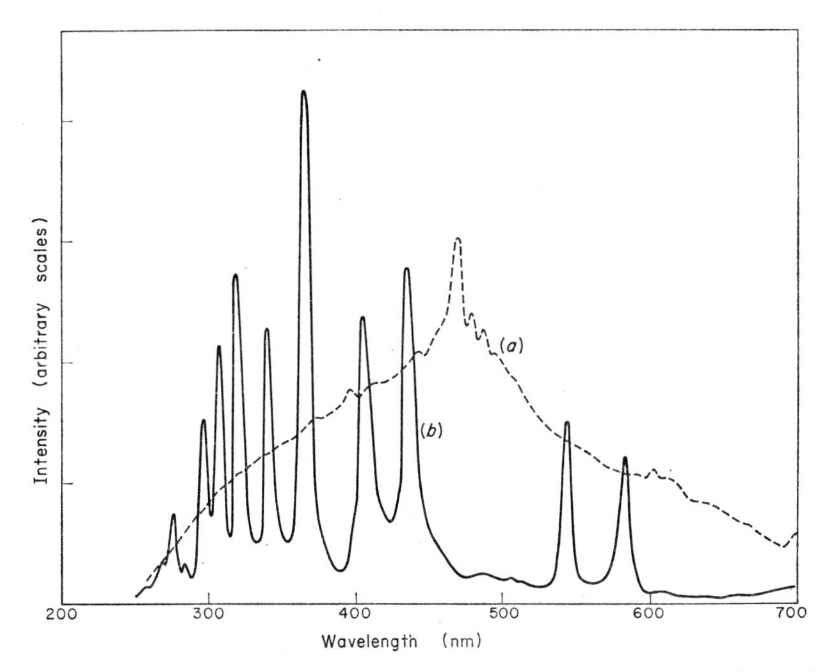

FIG. 2.2 The typical emission spectrum (quanta per unit wavelength interval) from (*a*) a high-pressure xenon arc (*b*) a high-pressure mercury arc.

2. Source Stabilization

In all fluorimetry it is necessary to keep the source intensity constant for the duration of the measurement. Intensity fluctuations may arise as a result of voltage or current changes in the power supply, changes in arc or discharge geometry, or a gradual decrease in operating efficiency. Although some sources, such as mercury lamps for fluorimeters, use unstabilized a.c. supplies, medium- or high-pressure arc lamps are most effective when operated from a current-stabilized d.c. supply. A useful current stabilizer with current regulation to $\pm 0{\cdot}1\%$ has been described [7]. For high-power lamps, smoothing problems are greatly simplified (by a factor of about 10 times) when a three-phase, rather than a single-phase supply is transformed and rectified. Intensity fluctuations resulting from movements of the arc or from power supply changes could be eliminated by using a photoresistive element to monitor a fraction of the light output from the lamp, and by feeding a correction signal back into the source current stabilizer. The measured fluorescence output signal can, of course, be modified to compensate for intensity changes when a source monitor is used.

3. *Filters*

Filters play an important part in the excitation and observation of luminescence. Their high transmission and large (or small) bandwidth may make their use essential in experiments where intensities are low, for example in fluorescence lifetime studies.

A wide choice of cut-off or band-pass filters is available for the near ultra-violet and visible regions: glass, gelatin, chemical filters and, less frequently, metal-dielectric and all dielectric (narrow band) interference filters and Christiansen scattering filters have been used in this region.

Although interference filters are very efficient in the visible they are expensive, and as the selected wavelength decreases, the peak transmission falls and the bandwidth of the filter increases. At wavelengths shorter than 300 nm the peak transmission is usually less than 20% and, in addition, such filters may transmit a small but significant amount of light (0·1% to 0·5%) up to much longer wavelengths.

Glass [8] and gelatin-based filters [9] are inexpensive and possess good transmission qualities, although some glass filters are luminescent and therefore are not suitable for fluorescence measurements. Band-pass filters of gelatin or glass are useful only for wavelengths greater than 310 nm. Special nickel oxide glasses are available for transmission in the ultra-violet region.

Chemical filters describe a wide range of inorganic [10] and organic [11] materials with useful transmission properties when used in solution, as vapour or in the solid phase [12]. Materials may sometimes be mixed in solution to provide the required filter properties [13]. Chemical filters contained in silica cells provide the only useful band-pass filters for the wavelength region below 300 nm, and transmission of several such filters has been described [2], [10]–[13]. Chemical and gelatin filters will not usually withstand high temperatures, and some chemical filters deteriorate as a result of photochemical decomposition with prolonged irradiation from the source.

A useful selection of complementary excitation and emission filter combinations has been described [14] for use with filter fluorimeters for the quantitative determination of a range of organic and inorganic substances.

4. *Monochromators*

The choice of a suitable dispersing element—usually a grating or prism monochromator—is of major importance since it not only defines the energy resolution of the measured spectrum, but also limits the maximum signal to noise ratio available.

Detailed theoretical and experimental comparisons have been made between the luminosities (the output flux from a source of unit luminance) of gratings and prisms in terms of their effective resolving powers and dimensions (the area of the grating or area of the prism base). These comparisons [15]–[18] reveal that for a given effective resolving power, defined by the monochromator slit width, the luminosity of a grating is superior to that of a prism at all wavelengths [15]. (See Fig. 2.3.) Although

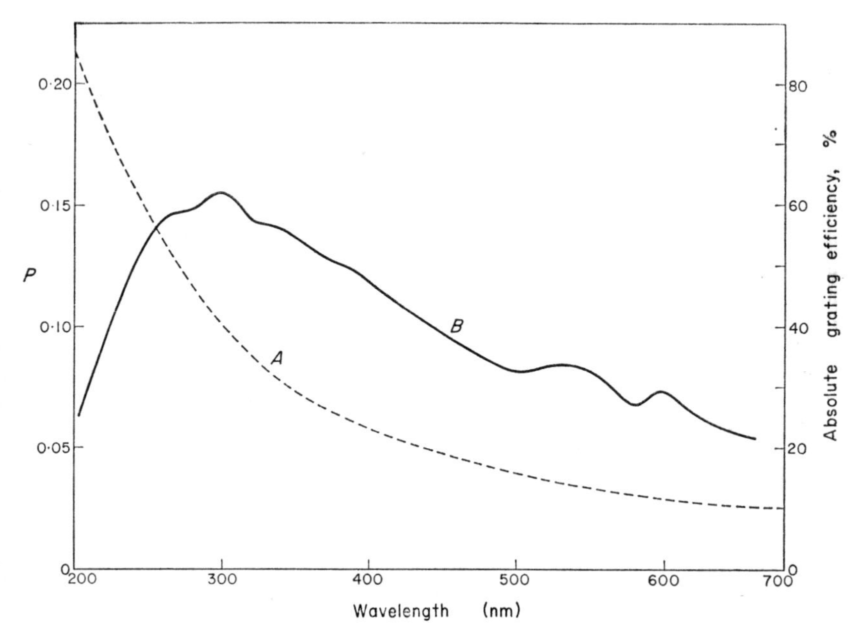

FIG. 2.3 Curve A: the variation, as a function of wavelength, of the ratio P, of the light flux given by a quartz prism to the flux given by a grating of the same size and with the same effective resolving power [15].

Curve B: the variation of absolute grating efficiency as a function of wavelength for a 600 lines per mm, 50 cm radius of curvature concave grating [19].

the grating is considered to be used at, or near, the maximum of the blaze, the variation in grating efficiency away from the blaze in the ultra-violet and visible region (Fig. 2.3) is insufficient to modify this conclusion.

In general, the light flux through the exit slit of such a dispersing system is proportional to the angular dispersion of the element. (A grating with twice the number of lines per millimetre, and hence twice the angular dispersion, will give twice the flux.) It is also proportional to the projected area of the element, to the transmission of a prism or the reflectivity of a grating, and it will increase with increase in bandwidth [15], [16]. Even in the short-wavelength region, where the angular dispersions of a prism and grating are similar, the decreasing transmission of the prism material may keep the prism-to-grating flux ratio small (Fig. 2.3). The highest grating efficiency will be achieved only when the blaze angle is appropriate to the use of the monochromator. For example, excitation and emission monochromators might be blazed for maximum efficiency at 350 nm and 500 nm respectively.

While a correction must be applied if a spectrum is to be automatically plotted (see 1, p. 17), a linear drive to a grating will provide spectra on a linear scale of wavelength. Unless a mechanical cam is used, a linear motor drive to a prism will produce a spectrum which is non-linear both in wavenumber and wavelength, and whose effective resolution (defined by a constant slit width) will vary markedly with wavelength.

A grating monochromator, used in the first order, suffers from the disadvantage that in the visible or red regions the grating will transmit light of shorter wavelength in the second or third order. These effects are eliminated when a glass filter is used, to attenuate higher diffracted orders, between 400 nm and 600 nm. Problems of this kind arise when the grating is used as an excitation and as an emission monochromator. Spectra recorded with a grating should be corrected for the grating response (Fig. 2.3, 2.5) otherwise anomalous spectral peaks will be recorded [19], [20], which result, in fact, from a polarization effect observed when diffraction of an order takes place along the surface of the grating [21]. Light diffracted from a grating is usually from 40% to 60% polarized by the grating rulings, and this factor may be important when polarization measurements are to be undertaken.

It is important that the monochromator should produce a minimum amount of radiation at the exit slit outside the chosen wavelength band. The criterion for minimum stray light is simply one of careful optical design. A good, expensive, prism or grating monochromator will have a stray intensity, at any wavelength, of about 0·001% of the peak transmitted intensity. The ultimate reduction in stray light will become available when two such monochromators are operated in tandem, that is, as a double monochromator. Inexpensive grating instruments are especially prone to stray light collection at the exit slit; this can sometimes amount to 0·1% of the transmitted light. The integrated stray light intensity at all other wavelengths may be sometimes as high as 5% to 10% of the total.

Wavelength calibration of monochromators in a fluorescence spectrometer is readily achieved by using the known emission lines from a low-pressure mercury lamp. With the spectrometer, it is important that the positions of the exciting source and fluorescent sample are accurately defined for substitution by the mercury lamp and scatterer, since the calibration will depend on the angle at which light enters through the entrance slit of each monochromator.

5. Sample

The total fluorescence intensity I_F (quanta per second) from a solution of c moles per litre of a material of extinction coefficient ε in a transparent cell *d* cm long, when excited by a parallel monochromatic light beam of intensity I_0 is

$$I_F = I_0\Phi(1 - 10^{-\varepsilon cd}) \qquad (2.1)$$

Φ is the fluorescence quantum efficiency of the solution. For dilute solutions, equation (1) becomes

$$I_F = 2{\cdot}303\, I_0\varphi\varepsilon cd \qquad (2.2)$$

If I_0 is normalized for all excitation wavelengths and Φ is assumed to be wavelength independent, then I_F is proportional to ε with an error of less than 4% for $\varepsilon cd < 0{\cdot}02$, and will give a profile identical to that of the absorption spectrum. For concentrations within the dilute solution range, I_F is proportional also to c, a relationship of great importance in filter fluorimetry.

When the solution is sufficiently concentrated for all the incident light to be absorbed at each wavelength, then $I_F = I_0\Phi$. In this case, after normalization of I_0, it can be established whether or not Φ is wavelength independent. Alternatively, if Φ is known to be constant, the variation in I_0 with wavelength may be measured.

The design of the sample compartment and geometry of the sample have been discussed in detail in previous reviews [22]–[25]. Frontal excitation and observation is most satisfactory both for solids (crystals) and for concentrated solutions. When dilute solutions (below 10^{-5} moles per litre), or sample holders prone to scattering, such as cylindrical low temperature dewars, are used, then right-angle viewing may be advantageous. With filters or monochromators mounted at right-angles as in Fig. 2.1, it is a simple matter to devise a sample mount which can be adjusted to provide both front surface and right-angle viewing.

A variety of physical and chemical effects at the sample can limit the sensitivity of analytical procedures, and modify, or possibly eliminate, fluorescence emission.

The material of the sample cell must be fluorescence-free. Pyrex glass or fused synthetic silica [26] are good cell materials. Some types of natural quartz and optical quality fused quartz are fluorescent [23].

The fluorescence signal is very susceptible to the presence of impurity species in the sample. Impurities in solvents, or those introduced by contamination with traces of vacuum grease, cleansing agents, chemical reagents, filter paper, etc., or as a result of photochemical decomposition in the sample are all important experimental hazards. For measurements in solution, spectroscopic grade purity solvents at least, should be used whenever possible.

Fluorescence quenching commonly occurs in the presence of dissolved oxygen. The oxygen concentration can be reduced to about 10^{-6} M for fluorescence measurements by bubbling with oxygen-free ('white spot') nitrogen gas. For delayed fluorescence or phosphorescence measurements in solution, the oxygen concentration can be reduced to less than 10^{-8} M by using a freeze-pump-thaw cycle of operations [53].

All measurements must be made at an accurately defined temperature since fluorescence deactivation processes are strongly temperature dependent. Local heating effects due to the source are largely excluded by the selectivity of the excitation filter or monochromator, but sometimes may have to be considered.

6. *Detectors*

A wide range of selective and non-selective detectors can be used in the region 200 nm to 800 nm [3], [27]. In luminescence studies, non-selective detectors such as thermopiles and bolometers, and also some selective detectors such as photoconductive and photovoltaic cells, are little used on account of their fragility, general lack of sensitivity or poor time response. The great majority of detectors used are of the photoemissive type, usually photomultiplier tubes.

A photomultiplier has a high cathode photoelectric quantum efficiency (up to 25% for a trialkali cathode) and can provide a current gain of from 10^7 to 10^9 for applied potentials of 2 kV to 3·5 kV. A typical 13 stage venetian-blind type photomultiplier will have an output proportional to the light intensity incident on the cathode for anode output currents of up to 30 μA or so. The maximum gain which can be obtained is usually restricted by the dark current signal from the photomultiplier. For a 13 stage tube, the dark current will probably be unimportant up to applied voltages of about 2 kV when the dark current will be approximately 1 μA and will then increase rapidly with increase in gain.

The dark current may be reduced considerably by cooling the photocathode to permit measurements of very low light intensities. In normal luminescence work, however, the output noise level will be defined primarily by light scattering effects, and photocathode cooling is probably unnecessary.

A limitation of photomultiplier detection is its rather rapid fluctuation in sensitivity as a function of wavelength, shown in Fig. 2.4 for different cathode types.

The long wavelength cut-off is defined by the work function of the cathode material, and usually lies between 650 nm and 850 nm. The profile through the visible and ultra-violet regions depends on the optical transmission properties of the semi-transparent cathode, with the short wavelength limit being set by the transmission of the end window (pyrex or quartz) [28], [29]. The falling sensitivity, in the red particularly, may effect a major alteration to the shape of an emission spectrum, and is eliminated by correction of the output signal for the photomultiplier sensitivity.

Although the photomultiplier output can be measured directly on a galvanometer of sensitivity about 100 mm per μA, the signal is frequently amplified by means of an a.c. or d.c. amplifier of gain from 10^1 to 10^5, and is fed into a recorder.

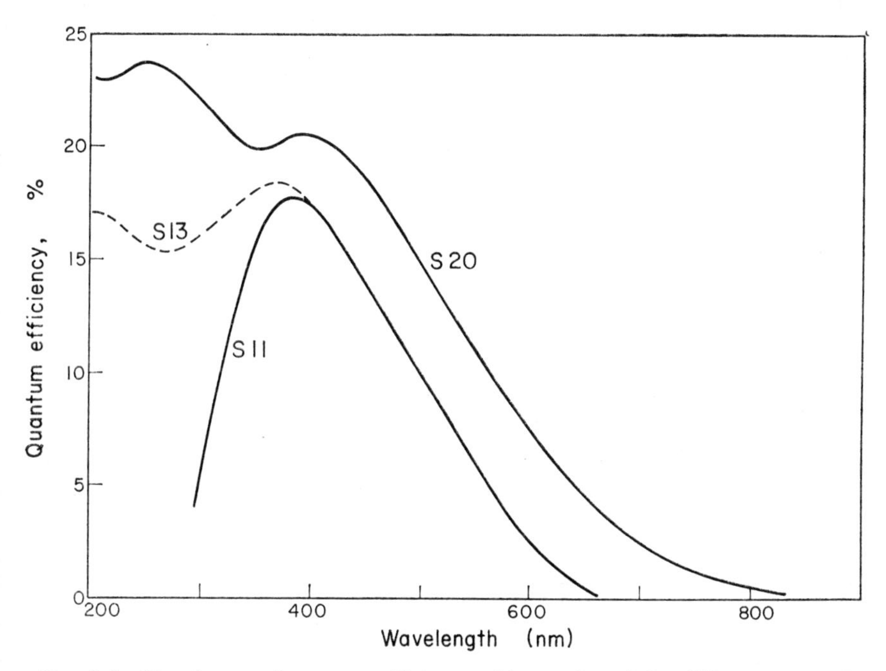

FIG. 2.4 The change of quantum efficiency with wavelength for different types of photomultiplier cathode: S2O, Sb—Na—K—Cs on fused silica; S13, Sb—Cs—O on fused silica; S11, Sb—Cs—O on glass [29].

Types of Instrumentation

FILTER FLUORIMETERS

A large number of filter fluorimeters are commercially available, and these differ considerably with respect to light sources, filters, detectors, and types of sample accepted. The relative merits of the construction and sensitivity of different instruments have been described fully [24], [25], [30]–[33] and details given of their use in chemical analysis [14].

The construction of a filter fluorimeter is fairly simple, and is outlined in Fig. 2.1. The integrated fluorescence intensity viewed by the detector through the emission filter is directly measured on a meter. Errors may be introduced by fluctuations in the source intensity, but these can be eliminated when a second detector is used to compensate for such changes. Alternatively, if a light-chopper is used and the detector views the exciting light and the fluorescence light alternately, then an a.c. amplifier and phase-sensitive detector may be used. In this second case, both source fluctuations and dark current effects are eliminated.

The output signal from the detector is proportional to concentration for dilute solutions (eqn 2.2). Actual calibration of the fluorimeter is established, normally, by measuring the fluorescence signal from a series of standards of known concentration. If an intense (high-power) source and sensitive detector are used, measurements can be extended down to concentrations of the order of 0·001 μg/ml for certain materials. The fluorimeter, of course, will only operate efficiently in a region of high detector sensitivity and where the detector response changes linearly with varying signal intensity.

Spectrofluorimeters

A logical development of the filter fluorimeter is to replace one or both of its filters with a monochromator.

Most commercial absorption spectrometers provide a fluorescence attachment which permits the measurement of fluorescence emission spectra and, sometimes, of excitation spectra [24]. Although such modifications represent primarily an extension to the capabilities of an absorption spectrometer, they play a useful part in quantitative fluorescence spectroscopy. To derive a maximum amount of information from a luminescent material, however, a two-monochromator system which can measure both emission and excitation spectra is essential.

1. Instrumental Corrections

The measured excitation and emission spectra will, of course, represent a combination of the true sample spectrum with various instrumental properties. No two light sources, monochromators or detectors possess an identical wavelength dependence. Clearly, therefore, no two uncorrected fluorescence spectrometers will produce identical spectra from measurements on the same sample, and sensible comparison of results from different workers is impossible. Even in one particular apparatus, changes may occur, in time, as the light source or detector efficiency alters.

Methods for determination of the several instrumental correction functions have been discussed [2], [34]–[36]. Measurement of a true fluorescence excitation spectrum, when the excitation monochromator is scanned through the sample absorption, with the emission monochromator fixed, is a straightforward procedure. It is achieved by viewing a fraction of the incident light used for excitation at the exit slit of the monochromator. The effective intensity spectrum of the lamp (Fig. 2.2) and any incidental intensity fluctuations occurring during the experiment, are measured either by a non-selective detector (usually a thermopile), a previously calibrated photomultiplier or a ‘quantum counter’ which absorbs all the incident light and has a constant quantum efficiency [37]. The monitor output is fed to a servomultiplier or a ratio recorder, or simply to a meter so that point by point corrections may be applied. In this way, the true excitation spectrum can be found and also the change of sample quantum efficiency with exciting wavelength studied, using dilute and concentrated solutions respectively (see 5, p. 23).

Although simple to provide, a quantum counter suffers from the disadvantages of limited useful wavelength and intensity operating ranges, and from possible deterioration with time. Correction in commercial instruments is usually applied by means of a monitoring thermopile, and gives spectra normalized for constant energy of excitation. This correction should be multiplied by the exciting wavelength in order to obtain the experimentally significant intensity normalized spectrum. Additional use of the thermopile signal to drive a slit servocontrol is unsatisfactory since the monochromator bandwidth continuously changes by an unknown amount throughout the spectrum.

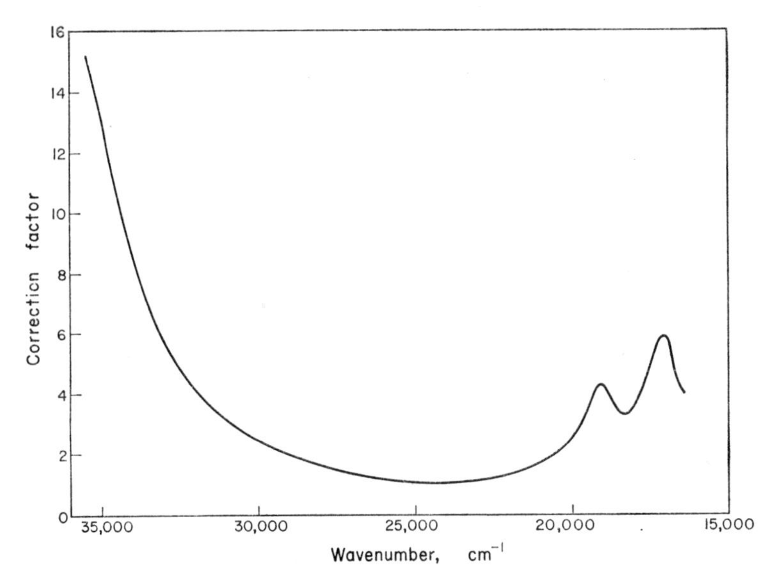

FIG. 2.5 A correction factor applied to a fluorescence emission spectrum to eliminate effects due to changing sensitivities of the grating monochromator and photomultiplier detector through the spectrum. The function is stored in an electrical cam [49].

When light scattering effects are ignored, the true fluorescence emission spectrum will be modified primarily by the grating and the detector response, which together may change by as much as 20 : 1 in the range 250 nm to 650 nm (see Figs. 2.3, 2.4). A measured correction curve, by which the detector output must be multiplied, is shown in Fig. 2.5, for a grating monochromator and photomultiplier in combination. All emission spectra should be plotted in terms of quanta per unit wave-number interval against wave-number (a linear energy scale) for greatest usefulness and physical significance [38].

The combined or separate responses of the monochromator and detector are established by calibration with a standard black body source of known colour temperature, a detector of known response, such as a thermopile, or by using a quantum counter [19], [20], [28], [34]–[36]. Even when more than one method is used for calibration, the reproducibility of such a correction curve (Fig. 2.5) is, at best, only about ±3%. At the high and low wavelength limits, the accuracy will be much worse than this. The correction function can be checked by measuring the spectrum of some 'standard' fluorescent material [39], [40].

2. Choice of Instrument

The wide range of modified absorption spectrophotometers and uncorrected fluorescence spectrometers available have been reviewed [23]–[25], [41], [42]. If only either excitation or emission spectra are to be measured then a single monochromator instrument will be sufficient providing its response can be found. The necessary second monochromator for such calibration is, of course, already available in a fluorescence spectrometer.

When a manual instrument only is available, correction of the spectra may be undertaken by using a curve-multiplying device [43], or, alternatively, by tedious point by point calculations. However, the tremendous usefulness, sensitivity and speed of operation obtained by using automatic, fully corrected (or 'absolute') fluorescence spectrometers have been generally appreciated, and instruments of this type are now commercially obtainable [44]–[46]. The number of accessories available makes such instruments invaluable in luminescence research.

Some absolute fluorescence spectrometers have been described [44], [45], [47]–[49] in which the method of correction for the instrumental response is specifically discussed. The correction usually is applied mechanically or, less frequently, electrically. Representation of the correction function on a mechanical cam may be inflexible (if the function changes with time) and may not operate efficiently over a large correction range or through very sharp variations in the function. An electrical cam [44], [49] or 'adjustable function generator' [50] is much more convenient in that it is commercially available and can be adjusted to provide any chosen function with visual representation of the function. Interpolation between the function fixed points is obtained by using a tapped helical potentiometer rotating synchronously with, for example, the monochromator wavelength drive. The correction to the output signal (Fig. 2.5 shows a 36 fixed-point electrical cam representation of a measured correction curve) may be applied by a servomultiplier driven by the function generator [49].

An expensive, but convenient method for obtaining corrected spectra has been described [51] which uses a programmed computer to plot out spectra and calculate relative quantum efficiencies.

PHOSPHORESCENCE

Under appropriate sample conditions (in a low-temperature glass, for example) fluorescence and phosphorescence emission spectra may be measured simultaneously in any fluorescence spectrometer so long as the exciting light is not chopped at high frequency for a.c. detection. However, the relatively long phosphorescence lifetime (10^{-4} s or longer for a detectable phosphorescence yield) permits the application of simple time resolved spectroscopy. Mechanical or optical shutters are used alternately to excite the sample and view the emission during a period of zero excitation. This procedure eliminates not only all fluorescence emission (lifetime less than 10^{-6} s), but also any scattered exciting light, and it is therefore possible to excite the sample with unfiltered radiation from a source of maximum intensity. In addition, the time dependence of the emission (phosphorescence or delayed fluorescence) can be measured directly on an oscilloscope within a time range determined by the chopping frequency.

The simplest type of phosphorimeter employs two slotted discs mounted on a common axle, the sample being placed between the discs. When the slots for excitation and observation are rotated out of line, such apparatus is convenient for delayed fluorescence measurements in solution at room temperature [52].

In a detailed review of phosphorescence and delayed fluorescence measurements [53] an elegant type of apparatus using two separate synchronous motors driven from a common power source has been described. The phase difference between the chopper discs can be altered simply by rotation of one of the motor bodies through a known angle. Thus the ratio of the phosphorescence to the total luminescence signal can be measured while the apparatus is kept running continuously.

All fluorescence spectrometers could be converted into spectrophosphorimeters by enclosing the sample in a slotted rotating can. A corrected instrument of this type is available commercially [46]. A similar, more versatile, instrument uses a Variac controlled variable speed motor which can alter the rotation of the slotted can from 200 rev/min to 7000 rev/min and permits lifetime determinations down to 10^{-4} s [47].

For efficient phosphorescence spectroscopy, it is essential to use an emission monochromator blazed in the visible-red region, and a red-sensitive (S20 response) photomultiplier. Lifetime measurements are greatly simplified when oscilloscope trace photography is replaced by the use of a pulse sampling oscilloscope and direct X–Y plot-out of the decay curve. Since the sensitivity will ultimately be determined by statistical fluctuations in the signal, the signal to noise ratio at the output from a spectrophosphorimeter could be considerably enhanced by using a signal averaging device.

Quantum Efficiency

The precise measurement of absolute fluorescence quantum efficiency is a difficult experimental technique involving corrections for various effects associated with cell reflectivity, solvent refractive index and polarization effects. Various methods used to evaluate luminescence efficiencies have been reviewed recently [54] and the most suitable materials for use as fluorescence standards have been discussed, with particular reference to inorganic compounds.

In practice, the determination of relative quantum efficiency is much simpler and more accurate than that of absolute efficiency. The unknown efficiency is measured by comparison with a standard compound whose absolute efficiency has been established previously under carefully defined experimental conditions. Such a comparison should be made using dilute solutions of the standard and the unknown under identical conditions of cell geometry, sample temperature and incident exciting intensity. When a corrected fluorescence spectrometer is used and the optical densities of the two solutions are the same, then from equation (2.2) the ratio of the fluorescence efficiencies is simply proportional to the ratio of the integrated areas under the fluorescence spectra. Good agreement has been obtained between independent absolute measurements on a series of organic solution standards [2], [55] and their relative fluorescence efficiencies [34].

Other relative efficiency determinations have been undertaken by using a light-scattering solution as a standard [56], [57], and by using a quantum counter solution to collect emitted fluorescence.

POLARIZATION

Observations of fluorescence polarization spectra or measurements of the degree of depolarization of fluorescence are particularly important in the study of solids and of large molecules such as proteins.

Depolarization of fluorescence is a measure of the rotational Brownian motion of the molecule within a viscous medium, and is directly related to the fluorescence lifetime [58]. Studies of this kind can be used in measurements of intermolecular energy transfer, or, in dilute solutions of complex molecules, to observe energy transfer between different parts of the same molecule [59].

A plot of the principal polarization of fluorescence as a function of exciting light frequency will indicate the relative orientations associated with the absorption and emission of light. Such measurements are of great importance in the identification of different overlapping electronic transitions, for example, in measurements of crystal spectra [60]. A versatile automatic recording polarization spectrofluorimeter has been described [61].

Fluorometric Measurements

Measurements of the time decay of fluorescence in the 10^{-9}–10^{-6} s region can be divided conveniently into two categories: (*a*) phase and modulation fluorometry, in which the excitation is periodic and a knowledge of the decay function is needed before an analysis of the results can be made; and (*b*) pulse fluorometry, where the time decay of the fluorescence intensity is observed directly, after a short initial pulse of excitation.

PHASE AND MODULATION FLUOROMETRY

When a fluorescent sample is excited with a sinusoidally modulated light signal, then the resulting fluorescence will also be modulated at the same frequency. However, the emitted light suffers a phase change (θ) and change in the modulation factor (m) with respect to that of the exciting light. If the fluorescence decay is a simple exponential function, then

$$\tan\theta = \omega\tau \tag{2.3}$$

and

$$m = \frac{1}{\sqrt{(1+\omega^2\tau^2)}} = \cos\theta \tag{2.4}$$

Equation 2.3 is the basic equation of the phase fluorometer. The fluorescence lifetime, τ, is determined from the measured phase difference between the exciting and emitted light at a known angular frequency, ω of the exciting light.

1. Phase Fluorometers

A schematic diagram of a typical phase fluorometer is shown in Fig. 2.6. The modulated light source is provided either by modulation of the applied

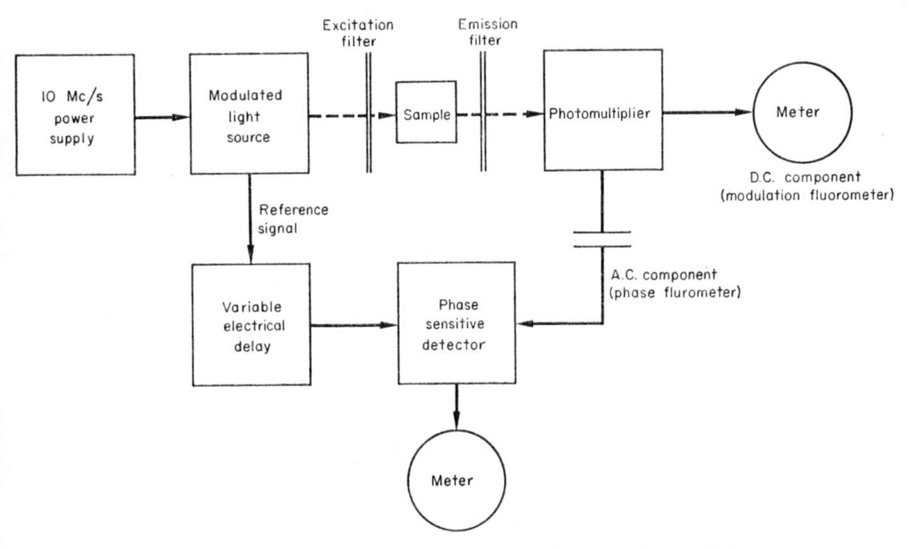

FIG. 2.6 Schematic outline of a fluorometer which can be used for measurements both of phase and of modulation changes produced by the sample [62].

potential across a discharge lamp or by passing a steady light beam through a device with variable optical density. In either case, since the accuracy of the result depends on exact knowledge of the modulation frequency, quartz crystal oscillators (resonant frequencies about 10^7 c/s) are generally used.

A.C. discharge lamps are carefully tuned to the output of the high-frequency power supply, and might dissipate about 30 W. The actual lamp construction can be very simple. For example, an effective source [62] can be made by sealing two tungsten wires about 1 mm apart in a Pyrex envelope (or molybdenum in silica) filled with hydrogen at 100 torr. The light output from the lamp follows the current profile closely and gives about 60% modulation.

For higher excitation intensities, the light from an intense unmodulated source (e.g. high-pressure mercury arc) is passed through an electro-optical modulator or an ultrasonic diffraction grating [63]. The various electro-optical modulators used depend on the Kerr effect [64] or the Pockels effect [65], [66] and may ultimately produce modulation factors approaching 100% [67]. The transmission properties of the modulator must, of course, be compatible with specimen excitation. The many different types of modulated light sources used in phase fluorometry have been reviewed [68].

The sample geometry must be arranged so as to collect a maximum amount of fluorescence emission with a minimum of scattered light (see 4, p. 22). Reliable measurements can be made only when two complementary filters (for excitation and emission wavelength selection) are available and are appropriate to the compound being studied. Light losses are normally too great to permit the use of monochromators in place of the filters. Front surface viewing is used frequently to minimize reabsorption effects.

The detector of the modulated fluorescence emission should be a high-gain, fast time response photomultiplier of the focused dynode type [69]. The output from the detector is fed to a phase sensitive device where it is compared in phase with a reference signal, whose phase is usually altered by means of an electrical or optical delay. Figure 2.6 illustrates a flouro-meter in which the reference signal, derived from the electrical power supply to the light modulator, is fed into the phase sensitive detector via an electrical delay line.

Considerable ingenuity has been used in the design of some of the many different types of phase sensitive devices used [68]. A simple, but sensitive, circuit has been described as part of a fluorometer to measure the lifetimes of small biological samples by viewing them through a microscope [70]. The sample and reference signals are fed to a centre tap transformer and when they are 90° out of phase a null output is obtained independently of the relative amplitudes of the signals and of any noise component in the sample signal [62]. The typical reproducibility for such measurements of phase angle using 10 Mc/s modulation would be to $\pm 0{\cdot}15°$ or approximately 10^{-10} s.

2. Modulation Fluorometers

The fluorescence lifetime (τ) can be obtained also from measurements of the modulation ratio, m (eqn 2.4). This is achieved, for example, by comparing the d.c. signals from the specimen and a reference scatterer, at identical amplitudes of their a.c. components compared on a narrow bandwidth detector at the modulation frequency. Such a technique has certain advantages over pure phase measurements and is of similar accuracy, but has been little used [71].

Since both phase and modulation fluorometry require essentially identical instrumentation, a logical, and indeed the most useful extension of these techniques has been the development of an instrument for the simultaneous measurement both of Θ and m. The two independent values so obtained may permit an exact description of the decay function, in particular cases when it is not a simple exponential [72]. This is an advantage if the instrument is to be used for energy transfer studies. So far, only one combined phase and modulation fluorometer has been described [62], although other phase fluorometers could be modified to combine both types of measurement [68].

3. *Experimental Limitations*

A disadvantage of phase and modulation fluorometry is that, in general, the decay function must be known to be a simple exponential if measurements can be made only at a single exciting frequency. For the analysis of particular non-exponential decays the form of the decay function must be assumed. If it is not correct, then the measured phase shift Θ and modulation ratio (m) could be very misleading.

Although many phase fluorometers have a claimed resolution of better than 10^{-10} s there are several potential sources of error for which corrections must be made, and which could make the final result much less accurate than this. The phase of the photomultiplier signal depends on the illuminated area of the photocathode and the stability of the applied potential. In addition, the emission lines in the source spectrum differ in phase, so that errors can arise in comparing the fluorescence of a specimen absorbing in one spectral region with that of a reference sample scattering heterochromatically. Finally, phase measurements of the fluorescence are particularly sensitive to the presence of scattered exciting light, which reduces the measured fluorescence lifetime.

The design and construction of a phase and modulation fluorometer demands a certain amount of technical ability, and unfortunately, no instruments of this type are commercially available at present. However, a phase fluorometer constructed from commercial sub-units and having a time resolution of less than 10^{-10} s has been described, and its accuracy and sources of error discussed [73].

PULSE FLUOROMETRY

The principles of pulsed measurement are simple. A pulsed (or chopped) light source is used intermittently to excite the specimen, and then the fluorescence emission produced is viewed with a high-gain, fast-response photomultiplier, and the signal is displayed and measured on an oscilloscope or by some other means (see Fig. 2.7).

Since the light pulse used for excitation (as viewed by the detector) is usually comparable in duration with the shortest lifetimes to be measured (about 5 ns or less), the fluorescence response actually observed, $F(t)$, at time t is given by

$$F(t) = \int_0^t I(t')f(t - t')\,dt' \tag{2.5}$$

$f(t)$ represents the true fluorescence response to be measured, which is modified by the instrumental response $I(t)$. $I(t)$ is determined both by the time response of the detection system (photomultiplier, cable, oscilloscope, etc.) and by the shape of the light pulse used for excitation, and is measured experimentally by observation of the light pulse under identical conditions to those used to observe $f(t)$.

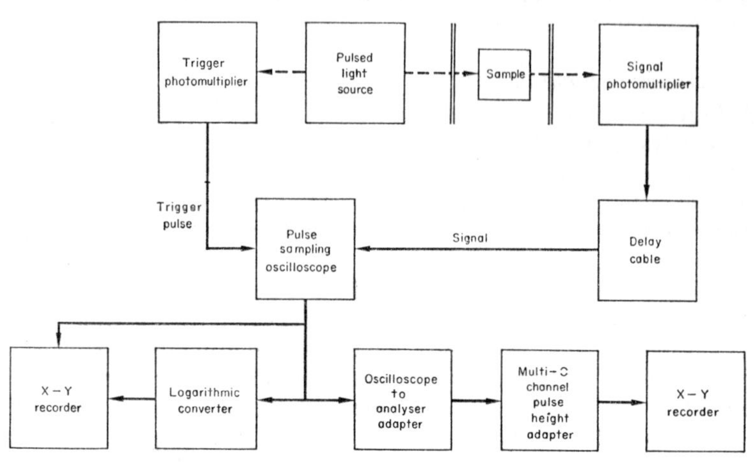

FIG. 2.7 Outline of a fluorometer using pulsed excitation, with two possible methods of recording the fluorescence decay curve.

However, the instrumental response is often smaller than the true fluorescence lifetime, usually for lifetimes greater than 10 ns, and so $I(t)$ may be neglected. When the fluorescence lifetime is short (of the order of 1 ns), then the true lifetime may be derived from the measured response $F(t)$ from equation (2.5) by various analytic or synthetic procedures [74].

1. Light Sources

Considerable effort has been devoted to developing intense light sources of short duration, a type of source useful not only for fluorescence excitation, but also for testing the time response of various photo detectors in the nanosecond region. Although various techniques have been used, such as light-chopping with an electro-optic shutter or rapidly rotating mirror, the simplest and most convenient source is a nanosecond discharge lamp [68].

In a lamp of this type, an effective capacitor, C (real or stray), in parallel with the lamp is slowly charged until the lamp breaks down. A current pulse of duration rC (where r is the conducting resistance of the lamp) will pass through the lamp and define the time profile of the emitted light if recombination and emission from the ionized molecules takes place sufficiently rapidly. For a 10^{-9} s pulse, since r is usually about 1 ohm, C must be less than 1000 pF. Consequently, for a breakdown potential of 10 kV, the maximum electrical energy available per pulse is approximately

0·05 J. The useful light energy per pulse will be very much less than this because of thermal dissipation and the fact that most of the emitted light will lie outside the region useful for excitation. A typical energy light output per nanosecond pulse would be much less than 1 erg (or less than 10^{11} visible photons). Attempts to increase the light output by increasing the breakdown potential will be restricted by the insulating properties of the condenser.

Lamp designs vary widely but can be of extreme simplicity; for example, breakdown in air between the centre and outer conductors of a short length of ordinary coaxial cable will produce a reasonably intense light pulse of a few nanoseconds duration. The problems inherent in pulsed light source construction have been discussed [75], [76] and several sources of this type are commercially available [77]–[80]. If the time restriction is relaxed slightly, then much higher intensities per pulse are available [81], [82].

For normal detection systems (not 'single photon' type apparatus) the low intensity of such pulsed sources excludes the possibility of specimen excitation through a monochromator. Two complementary filters are generally used for excitation and observation of the fluorescence.

2. Instrumentation

If the pulsed light source is carefully constructed then the instrument response will be determined primarily by the transit time spread (an approximately Gaussian function) of the photomultiplier used. Minimum transit time spread and maximum gain are provided by 14 or 15 stage focused dynode-type tubes, preferably with a quartz window and multi-alkali cathode (S20 response) [69]. The high gain of the detector permits direct display of the luminescence decay without the need for any intermediate amplification.

The photomultiplier output signal can be handled in several different ways [68]. The signal can be displayed directly on a fast (1000 Mc/s bandwidth) travelling wave oscilloscope. Although each sweep of the oscilloscope will record the complete fluorescence decay, some form of photographic integration is necessary to extract the lifetime information.

Higher sensitivity, better time resolution and improved signal to noise ratio can be obtained by using a fast pulse sampling oscilloscope [83]. Only a small portion of the total decay signal is sampled in each emission pulse, the complete decay curve being built up at a rate determined by the repetition rate of the light source. The information can be taken directly from the sampling oscilloscope, and electrically integrated and displayed on an X-Y recorder of appropriate sensitivity and time constant (Fig. 2.7). A logarithmic converter unit can be used to give a plot of the log of fluorescence intensity as a function of time. Absolute time measurements rely on the calibration and linearity of the oscilloscope time base and these are usually accurate to $\pm 3\%$. The calibration can be checked by using known lengths (delays) of coaxial cable to displace the oscilloscope signal.

The luminescence intensity in the time range 10^{-9}–10^{-6} s can usually be followed to 1%, or less, of its maximum value when equipment of this type is used. Any scattered exciting light can be distinguished from the fluorescence emission, since it appears as a sharp spike at the commencement of the decay. The apparatus is ideal for the observation of non-exponential decays, produced, for example, by crystal surface effects [84] or excimer emission [72]. A 'nanosecond spectral source system' is at present commercially available [79], but it is not a difficult procedure to assemble the components and set up and use a pulse fluorometer of this kind.

For very low intensity signals where the random photon noise fluctuations are large, alternative, but more expensive, systems are available for the signal display. Since the signal to noise ratio increases as the signal amplitude is added, the output from the sampling oscilloscope can be fed and added in digital form in the appropriate channels of a memory store, for example, in a multi-channel pulse height analyser operating in a multi-scale mode. In this way, the repetitive signals can be accumulated until the required statistical accuracy has been achieved. Such apparatus has been described and used to catalogue the fluorescence lifetimes of many aromatic molecules [85] including practically all the currently popular organic scintillators.

REFERENCES

[1] HERCULES, D. M. *Anal. Chem.* **38**, 29A (1966).
[2] CALVERT, J. G. and PITTS, J. N. *Photochemistry*, Wiley, New York (1966).
[3] KOLLER, L. R. *Ultra-violet Radiation*, Wiley, New York (1965).
[4] LEWIN, S. Z. *J. Chem. Educ.* **42**, A165 (1965).
[5] PAANANEN, R. *Appl. Phys. Lett.* **34**, 9 (1966).
[6] *Bulletin GET-1248J*, General Electric Co., Cleveland, Ohio.
[7] SCOULER, W. J. and MILLS, E. D. *Rev. sci. Instr.* **35**, 489 (1964).
[8] CHANCE-PILKINGTON LTD. St. Asaph, Flintshire. CORNING GLASS WORKS, Corning, New Jersey. JENA GLASWERK SCHOTT UND GEN., Mainz.
[9] ILFORD LTD., Ilford, Essex. KODAK LTD., London (Wratten filters).
[10] BOWEN, E. J. *Chemical Aspects of Light*, Oxford (1946).
[11] KASHA, M. *J. opt. Soc. Amer.* **38**, 929 (1948).
[12] PELLICORI, S. F., JOHNSON, C. A. and KING, F. T. *Appl. Optics*, **5**, 1916 (1966).
[13] BRAGA, C. L. and LUMB, M. D. *J. sci. Instr.* **43**, 341 (1966).
[14] GUNN, A. H. *Introduction to Fluorimetry*, Electronic Instruments Ltd., Richmond, Surrey (1963).
[15] JACQUINOT, P. *J. opt. Soc. Amer.* **44**, 761 (1954).
[16] JOHNS, H. E. and RAUTH, A. M. *Photochem. and Photobiol.* **4**, 673 (1965).
[17] JOHNS, H. E. and RAUTH, A. M. *Photochem. and Photobiol.* **4**, 693 (1965).
[18] *Technical Bulletin EB–128* (June 1965), Jarrel-Ash, Waltham, Mass.

[19] HAMMER, D. C., ARAKAWA, E. T. and BIRKHOFF, R. D. *Appl. Optics,* **3**, 79, 1964.
[20] CHRISTENSEN, R. L. and AMES, I. *J. opt. Soc. Amer.* **51**, 224 (1961).
[21] PALMER, C. H. *J. opt. Soc. Amer.* **42**, 269 (1952).
[22] BARTHOLOMEW, R. J. *Rev. pure and appl. Chem.* **8**, 265 (1958).
[23] PARKER, C. A. and REES, W. T. *Analyst,* **87**, 83 (1962).
[24] UDENFRIEND, S. *Fluorescence Assay in Biology and Medicine,* Academic Press, London (1962).
[25] HERCULES, D. M. *Fluorescence and Phosphorescence Analysis,* Interscience, New York (1966).
[26] THERMAL SYNDICATE LTD., Wallsend, Northumberland ('Spectrosil').
[27] BRADDICK, H. J. J. *Rep. prog. Phys.* **23**, 154 (1960).
[28] BIRKS, J. B. and MUNRO, I. H. *Br. J. appl. Phys.* **12**, 519 (1961).
[29] E.M.I. ELECTRONICS LTD., Hayes, Middlesex (Photomultiplier Tubes).
[30] BOWEN, E. J. and WOKES, F. *Fluorescence of Solutions,* Longmans, London (1953).
[31] WHITE, C. E. and WEISSLER, A. *Anal. Chem.* **34**, 814 (1962).
[32] WHITE, C. E. and WEISSLER, A. *Anal. Chem.* **36**, 116R (1964).
[33] LOTT, P. F. *J. chem. Educ.* **41**, A421 (1964).
[34] PARKER, C. A. and REES, W. T. *Analyst,* **85**, 587 (1960).
[35] WHITE, C. E., HO, M. and WEIMER, E. Q. *Anal. Chem.* **32**, 438 (1960).
[36] MELHUISH, W. H. *J. opt. Soc. Amer.* **52**, 1256 (1962).
[37] PARKER, C. A. *Nature,* **182**, 1002 (1958).
[38] BAKER, D. J. and BROWN, W. L. *Appl. Optics,* **5**, 1331 (1966).
[39] LIPPERT, E., NAGELE, W., SEIBOLD-BLANKENSTEIN, I., STAIGER, U. and VOSS, W. *Z. anal. Chem.* **170**, 1 (1959).
[40] MELHUISH, W. H. *J. phys. Chem.* **64**, 762 (1960).
[41] LOTT, P. F. *J. chem. Educ.* **41**, A327 (1964).
[42] KAYE, W. *Appl. Optics,* **2**, 1295 (1963).
[43] GEAKE, J. E. *J. sci. Instrum.* **43**, 60 (1966).
[44] SLAVIN, W., MOONEY, R. W. and PALUMBO, D. T. *J. opt. Soc. Amer.* **51**, 93 (1961); (Perkin-Elmer Ltd., Model 236 Spectrofluorimeter).
[45] TURNER, G. K. *Science,* **146**, 183 (1964); (G. K. Turner Associates, Model 210 'Spectro').
[46] AMERICAN INSTRUMENT CO. INC., Silver Spring, Maryland (Aminco Bowman Spectrophotofluorometer with corrected spectra attachment).
[47] HAUGEN, G. R. and MARCUS, R. J. *Appl. Optics,* **3**, 1049 (1964).
[48] ROSEN, P. and EDELMAN, G. M. *Rev. sci. Instr.* **36**, 809 (1965).
[49] HAMILTON, T. D. S. *J. sci. Instrum.* **43**, 49 (1966).
[50] PERKIN-ELMER LTD., Norwalk, Conn. ('Vernistat' A.C. or D.C. adjustable function generators).
[51] DRUSHEL, H. V., SOMMERS, A. L. and COX, R. C. *Anal. Chem.* **35**, 2166 (1963).
[52] BIRKS, J. B., MOORE, G. F. and MUNRO, I. H. *Spectrochim. Acta,* **22**, 323 (1966).

[53] PARKER, C. A. in *Advances in Photochemistry* (edited by W. A. Noyes, G. S. Hammond and J. N. Pitts), Vol. 2, p. 305, Interscience, New York (1964).
[54] LIPSETT, F. R. *Progress in Dielectrics*, Vol. 7 (edited by J. B. Birks), Heywood, London (1967).
[55] MELHUISH, W. H. *J. phys. Chem.* **65**, 229 (1961).
[56] WEBER, G. and TEALE, F. W. J. *Trans. Faraday Soc.* **53**, 646 (1957).
[57] HERCULES, D. M. and FRANKEL, H. *Science*, **131**, 1611 (1960).
[58] BIRKS, J. B. and DYSON, D. J. *Proc. R. Soc.* **A275**, 135 (1963).
[59] WEBER, G. *Biochem. J.* **75**, 345 (1960).
[60] *Optical Methods of Investigating Solid Bodies* (Edited by D. V. Skobel'tsyn), **25**, Consultants Bureau, New York (1965).
[61] AINSWORTH, S. and WINTER, E. *Appl. Optics*, **3**, 371 (1964).
[62] BIRKS, J. B. and DYSON, D. J. *J. sci. Instrum.* **38**, 282 (1961).
[63] METCALF, W. S., NATUSCH, D. F. S., PAGE, S. G., SHIPLEY, E. D. and WIGGINS, P. M. *J. sci. Instrum.* **42**, 603 (1965).
[64] STONE, J., LYNCH, G. and PONTINEN, R. *Appl. Optics*, **5**, 653 (1966).
[65] *Electro-Optic Light Modulators*, Baird-Atomic Inc., Cambridge, Mass. (Nov. 1965).
[66] KAMINOW, I. P. and TURNER, E. H. *Appl. Optics*, **5**, 1612 (1966).
[67] CHEN, F. S., GEUSIC, J. E., KURTZ, S. K., SKINNER, J. G. and WEMPLE, S. H. *J. appl. Phys.* **37**, 388 (1966).
[68] BIRKS, J. B. and MUNRO, I. H. *Progress in Reaction Kinetics*, Vol. 5 (edited by G. Porter), Pergamon Press, Oxford (1966).
[69] MULLARD LTD., London (Type 56 UVP). R.C.A. INC., Harrison, New Jersey (Type 6810 A). E.M.I. ELECTRONICS LTD. Hayes, Middlesex (Type 9594 B).
[70] VENETTA, B. D. *Rev. sci. Instrum.* **30**, 450 (1959).
[71] HAMILTON, T. D. S. *Proc. phys. Soc.* **B70**, 144 (1957).
[72] BIRKS, J. B., DYSON, D. J. and MUNRO, I. H. *Proc. R. Soc.* **A275**, 575 (1963).
[73] MULLER, A., LUMRY, R. and KOKUBUN, H. *Rev. sci. Instrum.* **36**, 1214 (1965).
[74] MUNRO, I. H. and RAMSAY, I. A. *J. sci. Instrum.* (*J. Phys. E.* Ser. 2, **1**, 147–8) (1968).
[75] YGUERABIDE, J. *Rev. sci. Instrum.* **36**, 1734 (1965).
[76] FRÜNGEL, F. B. A. *High Speed Pulse Technology*, Vol. 2, Academic Press, London (1965).
[77] ELECTRO-NUCLEAR LABS. INC., Mountain View, California (Nanosecond light pulse generator, Model 450).
[78] PEK LABS. INC., Palo Alto, California (Nanosecond light source 118).
[79] TRW INSTRUMENTS INC., El Segundo, California (Nanosecond spectral source system).
[80] 20TH CENTURY ELECTRONICS, Croydon, Surrey (Nanosecond light source, NLS–1).
[81] MACKEY, R. C., POLLACK, S. A. and WHITE, R. S. *Rev. sci. Instrum.* **36**, 1715 (1965).

[82] IMPULSPHYSIK, DR.-ING. F. FRÜNGEL, G.M.B.H., Hamburg (Fischer-Nanolite).
[83] HEWLETT PACKARD INC., Palo Alto, California (Model 185A or 140 with plug-ins).
TEKTRONIX INC., Beaverton, Oregon (Plus-in Model 1S1).
[84] BIRKS, J. B., KING, T. A. and MUNRO, I. H. *Proc. phys. Soc.* **80**, 355 (1962).
[85] BERLMAN, I. B. *Handbook of Fluorescence Spectra of Aromatic Molecules.* Academic Press, New York (1965).

Chapter 3

DISCHARGE LAMPS

R. P. WAYNE

THE radiation emitted from excited species is of considerable importance in technical and scientific applications which demand light sources of high efficiency or possessing a special spectral distribution. Such light sources almost always rely on electrical excitation of gaseous atoms, and the present chapter describes briefly some of these 'discharge lamps' which have been developed.

Many scientific investigations require a source of near-monochromatic radiation. For example, the physicist studying interference effects may wish to employ monochromatic light in order to observe single interference fringes, while the photochemist may wish to perform a photolytic reaction at a specific wavelength so that he may define closely the energetics of the processes. An incandescent radiant source—such as the tungsten filament lamp—suffers from the great disadvantage that it behaves more or less as a black body. For practical purposes, this means that, using conventional tungsten filament lamps, it is difficult to obtain significant intensities of radiation at wavelengths much shorter than the violet end of the visible spectrum. Again, filters or a monochromator may be employed to confine the radiation to a limited spectral region, but the narrower the pass-band, the lower the intensity of radiation attainable. In contrast, the radiation from gas-discharge lamps lies in a few fairly narrow bands whose wavelength depends mainly on the identity of the emitting gas, and, to a lesser extent, on the operating conditions of the lamp. Optical filters may readily be used with such a source to isolate one band of effectively monochromatic light, and the intensity of the band need not be much reduced by the filter. One further practical feature of the discharge lamp, and one which may benefit the user, is the particular geometric nature of the discharge. Lamps with highly localized, or with considerably extended, radiating regions may be required for different applications. An incandescent lamp is always restricted to the shape of its filament, which thus imposes certain limitations not found with discharge lamps; on the other hand the *constancy* of output from a discharge lamp is much lower than that from a tungsten filament lamp run under similar conditions.

Before extending the discussion of discharge lamps to specific examples, it is of interest to see how the efficiency of the tungsten filament bulb may be increased. Unless an incandescent lamp is operated at exceptionally high temperatures, most of the energy radiated lies in the infra-red region. The efficiency of an ordinary tungsten filament lamp is of the order of 10–15 lumens per watt (a lumen is a measure of 'visible' intensity which depends in part on the relative spectral sensitivity of the human eye). A much larger current than is normally permissible may be passed through a filament if a drastic reduction in lamp life can be tolerated. For example, a typical over-run Photoflood bulb might produce 33 lumens per watt for a five-hour lifetime. A recent development has gone some way towards improving the efficiency of the tungsten lamp, and has incidentally extended the useful short-wave limit of application. By enclosing a filament in a quartz envelope containing inert gas and a little iodine vapour, it is possible to operate tungsten lamps at 20–25 lumens per watt and maintain a life of 1000 hours (the accepted normal value for filament lamps); over-run 'quartz-iodine' lamps can produce up to 45 lumens per watt for a five-hour life. The role of the iodine appears to lie in maintaining a high pressure and in forming volatile tungsten iodides. A cycle of processes occurs in which the volatile iodides decompose on the hot tungsten filament but not elsewhere. Thus the lamp vessel is not blackened, and the lamp maintains its initial output throughout its life. For experimental purposes, these lamps may be a useful source of *continuous* radiation in the near ultra-violet. However, in order to reach higher efficiencies, discharge lamps in their different forms must be used.

Discharge lamps are often classified according to the pressure at which they operate: low, medium or high. A typical low-pressure discharge lamp is the sodium street lamp. The lamp contains a filling of about 1 cmHg of an inert-gas mixture (largely neon) and some sodium metal. Connection of the lamp to the electricity supply sets up a discharge in the inert gas, and the lamp becomes warm, thus vaporizing a little sodium. When this happens, the neon hands on its excess energy to the sodium, a process represented in eqn 3.1:

$$Ne^* + Na \rightarrow Ne + Na^* \tag{3.1}$$

Excitation occurs largely to the $^2P_{1/2,\ 3/2}$ states of sodium, and emission of the well-known sodium–D lines (at 589·6, 589·0 nm) arises in transitions to the ground $^2S_{1/2}$ state. Radiation involving transitions between the ground and (first) excited states is known as 'resonance radiation'.

A large fraction of the electrical energy put into the sodium lamp is radiated as the sodium–D lines, and since these lie within the visible region the visual efficiency of the sodium lamp is high (e.g., around 100 lumen/watt). This is not to say that the yellow light is satisfactory as illumination in many circumstances; it would be most irritating in the house, although the poor colour rendering makes for high contrast which is useful in street lighting. In many research applications the monochromatic nature of the light makes the sodium discharge an ideal source. For such purposes it may be desirable to use a 'hot-cathode' rather than a 'cold-cathode' discharge tube. In this type of tube an emissive filament is fitted to each end of the discharge tube, the two filaments are wired in series and connected to the supply. After the filaments have warmed up sufficiently, the link connecting them is broken (manually or automatically), thus placing the full supply voltage across the tube. A discharge should then be struck which maintains the temperature of the filaments.

A number of other elements may be substituted for sodium in hot- or cold-cathode discharge tubes, and lamps containing sodium, potassium, rubidium, caesium, calcium, zinc, mercury and cadmium are available commercially in low-power ratings (15 to 50 watts), as well as lamps containing the inert gases.

The low-pressure mercury 'resonance' lamp is of particular interest to the research worker. Properly speaking, the mercury resonance line is the permitted line at 184·9 nm resulting from the $^1S_0 \leftarrow {}^1P_1$ transition. However, the *absorption* of this line by mercury vapour is so strong that even at room temperatures the vapour pressure of mercury is sufficient to cause complete absorption within a millimetre. Lamps which are to produce the 184·9 nm line must be specially cooled and fabricated from selected silica whose transmission is satisfactory at 184·9 nm. It has become common practice, therefore, to refer to the line originating from the $^1S_0 \leftarrow {}^3P_1$ transition at 253·7 nm (which would be wave-mechanically forbidden if the term symbols written were adequate descriptions of the states) as the resonance line. Even this line is strongly reabsorbed by mercury vapour, as may be demonstrated dramatically in the 'mercury smoke' experiment. Radiation from a mercury resonance lamp is allowed to fall on a silicate or tungstate fluorescent screen, and a dish of mercury placed in the path of the radiation. Clouds of 'smoke' rising from the mercury can be seen as a shadow on the fluorescent screen.

About 95% of the total radiated energy from the low-pressure mercury lamp is concentrated in the one line at 253·7 nm, and the mercury resonance lamp provides an extremely useful source of monochromatic, photochemically-active radiation for the chemist. From the point of view of the illumination technologist, the ultra-violet radiation is not immediately of use. However, by coating the inside of a low-pressure mercury discharge lamp with a fluorescent powder, it is possible to convert much of the 253·7 nm line to visible radiation. In current practice, the phosphor used is a calcium halophosphate, with heavy metal activators whose nature determines the colour of emission (e.g., manganese gives a red colour); by judicious choice of phosphor the total light emitted from the tube may be made to approximate closely in colour-balance to daylight. Although both hot- and cold-cathode tubes may be used in this way, by far the most familiar fluorescent tube uses the hot-cathode technique with an automatic thermal-delay switching circuit.

The pressure ultimately developed in a discharge lamp depends both upon the structure of the lamp and upon the electrical conditions under which it is operated. In the example of a mercury discharge lamp, if the current is free to reach relatively high values, then the mercury initially vaporized will maintain a discharge in which power is dissipated, and the resultant heating will lead to further vaporization and concomitant increase in current passed. The pressure will therefore build up, and the current increase, until a limitation is imposed either by the electrical circuit, or by the amount of mercury available for vaporization. Commercial medium- and high-pressure mercury lamps operate at pressures from a few cmHg up to several hundred atmospheres.

Several changes are apparent when the pressure in a discharge lamp is increased. First, the nature of the discharge is itself altered. At low pressures the discharge is spread diffusely throughout the lamp, while at high pressures the discharge is confined, by and large, to the column of gas between the electrodes, and is, properly, an arc. Suitable design can thus provide high-power lamps whose radiating region is small. Many applications require a near-point source of high intrinsic brilliance.

Secondly, the spectral distribution of the radiation is affected by an increase in discharge pressure. All spectral lines are broadened at high pressures, and this is particularly apparent with the 253·7 nm resonance line. Further, the relatively cool mercury vapour near the envelope of the lamp absorbs the 253·7 nm sharply, and this leads to the phenomenon known as 'reversal'—the resonance line appears in absorption rather than emission. Not only are the lines broadened, but the relative intensity of each is changed. No longer does 95% of the radiation lie in the broadened and reversed resonance line. Instead, the energy is distributed more evenly over the various lines, the longer wavelength lines being to some extent favoured. The effect has its origin in the decreased rigidity at high pressures of the various selection rules for optical transitions. The strongest and most useful bands which may be isolated from the medium- or high-pressure lamp lie at 303, 313, 334, 366, 405, 436 and 546 nm. The preponderance of high intensity lines in the visible spectrum makes such lamps useful for illumination purposes; efficiencies of 45–50 lumens per watt are readily obtained. The light is, however, deficient in red, and several methods are employed in attempts to improve the colour rendering. The quartz arc tube may be enclosed in a second, glass, envelope coated with a phosphor emitting with a red bias to give a balanced white light. A further improvement—although with some reduction in efficiency—can be achieved by using a tungsten filament lamp as ballast in series with the discharge lamp, and enclosed together with it in the phosphor-coated outer envelope. The General Electric Company has recently developed a mercury vapour lamp which yields near-white radiation in a rather different way. Sodium iodide is present, together with mercury, in the lamp filling, and in operation the mercury pressure is about 400 cmHg, while the sodium iodide vapour pressure is about 2 cmHg. In the discharge region, sodium iodide is dissociated and yields thermally excited sodium. Near the cool envelope wall the sodium and iodine atoms recombine, thus avoiding attack of the silica by hot sodium vapour. The lamp radiates the mercury and sodium emission components, and gives near-white light at an efficiency of 90 lumens per watt.

Low-pressure mercury lamps for laboratory production of 254 nm radiation have a low intensity of emission per unit area, and the light cannot therefore be efficiently collected by lenses. They are best used either as a tube surrounded by the material to be irradiated, or coiled into a

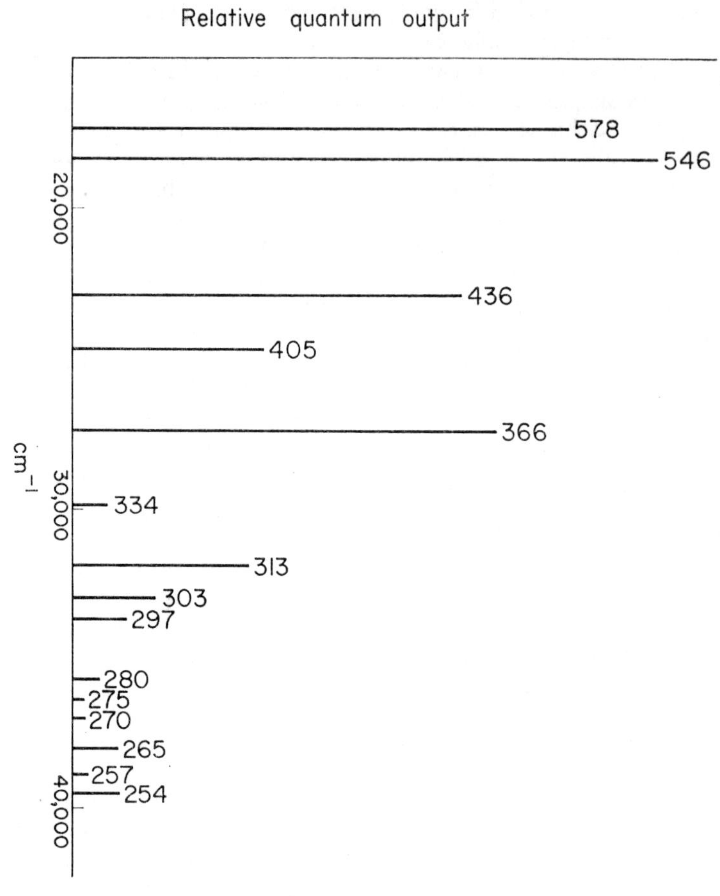

FIG. 3.1 Relative quantum intensities for the line emission of a representative medium-pressure mercury arc lamp, plotted against wave-number. Lines marked with wavelengths in nm.

helix or flat spiral form. The total output in all directions from an average lamp is of the order of 2×10^{-7} $\boldsymbol{Nh\nu}$ s/cm of arc length at 254 nm.

Figure 3.1 shows the relative quantum outputs for the lines emitted by a laboratory type medium-pressure mercury lamp operating without a glass envelope. The total ultra-violet output in units of $\boldsymbol{Nh\nu}$ at wavelengths shorter than 380 nm is of the order of 5×10^{-5} s/cm of arc length. With an ordinary photochemical set-up, using a lens and light filter, about 10^{-8}

Nhv cm²/s at the mercury line 366 nm may be expected from a 125-watt medium-pressure lamp.

About ten times the intensity can be obtained from a high-pressure (compact source) mercury arc, although the gain in the ultra-violet is less than for the visible, and the lines are broadened with an accompanying background. High-intensity capillary forms are useful where light is to be focused on to an instrument slit.

In principle, operation of discharge lamps at high pressure is possible for fillings other than mercury, although a limitation exists in that the filling must be sufficiently volatile for the required pressure to be developed at temperatures below the devitrification point of the envelope. Sodium (or any other alkali metal) fulfils this condition, but unfortunately attacks quartz at the high temperatures developed. One ingenious method of preventing attack has been described for the mercury-sodium lamp. Another approach has been made possible by the development of a translucent form of alumina. This is not attacked by hot sodium vapour, and an experimental lamp has been made in which a sodium vapour pressure of 20 cmHg is produced in operation at 400 watts. The broadened sodium spectrum gives near-white light at an efficiency of about 100 lumens per watt.

High-pressure inert gas arcs (in particular xenon arcs) have proved useful as sources of continuous radiation extending from the infra-red to wavelengths shorter than 2000 Å. The arc can be confined to a small column even at high powers, and xenon arc lamps have found applications as source lamps in high-dispersion absorption spectroscopy, in photochemical experiments (and, indeed, in film projection, where the intense white visible component of the emission is most valuable). So-called 'hydrogen' or 'deuterium' lamps are useful where a fairly level spectrally continuous emission between 400 and 200 nm is required, as for absorption spectroscopy. The continuum is due to electronic transitions from excited H_2 or D_2 molecules to the lowest triplet state which, being non-bonding, has no quantized vibrational levels.

'Flash' lamps, where by special circuitry large amounts of stored electrical energy can be suddenly discharged through a gas at low pressure, are used for work requiring high intensities of short duration. Depending on the gas-filling (krypton for most usual purposes) flash durations of 10^{-3} to 10^{-4} s are obtainable with a fairly level output from 440–260 nm of 10^{-3} *Nhv* per flash. Lamps can be designed for a flash duration as short as 10^{-9} s. Although their output is only about 10^{-10} *Nhv* per flash they have been used, with sensitive photomultiplier detectors, to measure fluorescence lifetimes in the nanosecond range. (See Chapter 2.)

No discussion of discharge lamps would be complete without mention of the laser, a light source which is finding an increasing number of applications. The lamps so far described depend for their emission upon *spontaneous* transitions from electronically excited species. The laser, on the other hand, makes use of *stimulated* emission to produce radiation. Since stimulated emission is the true counterpart of absorption, net emission is only observed in systems where the population of the excited state is greater than that of the ground state, a situation described as population inversion. In practice, a laser consists of the emitting material, in the form of a cylinder, enclosed between a pair of reflecting surfaces. The reflectors are arranged so that there is additive interference between waves travelling to and fro. As a result, all emission in the system is in phase, and the presence of radiation (derived from spontaneous emission) will trigger off stimulated emission in phase with that of the originating radiation. This radiation builds up in intensity as the result of the end reflections which maintain the phase relationship, and continues until the population inversion is destroyed. Radiation emerges from the laser through one of the end mirrors which is made partially transmitting.

The most striking feature of lasers is the 'coherence' of the radiation: all the wave-trains emitted are in phase. Ordinary sources of light produce non-coherent radiation, and the existence of means for the production of coherent radiation opens up a number of interesting possibilities for the physicist. However, the photochemist is not likely to be primarily concerned with the property of coherence; the extreme monochromaticity and high intrinsic brilliance of laser sources are the features likely to prove useful to him. The brilliance of a laser beam may exceed by orders of magnitude that of the sun. On the other hand, the total energy obtainable from a laser cannot exceed that of the exciting source, and, indeed, the conversion efficiency is frequently as low as 0·01%, although some newly-developed types may approach 10%. The fundamental frequency depends on the material from which the laser is constructed; a small amount of tuning can be effected in suitable cases by alteration of the temperature or by application of a magnetic field. In all instances the band emitted is extremely narrow, and for special gas lasers it can be a million times sharper than the best older line sources. The accurately parallel nature of a laser beam permits it to be focused on to a very small spot, with the production there of very high temperatures and radiation densities.

Power input to achieve population inversion is applied in different ways. Gas lasers, of which the helium-neon type is the commonest, are excited by electric discharge from external or internal electrodes. With continuous excitation, powers of 1–25 mW are obtainable at a wavelength of 633 nm. Special stabilization techniques enable the frequency to be kept to within a few parts in 10^9. Image formation by holography is an interesting use for this laser. The argon laser is capable of producing a continuous output of 50 W at the wavelengths 514·5 and 488 nm., but is expensive to construct and difficult to operate. Carbon dioxide lasers have the highest efficiencies and outputs, reaching 500 W at the infra-red wavelength of 10,600 nm, and are capable of dramatic effects such as cutting through steel sheets, or welding metals. The latest developments centre round methods of achieving momentary population inversion by very rapid excitation pulses. 10–500 kW emissions of 10^{-8} s duration have been so obtained from lasers containing neon (at 540 nm) and nitrogen (at 337 nm).

Solid state lasers are excited by a nearby flash tube. The ruby laser is described in Chapter 5. In a typical commercial ruby laser the lamp flash input is 2500 joules, and the laser output is 25 joules delivered in 5×10^{-4} s; the power delivered is thus 50 kW. Much higher powers may be achieved by the technique of Q-switching. If by some means (in effect by decreasing the amount of light reflected at each traverse) the Q of the laser optical system is reduced below that required for laser operation, the population inversion may be allowed to build up under the influence of one or more flashes. If now the Q is restored to its original value, a laser flash of extremely short duration is obtained. For a very modest ruby laser using single-flash excitation, a stored energy of 1 joule may be released in 10^{-8} s, which corresponds to a power of 100 mW. Larger lasers have been described which are capable of delivering powers reaching billions of watts.

The lasers commercially available may have a repetition rate in the range 1–10 flashes per second. Faster repetition is possible at reduced power, and continuously operating solid state lasers have been designed in which a high power mercury discharge lamp is used to maintain a population inversion achieved initially by means of a flash tube.

It has not been the intention in this chapter to provide a catalogue of all the different discharge lamps that have been made. Instead, an attempt has been made to describe the different ways in which an electrical discharge in a gas can be used to produce radiation. The reader who intends to embark upon specific experimentation is advised to consult the manufacturers' data sheets (for example, Philips, A.E.I., G.E.C., Hanovia) as well as Calvert and Pitts, *Photochemistry*, p. 687, Wiley, New York (1966).

Chapter 4

LUMINESCENCE IN THE GAS PHASE

R. P. WAYNE

PHOTOEMISSIVE processes in gaseous systems differ from those in condensed phases mainly as a result of the relatively smaller rate of collision, which enables radiation to compete more successfully with deactivation processes. It is possible, by making observations of the emission from species of short radiative lifetime at low pressures, to obtain information about the excited species as it is first formed in the exciting process. Indeed, in the case of a chemiluminescent process (i.e., where the excess energy of a chemical reaction leads to excitation), it may be possible to infer the nature of the transition state in the reaction. Such investigations are, of course, of importance in testing theories of reaction kinetics. Additional kinetic interest attaches to certain gas phase chemiluminescent processes since they may be used to measure concentrations of various atomic species. Studies of fluorescence and chemiluminescence are directly complementary to each other, since an investigation of emission from, and quenching of, species excited by absorption of light may frequently yield valuable data about the species excited by chemical reaction.

Emphasis has been placed on the number of collisions suffered by an emitting species in gaseous and condensed systems. A consequence of the difference is that the spectroscopic nature of the emission may be distinct in the two phases. A brief discussion follows, therefore, of the kind of emission spectra observed in the gas phase.

Spectroscopic Nature of Gas-phase Luminescence

The most apparent difference between emission spectra in gaseous and condensed phases arises from the almost complete relaxation of vibrationally excited levels in upper electronic states before emission in condensed systems. Thus if any vibrational structure is observed at all in condensed systems it consists of transitions from the lowest vibrational level of the excited state. Absorption necessarily occurs largely from the lowest vibrational level of the ground electronic level at normal temperatures, and since the vibrational spacing and interatomic distances are frequently similar for the upper and lower states of the relatively large molecules studied in condensed phases, the emission and absorption spectra appear to be mirror images (see Chapter 6). While such effects occur also in many gaseous systems, yet the relationship between absorption and emission may frequently be of a more involved nature. The equilibrium interatomic distances (r_e', r_e'') for the upper and lower electronic states involved may be significantly different from each other. Such a situation is represented for the ground ($^3\Sigma_g^-$) and upper state of the Schumann–Runge system ($^3\Sigma_u^-$) of oxygen in Fig. 4.1. The operation

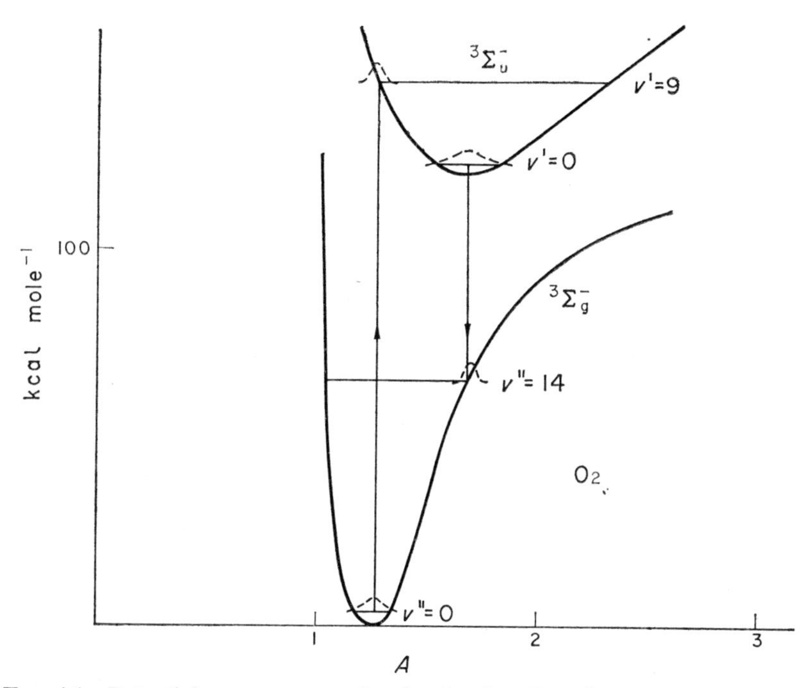

FIG. 4.1 Potential energy curves for O_2 showing Franck–Condon transitions for absorption and emission. Dotted curves are wave-mechanical representations of internuclear vibrations.

of the Franck–Condon principle ('vertical transitions') suggests that the strongest vibrational bands in absorption will be from $v''=0$ to v' levels from 7 to 11 (i.e., at about 185 nm). On the other hand, emission, if it occurs from the $v'=0$ level, will be strongest to v'' levels around 14 (i.e., at about 340 nm). Absorption and emission maxima therefore lie at quite different wavelengths in this kind of molecule. Weaker emission bands from levels of $v'>0$ are observed in the gas phase, although such behaviour is virtually unknown for liquid-phase emission.

In condensed phases, the interaction of solvent with the upper and lower electronic levels may lead to a shift of a given vibrational level observed in emission and absorption. (For a fuller discussion, see Chapter 6.)

Any shift resulting from intermolecular interactions is always small in gaseous systems, and at low pressures it is, to all intents and purposes, non-existent.

Discontinuities in a gas-phase chemiluminescence or fluorescence spectrum may suggest the occurrence of *predissociation*, i.e., a non-radiative transition from one electronic energy level to another of appropriate properties at a point where their total energies, electronic + vibrational, happen to coincide. It has long been realized that the breaking off of a sequence of vibrational levels in emission is a more reliable test for predissociation than is the corresponding blurring of fine structure in absorption. The behaviour of nitrogen dioxide nicely illustrates the relationship between absorption, emission and predissociation. Absorption excites the molecule, and fluorescence, quenching and chemical dissociation may occur, although the relative importance of the processes depends on the energy of the excited molecule.

$$NO_2 + h\nu \rightarrow NO_2^* \quad \text{absorption} \tag{4.1}$$
$$NO_2^* + M \rightarrow NO_2 + M \quad \text{quenching} \tag{4.2}$$
$$NO_2^* \rightarrow NO_2 + h\nu \quad \text{fluorescence} \tag{4.3}$$
$$NO_2^* \rightarrow NO_2' \rightarrow NO_2 + O \quad \text{(pre)dissociation} \tag{4.4}$$

The chemiluminescent process

$$O + NO + M \rightarrow NO_2^* + M \tag{4.5}$$
$$NO_2^* \rightarrow NO_2 + h\nu \tag{4.6}$$

is also thought to give rise to the same electronically excited state as leads to fluorescence. The absorption spectrum of NO_2 shows a region of diffuseness at wavelengths somewhat shorter than 400 nm, suggesting the onset of predissociation. The chemiluminescent emission from processes (4.5) and (4.6) shows a short-wavelength limit coinciding exactly with the onset of diffuseness in absorption. At the same time, the quantum yield for *fluorescence* drops sharply as the wavelength of the exciting light is made shorter than 400 nm. Confirmation that predissociation occurs at this wavelength is obtained from the primary quantum yield of the dissociative process (4.4). At 436 nm the quantum yield is effectively zero, while at 366 nm it is unity; at 405 nm—at which wavelength fluorescence is weak—the quantum yield of (4.4) is 0·4.

The occurrence of chemiluminescent and fluorescent emission from electronically excited nitrogen dioxide is of considerable interest, and will be discussed in greater detail in subsequent sections of this chapter.

Finally, in this account of the nature of luminescence in gaseous systems, the occurrence of emission from states in which mode of excitation is not electronic should be noted. At low pressures, collisional deactivation may be sufficiently reduced to allow of emission from states of long radiative lifetime. Radiation from vibrationally excited species may therefore be observed in gaseous systems, and emission from vibrationally 'hot' molecules does in fact occur under a wide variety of conditions. The excitation obtained *thermally*, in, say, a shock tube, lies outside the province of this chapter. On the other hand, chemical activation in processes such as

$$H + Cl_2 \rightarrow HCl^\dagger + Cl \tag{4.7}$$

(where $HCl^\dagger$ represents a vibrationally excited HCl molecule in its ground electronic state) followed by emission

$$HCl^\dagger \rightarrow HCl + h\nu \tag{4.8}$$

clearly fits into the ordinary understanding of the term 'chemiluminescence'. As will be seen later, reactions of this kind have proved particularly susceptible to interpretation in terms of the transition state of the reaction.

Fluorescence Quenching

A knowledge of the relative importance of radiative and quenching processes for a given electronic state is often of prime importance in the elucidation of chemiluminescent reaction mechanisms. Fluorescence quenching studies give this data *if it can be assumed that the excited states in the fluorescent and chemiluminescent processes are identical.* Alternatively, a comparison of radiative and quenching rate constants measured in the two processes may help to establish whether or not the excited states are the same.

The simplest scheme that can be written for a fluorescence process with quenching is

$$A + h\nu \xrightarrow{k_i} A^* \text{ absorption} \tag{4.9}$$

$$A^* + M \xrightarrow{k_q} A + M \text{ quenching} \tag{4.10}$$

$$A^* \xrightarrow{k_r} A + h\nu \text{ fluorescence} \tag{4.11}$$

where A, A^* are the unexcited and excited states of the fluorescing species, and M is the quenching gas (which may, of course, be A).

If the scheme represents the only fates which may befall A^*, then we may write $k_i[A] = I_{abs}$, the absorbed light intensity. A stationary state may be assumed for A^*, so that the emitted light intensity, I_{fluor}, is given by

$$I_{fluor} = k_r[A^*] = \frac{k_r}{k_r + k_q[M]} \cdot I_{abs}$$

Inversion of this expression gives the 'Stern–Volmer' relation:

$$\frac{1}{I_{fluor}} = \frac{1}{I_{abs}} \left(1 + \frac{k_q[M]}{k_r}\right)$$

Hence a plot of the reciprocal of the emitted intensity against the gas pressure should be linear, and the ratio of slope to intercept will be k_q/k_r, no matter what units are used to measure the intensity of fluorescence.

That ratio k_q/k_r is often the information needed in a problem. There are, however, instances where the absolute value of k_q is required, and to find this it is necessary to obtain a value for the radiative lifetime.

Several experimental techniques exist for the determination of radiative lifetimes. In the first place, the lifetime may be calculated from the integrated extinction coefficient for the same transition measured in absorption, account being taken of the ν^3 relation between *spontaneous* emission and (stimulated) absorption lifetimes at a frequency ν. Secondly, it may be possible to follow the first-order decay of fluorescence intensity with time (under conditions where quenching is negligible or can be allowed for) either by using a pseudo-steady state arrangement involving chopped light beams, or by observing the intensity as a function of time after excitation by a short-duration flash tube. The pseudo-steady state technique is useful for radiative lifetimes down to 10^{-6} s, while the flash technique when used in conjunction with photomultiplier-oscilloscope detection systems is used down to about 10^{-9} s. A third method employed to measure radiative lifetimes in fluorescent systems uses a beam of light modulated at a frequency, n, in the range 5–12 Mc s^{-1}. Because of the finite delay between excitation and fluorescence, the modulation of the fluorescent radiation differs in phase from that of the exciting radiation by an angle φ, and it may be shown that the radiative lifetime, τ, is given by

$$\frac{1}{k_r} = \tau = \frac{1}{2\pi n} \tan \varphi$$

Values of τ down to about 4×10^{-10} s can be measured by this method.

Some discrepancy between the radiative lifetimes measured directly from decay experiments and those calculated from absorption data may arise, in the case of small molecules, as a result of *reversible* electronic-vibrational energy exchange. For example, in the case of nitrogen dioxide, reversible 'internal conversion' between electronic energy of the excited state and vibrational energy of the ground state may occur and prolong the apparent radiative lifetime. With larger molecules the exchange is no longer reversible, only electronic to vibrational exchange being important, and the radiation delay does not occur.

Intermolecular energy exchange is perhaps a more familiar phenomenon than intramolecular exchange, and the next section deals briefly with the effects of energy exchange on radiative phenomena.

Energy Exchange

Franck first suggested that electronic energy could be transferred from an excited atom of one element to a ground state atom of some other element, and that the efficiency of the process would be greatest when there was near-resonance between the energy levels of the 'donor' and 'acceptor' atoms. The observation of *sensitized fluorescence* was first made in a system containing a mixture of mercury and thallium vapour, irradiated with the 253·7 nm line of mercury (which is not absorbed by thallium). Emission was observed from excited states of thallium. The 253·7 nm line excites mercury from the $^1S_0$ ground state to the $^3P_1$ level. Thus the steps in the emission from thallium appear to be

$$Hg(^1S_0) + h\nu\ 253{\cdot}7 \rightarrow Hg^*(^3P_1) \quad (4.12)$$
$$Hg^*(^3P_1) + Tl \rightarrow Hg(^1S_0) + Tl^* \quad (4.13)$$
$$Tl^* \rightarrow Tl + h\nu \quad (4.14)$$

Later experiments indicated that a resonant exchange was more efficient than a non-resonant process. The fluorescence of sodium sensitized by mercury shows a strong doublet at 442·3 nm and 442 nm (corresponding to the $Na9(^2S) \rightarrow 3(^2P)$ transition) and a weak doublet at 568·8 nm and 568·3 nm (which corresponds to the $Na4(^2D) \rightarrow 3(^2P)$ transition). In normal emission processes the longer wavelength doublet is the stronger, and the anomalous behaviour of the sensitized fluorescence is ascribed to almost exact resonance of the $^3P_1$ Hg and 9^2S Na levels (4·86 eV and 4·88 eV) leading to the highly efficient energy exchange. If further evidence were required to show that resonance favours energy exchange, it might be adduced from the effect of nitrogen on the mercury-sodium system. Nitrogen can quench the $Hg(^3P_1)$ level to the metastable $Hg(^3P_0)$ state, which lies at an energy of 4·64 eV. The Na $7(^2S)$ state lies only 0·07 eV away, at 4·71 eV, and in the presence of nitrogen the intensity of the Na $7(^2S) \rightarrow 3(^2P)$ transition is considerably enhanced.

Intermolecular electronic excitation by exchange from other modes of excitation is also well established. Sodium vapour introduced in small quantities into a bunsen flame behaves as a nearly-total emitter and absorber for its resonance radiation, and the intensity of that radiation which escapes coincides rather well with the flame temperature. However, the relative importance of excitation from chemiluminescent recombination involving atomic sodium, and true exchange processes involving translational or vibrational energy, is not clear. A more definite case of vibrational energy leading to electronic excitation is reported for shock waves in nitrogen to which a little sodium salt has been added. The temperature measured from the intensity of sodium resonance emission is virtually identical to the vibrational temperature calculated on the basis of the shock velocity. That is, the process

$$N_2^\dagger + Na \rightarrow N_2 + Na^* \quad (4.15)$$

occurs without loss of energy ($N_2^\dagger$ is vibrationally excited nitrogen; Na^* is electronically excited sodium). The inverse process—vibrational excitation by conversion of electronic energy—has also been investigated by a study of luminescent phenomena. For example, the irradiation, at $\lambda =$ 253·7 nm, of carbon monoxide containing mercury vapour, leads to the appearance in emission of the first overtone bands of the carbon monoxide vibrational spectrum at wavelengths around 2·5μ. The process may be represented by:

$$Hg(^1S_0) + h\nu\ 253{\cdot}7 \rightarrow Hg^*(^3P_1) \quad (4.16)$$
$$Hg^*(^3P_1) + CO(v''=0) \rightarrow Hg(^1S_0) + CO^\dagger\ (v''=0) \quad (4.17)$$

The relative intensities of the transitions from the different vibrational levels suggest that the electronic energy of the mercury is used to produce high levels of vibrational excitation in the carbon monoxide.

Chemiluminescence

The energy released in an exothermic chemical reaction does not necessarily appear initially in the equilibrium distribution about the various possible modes of excitation of the products. However, in many cases the rate of equilibration is so much faster than the rate of radiation that to a first approximation the reactions may be said not to emit light. Two factors favour chemiluminescence from the reactions of small molecules in the gas phase. First, the number of degrees of freedom in a small molecule is strictly limited, so that it may be difficult for energy which has accumulated in one mode of excitation to reach its equilibrium distribution. Secondly, a relatively small rate of intermolecular encounter in the gas phase prevents rapid quenching of excited species.

One of the first gaseous chemiluminescent phenomena to be observed was the 'air afterglow' seen by Lord Rayleigh in the electric discharge products of air. At the time, Rayleigh thought that the chemiluminescence arose from the reaction of nitric oxide and ozone formed in the discharge. Although this reaction is indeed chemiluminescent (and will be discussed later), it is now known that the air afterglow involves the recombination of atomic oxygen and nitric oxide:

$$O + NO + M \xrightarrow{k_c} NO_2^* + M \quad (4.5)$$

$$NO_2^* \xrightarrow{k_r} NO_2 + h\nu \quad (4.6)$$

The emission is yellow-green in colour, and shows bands which suggest that it arises from the same transition of NO_2 as that responsible for the visible absorption and the fluorescence of NO_2 (cf. the discussion of fluorescence quenching).

Some justification for the inclusion of the 'third body', M, in reaction (4.5) must be given. The air afterglow has frequently been studied over the last decade or so in discharge-flow systems. An electric discharge in molecular oxygen can be made to produce relatively high concentrations of atomic oxygen, and in a flow apparatus reactant gases can be added to the stream of atomic and molecular oxygen after the discharge. It is found that the intensity of chemiluminescence, I_{chem}, can be related to the concentrations of atomic oxygen and nitric oxide by the equation

$$I_{chem} = I_0[O]\,[NO]$$

I_0 is a constant which is independent of the *concentration* of M, although it is dependent on the *nature* of M. If M is able to quench excited NO_2 and also to assist in the recombination (4.5), these facts may be interpreted:

$$NO_2^* + M \xrightarrow{k_q} NO_2 + M \quad (4.18)$$

Solving the steady state equations for NO_2^* gives

$$I_{chem}=\frac{k_c k_r[O]\,[NO]\,[M]}{k_r+k_q[M]}$$

At the pressures normally encountered in discharge flow systems ($\sim$10–10^{-1} mmHg), $k_q[M]\gg k_r$, so that to a first approximation

$$I_{chem}=\frac{k_c k_r}{k_q}\,[O]\,[NO]$$

That is to say, I_{chem} does not show any kinetic dependence on [M], although k_q will depend on the chemical identity of M.

Since the air afterglow and fluorescence both involve the same excited state of nitrogen dioxide, it is possible to use values of k_r/k_q obtained in fluorescence quenching experiments to predict the pressures at which k_r can no longer be neglected and at which the intensity will become dependent on [M]. The pressures involved are about 50 microns, and recent experiments using special low-pressure flow systems have indeed shown that the emission kinetics ultimately become third-order at pressures below 50 microns.

Measurement of the absolute emitted light intensity can give a value for k_c, the rate constant of the recombination reaction, if in the value of k_r/k_q allowance is made for the reversible crossing described in the discussion of radiative lifetimes. Comparison of the absolute intensity with the *total* rate of reaction shows that most of the recombination in fact goes to form *unexcited* nitrogen dioxide in the reaction

$$O+NO+M\rightarrow NO_2+M \tag{4.5a}$$

and that only a relatively small fraction of the newly-formed nitrogen dioxide molecules are excited.

At concentrations of NO, small compared with [O], reaction (4.5) makes virtually no contribution to the decay of atomic oxygen. Rather less obviously, under these conditions nitric oxide is not consumed either. The reaction

$$O+NO_2\rightarrow NO+O_2 \tag{4.19}$$

is very rapid, having a rate constant of about $1{\cdot}5\times10^9$ litre mole^{-1} s^{-1} at room temperature. Thus, in the presence of excess atomic oxygen, nitric oxide is regenerated from every nitrogen dioxide molecule formed.

The chemiluminescence of the air afterglow, coupled with the rapidity of reaction (4.19), furnishes an elegant method for the determination of atomic oxygen concentrations in discharge-flow systems. Such determinations are naturally of great value in investigations of the reactivity of atomic oxygen, and the technique to be described has found very wide application.

If nitrogen dioxide, rather than nitric oxide, is added to a stream of atomic oxygen, then nitric oxide is produced in reaction (4.19) and the emission of the air afterglow is observed. Over a certain range, increase in $[NO_2]$ increases the intensity of emission, but when the concentration of NO_2 added is sufficient to react with more than half the atomic oxygen the intensity decreases with $[NO_2]$, and is finally extinguished when all the oxygen is consumed, i.e., when the concentration of NO_2 is equal to the

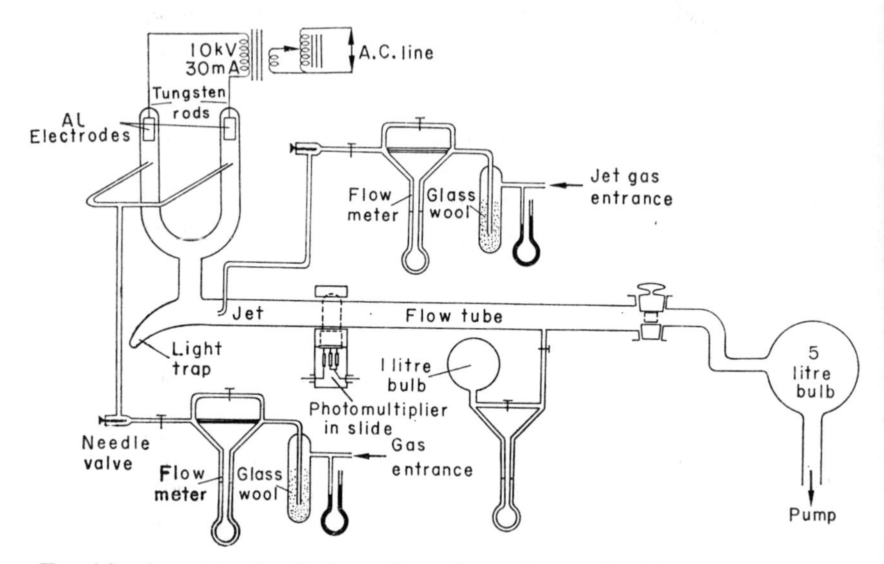

FIG. 4.2 Apparatus for discharge flow of a gas (e.g., O_2), with an inlet jet for an added gas (e.g., NO_2) and photomultiplier sliding on the flow tube.

concentration of atomic oxygen originally present. The end-point of this gas-phase 'titration' can be determined by the eye, or better, with a photomultiplier detector; the measured concentration of added NO_2 is then the same as the value of [O] in the absence of NO_2.

A somewhat similar technique has been devised for the measurement of atomic hydrogen concentration. The recombination reaction

$$H + NO + M \rightarrow HNO + M \tag{4.20}$$

yields some electronically excited HNO radicals which emit radiation in the red and near infra-red regions of the spectrum. The reaction

$$H + NO_2 \rightarrow NO + OH \tag{4.21}$$

is even more rapid than the corresponding reaction with atomic oxygen (4.19), and has a rate constant of $2{\cdot}9 \times 10^{10}$ litre mole^{-1} s^{-1} at room temperature, and atomic hydrogen may therefore be titrated with nitrogen dioxide. However, the stoichiometry implied by eqn (4.21) is invalidated as a result of secondary reactions of the hydroxyl radicals formed: at low *molecular* hydrogen concentrations three nitrogen dioxide molecules are consumed for every two hydrogen atoms initially present.

Nitrogen through which an electric discharge has been passed (i.e., that containing so-called 'active nitrogen') is also known to emit radiation. The radiation, which this time is yellowish in colour, is that of the First Positive Band system ($B^3\Pi_u \rightarrow X^1\Sigma_g^+$) of nitrogen. The recombination reaction

$$N + N + M \rightarrow N_2^* + M \qquad (4.22)$$

is the ultimate source of excitation, although the $B^3\Pi_u$ state of nitrogen does not correlate with two ground state (^4S) nitrogen atoms, but rather with one ^4S atom and one ^2D atom. The $B^3\Pi_u$ state must be populated, therefore, via another state (e.g., the $A^3\Sigma_u^+$) which crosses it and which *does* correlate with two ^4S atoms. Absolute intensity measurements indicate that about one-third of the newly-formed nitrogen molecules pass through the $B^3\Pi_u$ state, so the crossing process must be remarkably efficient.

The yellow nitrogen afterglow may be used as an indicator for the end-point of a titration reaction for the atoms. Atomic nitrogen reacts so rapidly with nitric oxide that under ordinary discharge-flow conditions the process

$$N + NO \rightarrow N_2 + O \qquad (4.23)$$

may be regarded as stoichiometric (the rate constant at room temperature is about $1{\cdot}3 \times 10^{10}$ litre mole^{-1} s^{-1}). Addition of nitric oxide to a gas stream containing atomic nitrogen therefore reduces the intensity of the yellow emission, until [NO] = [N] when the glow is extinguished. If *excess* NO is added, the green air afterglow is excited by the reaction of O formed in (4.23) with the NO. The reaction of atomic nitrogen with nitric oxide does, in fact, constitute a convenient technique for the preparation of atomic oxygen, in known concentration and in the absence of molecular oxygen.

In addition to the kinetic evidence, the participation of the third body, M, in the recombination reaction may be adduced spectroscopically. The emission observed always originates from vibrational levels of the $B^3\Pi_u$ state lower in energy than that afforded by the recombination of two ^4S nitrogen atoms. Although almost all the known chemiluminescent *recombination* reactions involve a third body, there are, nevertheless, many examples of exothermic bimolecular reactions in the gas phase which give rise to chemiluminescence.

One such reaction is the oxidation of nitric oxide by ozone

$$NO + O_3 \rightarrow NO_2 + O_2 + 49 \text{ kcal mole}^{-1} \qquad (4.24)$$

It is particularly interesting in that it provides an example of the way in which spectroscopic and kinetic data can complement each other and help to elucidate the reaction mechanism. Chemiluminescence is observed over a wide spectral range which starts at about 590 nm and extends to at least 3 microns. No emission from electronically excited O_2 is observed, and since the exothermicity of the reaction (49 kcal mole^{-1}) is only marginally greater than the energy of the shortest wavelength observed (48·4 kcal mole^{-1}) it would appear that the emission involves a transition from an excited state of NO_2 to its ground, $^2A_1$, state. This conclusion is supported by the presence of spectroscopic features corresponding to vibrational spacings in the $^2A_1$ state, and further evidence suggests that the upper state is the $^2B_1$, the same state responsible for the fluorescence, air-afterglow and visible absorption spectra. The difference between the $NO + O_3$ chemiluminescence spectrum and the other spectra mentioned arises from the population of only the lower vibrational levels of the $^2B_1$ in the $NO + O_3$ reaction; it seems therefore that there is no evidence for anomalously high vibrational excitation of the product NO_2 molecules in the reaction.

Comparison of the absolute intensity of emission with the rate of reaction leaves no doubt that the reaction proceeds via two paths, of which the one leading to electronic excitation of NO_2 contributes less than 10% to the total reaction at room temperatures. It has been shown that the pre-exponential factors of the rate constants for the reaction to the ground and excited states of NO_2 are similar (about 5×10^8 litre mole^{-1} s^{-1}) and that the difference in rate is almost entirely a consequence of differing activation energies for the two processes (2400 kcal mole^{-1} and 4200 kcal mole^{-1} respectively). The data is consistent with a transition state which initially forms the linear $^2\Pi$ state of NO_2. Bending the NO_2 to give a bond angle of 134° yields either the $^2A_1$ or $^2B_1$ states, depending upon the orbital filled by the odd electron. Thus the pre-exponential factors may be expected to be identical, whatever the final state of NO_2 after bonding. On the other hand, the number of $^2\Pi$ molecules reaching the $^2A_1$ and $^2B_1$ configurations will depend in the usual way on the energy barriers to entering these states, and this difference is reflected in the disparate activation energies for the overall reaction leading to NO_2 ($^2A_1$) and NO_2 ($^2B_1$).

Another highly exothermic reaction of ozone is the process

$$H + O_3 \rightarrow OH + O + 80 \text{ kcal mole}^{-1} \qquad (4.25)$$

and radiation is emitted from the system. In this case, however, the chemiluminescence is due to *vibrational* excitation of the hydroxyl radical rather than the formation of electronically-excited products. The observed spectrum extends from the infra-red into the long wavelength end of the visible spectrum; the bands in the visible and near infra-red regions result from 'forbidden' overtone transitions in which Δv is 4 or 5. The overtone vibrational emission spectrum of OH is observed as one of the features of the glow observed from the night sky. Reaction (4.25) seems a feasible source of excited OH in the upper atmosphere, especially since the distribution of intensities for the various bands is similar in the nightglow and in laboratory emission experiments. On the other hand, although there is sufficient ozone, the concentration of atomic hydrogen in the upper atmosphere is apparently far too low to account for the observed intensity of OH bands in the nightglow. It has, however, been shown that traces of atomic hydrogen can give relatively large intensities of the OH emission if an excess of atomic *oxygen* is present. The very rapid reaction

$$O + OH \rightarrow O_2 + H \tag{4.26}$$

regenerates atomic hydrogen every time this is consumed in (4.25), and the combined processes (4.25) and (4.26) amount to a hydrogen-catalysed reaction of atomic oxygen with ozone. Atomic oxygen is a major constituent of the upper atmosphere, and it seems that the observed OH emission intensities and hydrogen atom concentrations in the night sky may be reconciled by invoking reaction (4.26). (It might be mentioned at this stage that the presence of atomic oxygen in the upper atmosphere has been demonstrated in a most convincing way by discharging nitric oxide from a rocket at an appropriate altitude: an easily observed glow develops which has its origins to a large extent in the 'air afterglow' reaction, O + NO + M.)

One of the most tempting problems which might be solved by a study of gas phase chemiluminescence is the elucidation of reaction paths by analysis of the energy distribution in the reaction products. It is clear that systems which yield vibrationally, rather than electronically, excited species are potentially more tractable to this kind of treatment, although even here the system needs to be of the utmost chemical simplicity. Atomic hydrogen reacts with the halogens to give vibrationally excited hydrogen halide molecules, and J. C. Polanyi has recently measured the energy distribution in the product molecules. The work may perhaps be said to have its origins in investigations made by his father M. Polanyi in the 1930s. A great deal of interest was focused at that time on the 'atomic flame' reactions of sodium with the halogens. Chemiluminescent emission of sodium resonance radiation was observed, and the processes leading to excitation were suggested to be

$$Na + X_2 \rightarrow NaX + X \tag{4.27}$$

$$X + Na_2 \rightarrow NaX^{\dagger} + Na \tag{4.28}$$

$$NaX^{\dagger} + Na \rightarrow NaX + Na^* \tag{4.29}$$

where $NaX^{\dagger}$ represents a molecule of *vibrationally* excited sodium halide and Na^* is an *electronically* excited, ^2P, sodium atom.

Although the mechanism of the energy transfer reaction (4.29) was not clear at the time, it was realized that it must be relatively efficient. It was the efficiency of reaction (4.28), however, which enabled Polanyi to predict the general features of the energy surface describing the process. Where iodine is the halogen employed, the exothermicity of (4.28) is only 1 kcal mole^{-1} greater than the energy needed to excite sodium resonance radiation. It follows, therefore, that not only must (4.29) occur without loss of energy, but also all the energy liberated in (4.28) must go into vibration of the newly formed Na—I bond. This in turn means that the transition state must be a co-linear system, and, further, that there is virtually no repulsion between the newly-formed particles in the final state of the activated complex; if either of these conditions is not satisfied, rotational and translational excitation will result. A calculation of the actual energy surface for the reaction with atomic chlorine confirmed the absence of repulsion in the final electronic state of the activated complex, and conformed to the type of surface predicted.

The development of molecular beam techniques has permitted the elucidation of the excitation mechanism for reaction (4.29). A triple beam experiment has been devised, in the first part of which molecular beams of bromine and atomic potassium cross each other. Vibrationally excited KBr is formed

$$K + Br_2 \rightarrow KBr^{\dagger} + Br \qquad (4.30)$$

and is then collimated into a further beam. The beam of KBr† now meets a beam of atomic sodium, and the emitted radiation is observed. The chemiluminescence is that of the *potassium* resonance lines, so that the energy transfer process must be written

$$KBr^{\dagger} + Na \rightarrow K^{*} + NaBr \qquad (4.31)$$

The preceding description of the now-classical experiments on vibrationally-excited sodium halides serves to explain the objects of the current series of investigations being carried out on vibrationally-excited hydrogen halides. These latter experiments are, of course, much more sophisticated in that they do not rely, for information about vibrational excitation, on circumstantial evidence adduced from the observation of electronically excited species. It is possible with these systems to investigate the relative population of the several vibrational levels, and to determine the rotational excitation for each. For the reaction

$$H + Cl_2 \rightarrow HCl^{\dagger} + Cl \qquad (4.7)$$

vibrationally excited HCl† is formed with up to 6 vibrational quanta, and the vibration-rotation spectra for transitions of $\Delta v = 0$, 1 and 2 have been recorded.

The vibrational energy distribution of the newly-formed HCl† bears no relation to an equilibrium distribution, and the experiments show that the reaction rate is greatest into the lowest vibrational levels. On the other hand, the rotational energy does appear to be consistent with thermal equilibrium (at least for levels up to $J = 7$) at a temperature not more than 100°C above that of the reaction vessel. These results may be used to test hypothetical reaction surfaces and to distinguish between rival mechanisms. However, the importance of looking at the *initial* energy distribution must be emphasized. In preliminary experiments on the $H + Cl_2$ reaction it appeared that the vibrational energy followed a Boltzmann distribution, albeit at a high equivalent temperature. It was only performance of the investigations at the very low pressure of about 10^{-2} mmHg (experiments which necessarily required a particularly sensitive infra-red detection system) that revealed the departure from equilibrium.

It is hoped that this chapter has shown that a study of luminescence in the gas phase can help in the elucidation of reaction mechanisms, by application of the kinetic data which can be obtained and of the information about excited species produced in the course of a reaction. The author believes that much progress will be made in the future by investigating the formation of excited species in chemical processes, and by studying the reactivity of such species. Emission of radiation may always be anticipated from excited species if the experimental conditions can be suitably adjusted, and the importance of luminescence is thus seen.

BIBLIOGRAPHY

Annual Reports of the Chemical Society of London, **62**, 17 (1965).
Chemistry in Britain, **2**, 287 (1966).

Chapter 5

LUMINESCENCE OF INORGANIC SUBSTANCES

W. A. RUNCIMAN

OBSERVATIONS on inorganic luminescence probably date from about 1600, when V. Casciorolo, a cobbler of Bologna, calcined heavy spar and obtained a product which, after exposure to sunlight, phosphoresced in the dark. Two hundred and fifty years elapsed before the subject was scientifically examined, but during this century it has grown to a vast extent because of its practical uses and theoretical interest.

Most inorganic substances are not luminescent, and the apparently simple question as to why they are not cannot yet be fully answered. Quantum mechanical theory is well developed for interpreting atomic energy levels and the radiative transitions between them. However, the treatment of non-radiative effects, i.e., the conversion of electronic energy into interatomic vibrations (called 'phonons' when they run as waves in crystals), is very intractable. Certain generalizations only can be set out. Electronic transitions of a 'forbidden nature, and therefore of long radiative life, are to be expected to compete less successfully against energy conversion to heat. Electron-transfer reactions, especially those involving covalently bound groups of atoms, may act in a similar way. Absorption of light by a crystal of a single substance may be 'non-localized', i.e., the excitation energy may be visualized as rapidly passing from atom to atom. It is not surprising that inorganic luminescence is usually associated with dilute concentrations of ions more or less isolated in a relatively inert matrix; these ions are referred to as 'activators', and their energy states are treated by basic atomic theory modified by the crystal field of the matrix or 'host'. The activator ion, with the somewhat distorted arrangement of lattice ions round it, is often called a 'centre'. Host lattices are usually metallic oxides, phosphates, silicates, etc., where the large size of the oxygen ion almost dominates the lattice structure, or certain sulphides, and alkali or alkaline earth halides. Glasses can also act as hosts; here the energy levels are broadened because of the variable environments of the activating ions. Solid inorganic luminescent materials are called 'phosphors', although their emission is usually termed 'fluorescence'. The description 'phosphorescence' in this field is used for emission from those phosphors where electrons are slowly released from 'traps'. Great care to ensure very high purity is necessary in phosphor preparation; minute concentrations of certain ions, such as iron, may weaken the power to radiate. Where 'traps' are concerned the perfection of the crystal structure—presence of ion vacancies, dislocations, etc.—as determined by stresses or heat treatment, has an important effect on the luminescence properties. This chapter describes a limited number of phosphor systems to illustrate the way their behaviour is interpreted.

Trivalent Lanthanide Activators

These ions have an incomplete $4f$ shell of electrons which is shielded by complete $5s$ and $5p$ shells, so that the energy levels appropriate to the free ions are not greatly perturbed in a crystal lattice such as a suitable phosphate. The trivalent europium ion Eu^{3+} has six electrons in the $4f$ shell, and by Hund's rule the ground state is denoted $^7F_0$, which indicates that none of these electrons is spin-paired. Within the $4f$ shell there can be many other arrangements of electrons; these may be of higher energy, with energy differences determined by 'spin-orbit' coupling, forming a set of multiplets. These extend over an energy range of several thousand cm^{-1} ($\sim$10 kcal $mole^{-1}$). The details of the levels are very complex, but calculations of the electron-electron and spin-orbit interactions can be made by using the group theoretical methods of Racah. Luminescence, however, only involves the lowest excited multiplet denoted $^5D_{0-4}$. In this state two f electrons are spin-paired. Figure 5.1 shows the energy levels and the observed transitions between the ground multiplet and the excited one. The 0–0 transitions (shown dotted) are formally forbidden and are usually not observed. However when the $^5D_0 \rightarrow {}^7F_0$ transition is observed strongly as in $Sr_2\,TiO_4$ (Eu^{3+}), it indicates that a first-order electric field term is allowable in the harmonic expansion of the crystal field. Similarly other hypersensitive lines, such as $^7F_0 \rightarrow {}^5D_2$ in Eu^{3+}, whose intensities depend strongly on the environment, have $\Delta J = 2$. Therefore these lines also give information concerning the environment of the ion when it is in crystal or solution. In any event the crystal field splits each J level into a maximum of $2J+1$ sub-levels for an even number of electrons and $J+\frac{1}{2}$ sub-levels for an odd number of electrons. In the latter case the two-fold Kramers degeneracy can be removed by a magnetic field. In the Zeeman effect, the application of a magnetic field removes all degeneracy, even for a free ion, and, for a field of 30 kilogauss, causes splittings of a few cm^{-1}. The crystal field splittings which are usually larger, $\sim$100 cm^{-1}, are examples of the Stark effect. Thus a detailed examination of crystal field and Zeeman effects can give much information on the chemical environment of the rare-earth ion. Europium-activated phosphors, in particular YVO_4 (Eu^{3+}) are now being used as the red component for colour television tubes and also for colour-correction in mercury discharge lamps. Similar remarks apply to the transuranic ions with an incomplete shell of $5f$ electrons.

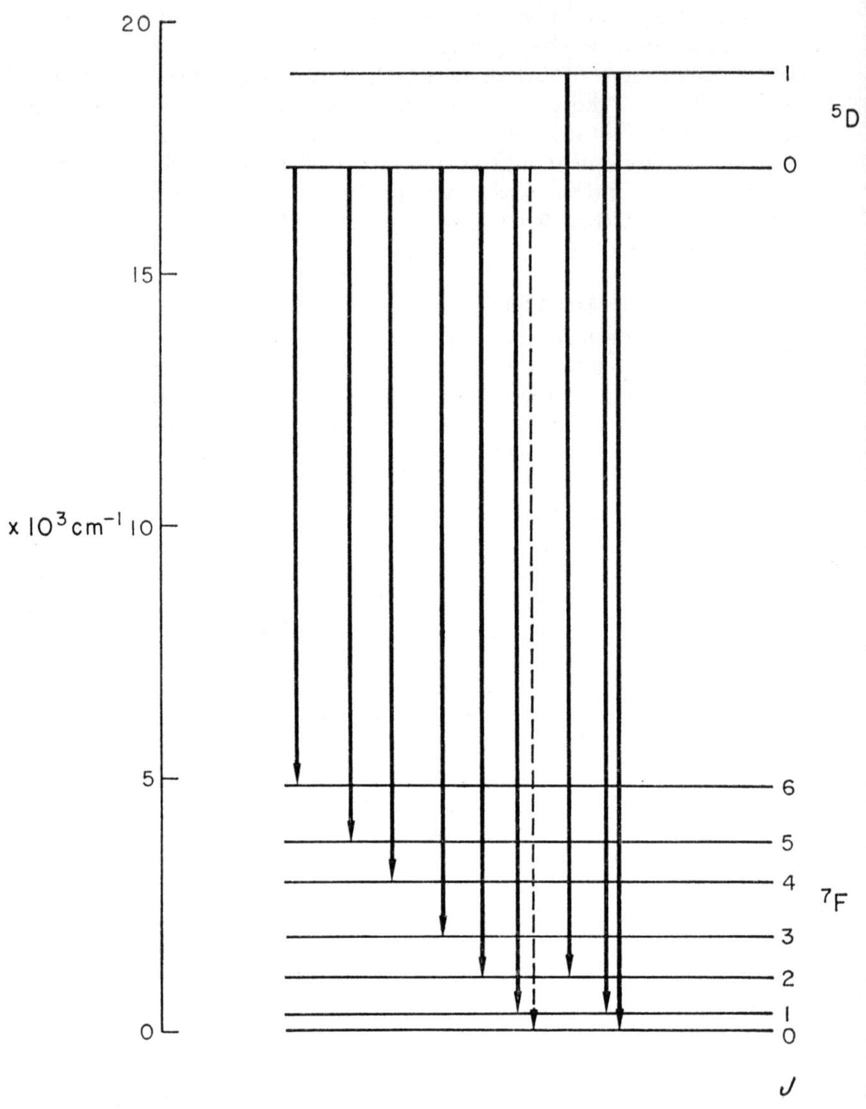

FIG. 5.1 Energy levels and fluorescent transitions in Eu^{3+}.

Transition Element Activators

Divalent manganese is the best-known luminescent activator, fluorescent lamp phosphors consisting of manganese-activated calcium halophosphate. The electronic transition occurs within the $3d^5$ electron configuration. Likewise ruby consists of chromium-activated aluminium oxide, Al_2O_3 (Cr^{3+}), which has a $3d^3$ electron configuration. It is therefore of interest to find out why manganese emits a broad emission band whereas ruby has a two-line spectrum, the R lines being at 692 and 693·4 nm at the temperature of liquid nitrogen or lower. Certainly the chromium ion is more compact, being trivalent, but the electrons involved are outer electrons exposed to the full effects of the neighbouring ions. The answer lies in the detailed nature of the upper and lower electronic states. The environment of the Cr^{3+} ion consists of six negative ions in approximately octahedral co-ordination. The effect of an octahedral crystal field on the electron states of one d-electron is to split them up into three t_2 states and two e states, separated by an energy Δ (sometimes referred to as 10 Dq), Fig. 5.2(a). The t_2 orbitals avoid the neighbouring ions, transforming like yz, zx and xy, whereas the e orbitals transform like $x^2 - y^2$ or $2z^2 - x^2 - y^2$, (Fig. 5.2(b)).

The energy levels of an ion with configuration d^n can be calculated in

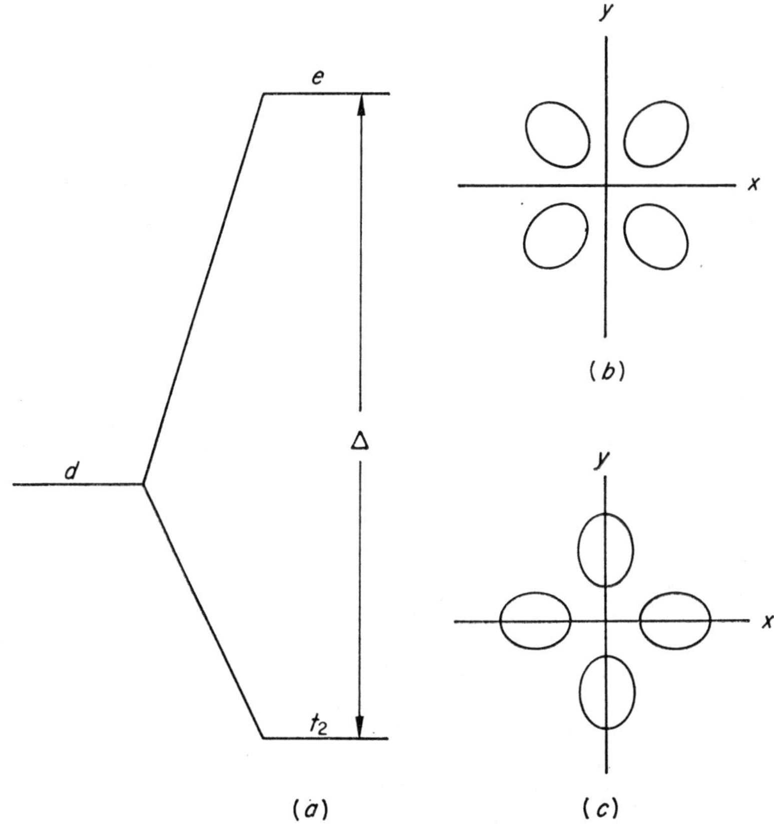

FIG. 5.2

(a) Effect of an octahedral crystal field of d-electron states.
(b) The charge distribution of a t_2 orbital, xy.
(c) The charge distribution of an e orbital, $x^2 - y^2$.

terms of the crystalline field strength Δ and the Racah parameters B and C which define the electron-electron interaction. In some cases the spin-orbit interaction parameter, ζ_{3d}, also needs to be included. If the ratio C/B is fixed, and it is usually about 4·5, then the energy levels can be plotted in units of B as a function of Δ. These are Tanabe–Sugano diagrams and are shown for d^3 and d^5 in Fig. 5.3. The left-hand side of the diagrams shows the energy levels for weak crystal fields, and the right-hand side

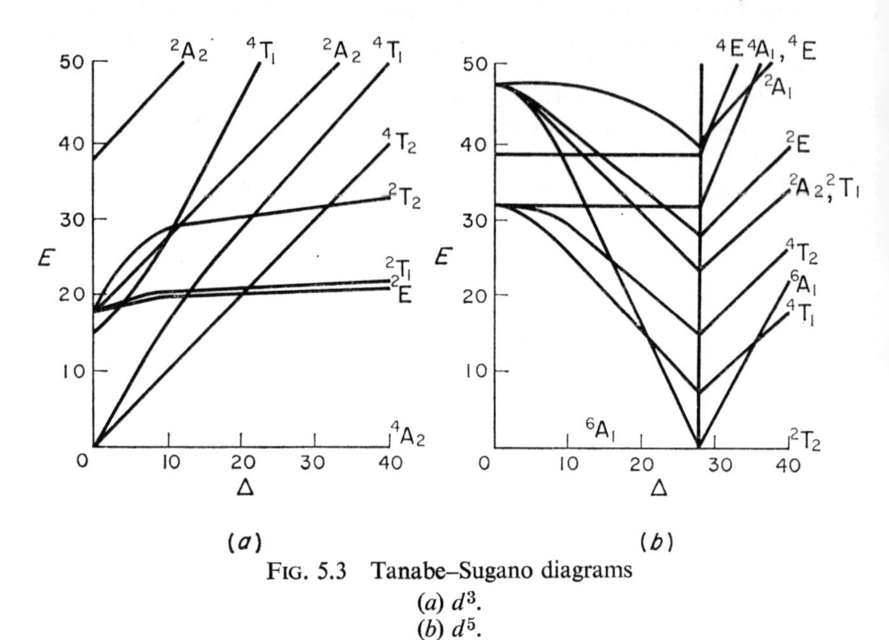

FIG. 5.3 Tanabe–Sugano diagrams
(a) d^3.
(b) d^5.
The energy E and the crystal field parameter Δ are expressed in units of the Racah parameter B. For simplicity some of the higher levels have been omitted.

those for strong crystal fields. In some cases, as for d^5, there is a transition in the nature of the ground state for strong crystal fields, but they need not concern us here, since, in practice, Δ is about $10B$ for most luminescent substances. For Cr^{3+}, Δ is about $30B$ and corresponds more to the strong field case. However for any detailed calculation the full intermediate field theory must be used and it is permissible to start with basic wave-functions derived from either the weak or the strong field models. The energy levels are described by labels of the form $^{2S+1}\Gamma$ where the spin S is written as the multiplicity $2S+1$, as for Russell–Saunders coupling, and Γ refers to an irreducible representation of the octahedral group. L is no longer a good quantum number since the crystal field quenches some of the orbital angular momentum. Hence we use Γ which is written A_1 or A_2 for singly-degenerate levels, E for doubly-degenerate levels and T_1 or T_2 for triply-degenerate levels. (Some authors use F_1 or F_2 for triply degenerate levels.) The total degeneracy is the product of this degeneracy with the spin multiplicity $2S+1$. For both Mn^{2+} and Cr^{3+} the ground states have only spin degeneracy, and the splitting of the sub-levels through spin-orbit and trigonal crystal field interactions is small (0·38 cm^{-1} for ruby). For orbitally degenerate states, the splitting due to spin-orbit interactions and to crystal fields of less than cubic symmetry can rise to several hundred cm^{-1}. We can now compare the nature of the luminescent transitions. For Mn^{2+} the transition will be $^4T_1 \rightarrow {}^6A_1$ and for Cr^{3+} it will be $^2E \rightarrow {}^4A_2$. The essential point seems to be the slope of the line for the upper level in the region of interest in the Tanabe–Sugano diagram. Lines with levels of large slope are more sensitive to fluctuations in crystal field strength. Broad bands occur when, as for Mn^{2+}, there is a large slope, and sharp lines when the slope is small as for Cr^{3+}. In other words, both the ^2E and $^4A_2$ states belong to the same strong field configuration t_2^3 of d^3, whereas the $^4T_1$ and $^6A_1$ states of Mn^{2+} belong to different strong field configurations of d^5.

Since the ^2E state consists of a half-filled shell of t_2 electrons, the first-order spin-orbit and trigonal field splittings are zero. However, higher order terms yield a total splitting of about 30 cm^{-1}, and hence two fluorescent lines, called R_1 and R_2, are observed. At liquid helium temperature the upper of the two levels is depopulated in accordance with a Boltzmann distribution, and R_1 becomes much stronger than R_2.

Ruby is used as the active material in many solid-state lasers. A xenon flash lamp in an elliptical reflecting cavity is used to pump energy into the broad absorption bands due to the $^4T_2$ and $^4T_1$ upper states. These upper states decay non-radiatively to the lowest excited states, ^2E. If population inversion is obtained, then light amplification can be observed due to stimulated transitions from the upper to lower states. This results in an intense, narrow beam of coherent radiation being emitted through a partially reflecting end-mirror on the cylindrical ruby rod. The other end of the ruby either has a totally reflecting mirror or a wedge-shaped end to give total internal reflection. A typical pulse might emit about a megawatt of power for ten microseconds, although special Q-switching techniques enable far higher power to be obtained for shorter times of the order of nanoseconds.

Where manganese is used as an activator, it is frequently found that the emission is enhanced by the addition of a 'sensitizer' which in some fluorescent lamps is antimony, but may also be arsenic, tin, lead or bismuth. These elements all are activators with emission bands at shorter wavelengths than the manganese. Energy is transferred from these ions to the manganese ions by dipole-dipole or exchange interactions (see Chapter 8). The process is referred to as sensitized luminescence when all the emission is in the manganese band, but more appropriately as double activation if the concentrations of the two activators are such that both emit.

Divalent Rare-earth Ions

The divalent rare-earth ions are luminescent activators in which both *f* and *d* electrons are involved, and hence combine features of the last two sections. The ground state belongs to the f^n configuration and the electronic transition occurs from an upper $f^{n-1}d$ state or in some cases from another f^n state. In either case the most efficient absorption process will be by an allowed transition to upper $f^{n-1}d$ states. As an example, consider divalent europium in calcium fluoride. The mineral fluorite often occurs in this fluorescent form, and indeed the term 'fluorescence' is derived from 'fluorite' on this account. The fluorescence consists of a narrow line at 413 nm and an associated broad band at longer wavelengths. Since the

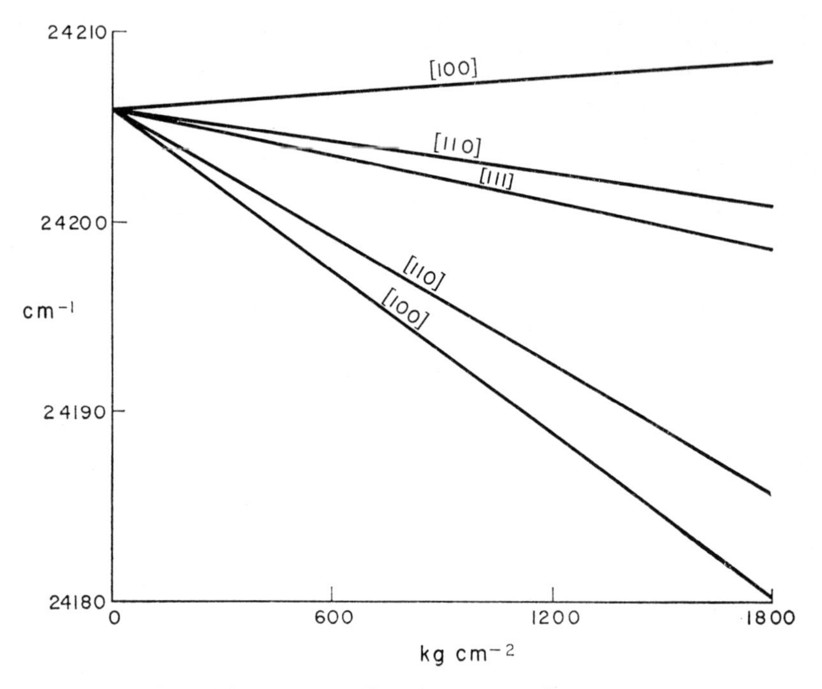

FIG. 5.4 Pressure effects in $CaF_2(Eu)^{2+}$ at 4·2°K.

europium ion can replace a calcium ion of the same valency, each europium is surrounded by eight fluorine ions in a site of cubic symmetry, O_h. The absorption spectrum consists of a narrow line at the same wavelength as the fluorescence, and broad bands at shorter wavelengths. The resonance line at 413 nm is due to an allowed electric dipole transition from an f^7 ground state to an f^6d upper state. Since f^7 is a half-filled shell, the ground state will be $^8S_{7/2}$. From paramagnetic resonance data, it is known that this ground state is split into three levels by crystal field and spin-orbit interactions with a total width of 0·2 cm^{-1}. The optical line-width is about 1 cm^{-1} so this fine structure is not observed. However, in a magnetic field the ground state splits into eight components fairly regularly spaced by an amount expected from a spin-only state with $g = 2$. The clearest indication of the degeneracy of the upper state comes from the spectrum obtained when a uniaxial stress is applied to the crystal (Fig. 5.4). Since the ground state which is spin-degenerate will not be affected by stress, the large splitting found, of about 30 cm^{-1}, must be due to the upper state. As the stress will not remove Kramers degeneracy, the twofold splitting of the upper state implies a total fourfold degeneracy. Hence the upper state belongs to the Γ^8 representation of the cubic group O_h. Unfortunately the f^6d configuration is very complex, so the detailed description of this upper state is not known. The broad fluorescence band is due to transitions in which part of the energy goes into exciting phonons (i.e., vibrations) either of the localized centre or of the lattice. This type of process is discussed more fully below.

Uranium Fluorescence

The yellow-green fluorescence of the uranyl ion (UO_2^{2+}) in many compounds is well known. Uranyl nitrate, sulphate, and phosphate crystals, for example, fluoresce strongly, and many substances are made fluorescent when doped with uranium. The fluorescence of hexavalent uranium in sodium fluoride is the basis for a very sensitive fluorometric method of determining uranium concentrations. Uranium fluorescence in all cases consists of a series of lines in groups about 900 cm^{-1} apart. The uranium ion may or may not be in a uranyl group. For instance in uranium-activated calcium fluoride the hexavalent uranium ion is thought to have four O^{2-} and four F^- nearest neighbours. However, for simplicity only uranyl groups will be considered in the following discussion, as similar considerations apply in the other cases. At low temperatures the highest energy line is a resonance line and the other lines are due to some of the energy being absorbed by vibrations within the UO_2 group with frequencies typically $A=860$ cm^{-1}, $B=940$ cm^{-1} and $C=220$ cm^{-1}. Since up to six A or B vibrations may be excited, very regular series are formed. The vibrations are those of a linear triatomic molecule, the A vibration being the symmetric stretching mode, the B vibration being the antisymmetric linear mode and the C vibration being the bending mode. These assignments have been checked by the isotope shifts due to O^{18} substitution of the normal O^{16} isotope.

The electronic energy levels of the uranyl ion are not well understood. The uranyl complex has much covalent binding and behaves very like a fluorescent organic molecule such as anthracene. The first excited state is probably a singlet state like the ground state, as no Zeeman splitting of the fluorescent transition has been observed. However, absorption to higher levels can lead to an observable Zeeman effect, and the upper levels concerned are appropriately known as the 'magnetic' series. One possible model for the excited levels is a charge transfer process in which a bonding electron becomes a $5f$ electron on the uranium ion.

Heavy Ion Fluorescence

The isoelectronic series of ions Tl^+, Pb^{2+} and Bi^{3+} are all efficient activators. The ground state is the $^1S_0$ state of the $6s^2$ configuration, and the upper state is the $^3P_1$ state of the $6s\ 6p$ configuration. Thallium-activated potassium chloride is one of the most studied materials, and calculations have been made of the appropriate configuration co-ordinate diagram. The absorption and emission bands approximate to Gaussian form because the number of vibrational quanta excited during absorption or emission is large, being around forty. The main absorption bands are centred on 196 nm and 249 nm with emission bands at 305 nm and 475 nm resulting from absorption in either band. At helium temperature the emission in the 305 nm band is observed to be polarized if the radiation absorbed in the 249 nm band is polarized along a cube axis. This polarization can be understood in terms of a Jahn–Teller distortion of the centre. A Jahn–Teller distortion is expected in a symmetrical orbitally-degenerate centre since a reduction of the symmetry will result in a lower energy of the system. It is somewhat surprising, however, that polarization is observed, since depolarization might be expected because of the many vibrational quanta also participating in the process.

Bismuth-activated calcium oxide has absorption and emission bands with well-defined vibrational series. The number of vibrational quanta involved is from zero to eight and this is due to the more tightly bound nature of the trivalent ion as compared with monovalent thallium. The absorption series may be expressed as the series in units of cm^{-1}: $\nu = 27{,}230 + 430v_1 + 215v_2$ where v_1 ranges from 0 to 8 and v_2 is usually 0 or 1. This absorption is believed to be due to the $^1S_0 \rightarrow {}^3P_1$ transition and is allowed apart from a spin selection rule, $\Delta S = 0$, which does not apply to heavy ions having large spin-orbit coupling. The absorption line for which no phonons are created is known as the zero-phonon line and is considerably sharper than the other lines in the series.

The emission series can be similarly written in the form:

$$\nu = 25{,}717 - 493v_1 - 195v_2$$

Since the zero-phonon line has a different wave-number, $\Delta\nu = 1513\ cm^{-1}$, this must be due to a different electronic transition, and it is now suggested that this is the $^3P_0 \rightarrow {}^1S_0$ transition. This is a highly forbidden transition, being from $J' = 0$ to $J = 0$, and it is not observed in absorption. However, charge compensation of the Bi^{3+} ion in a Ca^{2+} site may well involve a neighbouring Ca^{2+} vacancy. The resultant asymmetric crystal field can lead to a first-order crystal-field term, i.e., an electric field at the Bi^{3+} ion, which will allow this transition to occur in fluorescence. Of course in a point-ion approximation a charged ion is bound to be in a potential minimum with zero electric field; but because of the finite size of the ion and the presence of exchange forces this very simple model cannot be used to eliminate linear terms when these are allowed by group theory for a particular centre symmetry.

Vibrational series can also occur in defect centres produced by radiation damage in alkali halides and other crystals. In general the relative strength of the zero-phonon and other lines in a crystal at low temperatures is given by the relation:

$$I(q) = e^{-S} \cdot \frac{S^q}{q!}$$

where $I(q)$ is the intensity of the line due to q associated phonons, assuming that we may regard all phonons as of the same type. S is the mean number of phonons and is a measure of the strength of the coupling between the electronic and phonon systems. For $q=0$ this reduces to e^{-S}, the fraction of intensity in the zero-phonon line.

Zinc Sulphide Phosphors

Zinc sulphide is a phosphor frequently used in cathode-ray tubes. The green luminescence commonly seen is due to copper-activated zinc sulphide, and at high copper concentrations this phosphor is also used for electro-luminescence, i.e., luminescence caused by an electric field, usually alternating, applied across a phosphor layer.

It has proved a very difficult task to find out the nature of the different luminescent centres in zinc sulphide. The problem cannot be treated in terms of activator levels alone; the lattice is greatly involved. Three main types of model, shown in Fig. 5.5, have been suggested, and all have a

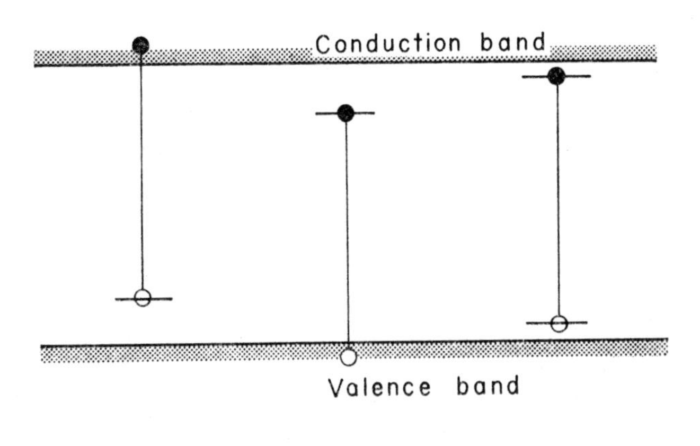

FIG. 5.5 Luminescent transitions in ZnS crystals.
(*a*) Schön-Klasens model.
(*b*) Lambe-Klick model.
(*c*) Localized transitions.

certain validity. In the Schön–Klasens model the upper state consists of an electron in the conduction band which can make a transition to a localized state. This is thought to be the case for the green copper emission due to copper on substitutional zinc sites. On the other hand blue copper emission may be due to interstitial copper in which the excited state is localized and an electron fills a hole in the valence band in the ground state. Both these bands move to higher energies as the temperature is reduced from room temperature to 77°K or lower. This is expected on any model involving the energy gap, E_g, which equals the energy required to raise an electron from the valence to conduction band. The green copper luminescence occurs in crystals which have chlorine or aluminium present as charge compensators for the substitution of monovalent copper for divalent zinc. Zinc sulphide exists in both a cubic (blende) and hexagonal (wurtzite) form, the co-ordination of the zinc and sulphur ions being four in each case. In the hexagonal form it has been found that the luminescence is polarized perpendicular to the hexagonal *c*-axis for both the blue and the green luminescence. It thus appears that the charge compensator for the copper ion is not sufficiently close to reduce the symmetry below that of a normal substitutional site in the case of the green luminescent centre. The distinction between models (*a*) and (*b*) is quite subtle, and has been made on the basis of thermoluminescence and photoconductivity measurements.

In a centre such as that in model (*a*) the electrons freed by excitation may be caught in 'traps'. These consist of impurities or crystal imperfections which bind electrons with binding energies less than the energy gap. Information about the depths of such traps in an excited phosphor can be obtained from the maxima in 'glow curves' obtained by plotting emission intensity against temperature when the material is slowly heated. Each maximum corresponds to the point where thermal emptying of a trap with injection of electrons into the conduction band becomes very rapid. The mobile electrons then give rise to photoconductivity and to luminescence. The preparation of phosphors with definite trap depths is a tricky and empirical matter; special heat treatments and addition of traces of impurities such as cobalt are commonly used.

Of quite a different nature is the self-activated luminescence which is obtained when zinc sulphide with added chloride is heated in air at 950°C, without any copper (Fig. 5.6). In this case the peak energy is reduced as the temperature is lowered. This and some other luminescence centres in zinc sulphide are thought to be caused by electronic transitions in localized centres. In this case the centre may be due to a zinc vacancy with partial charge compensation, a monovalent chlorine ion occupying a normal sulphur ion site.

In cadmium sulphide there is a lot of structure in the sharp emission lines which occur at energies near the band gap. This so-called edge emission does contain some luminescence characteristic of the pure crystal, but most often appears in lines associated with impurity levels in the crystal. Similar effects are also found in selenide and telluride phosphors. These transitions can then be thought of as being due to the luminescent decay of bound excitons. An exciton may be regarded as a combination of an electron in the conduction band and a hole in the valence band. If these move together because of their mutual electrostatic attraction, but quite freely in the crystal, they form an exciton. This becomes bound if either the hole or the electron is caught at some trapping centre. Bound excitons will tend to have short lifetimes, because of the influence of the trapping site, where the electrons combine, either radiatively or non-radiatively, with the holes.

Zinc sulphide is used for self-activated phosphors as in instrument and

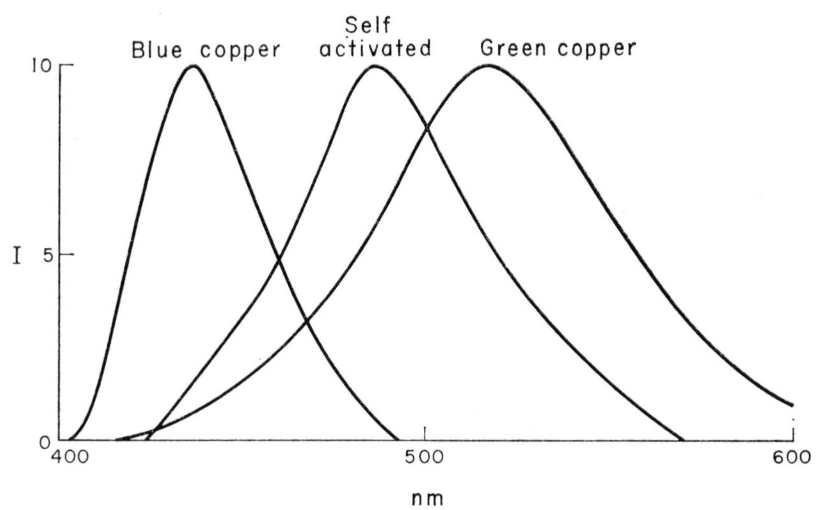

FIG. 5.6 Luminescence spectra of ZnS phosphors, at 4·2°K.
(*a*) green copper luminescence.
(*b*) blue copper luminescence.
(*c*) self-activated luminescence.

watch dials. Radioactive α-particle emitters are incorporated, and the life of the material is limited by radiation damage. For some purposes excitation by tritium is used, because of its freedom from this defect and its relatively harmless nature.

In semiconductor crystalline compounds, such as GaAs, edge emission may be excited by electron injection across a n–p boundary. If the material is regarded as Ga^+As^- (without specifying the actual numerical charge values), n-type crystals have some uncharged gallium due to the presence of impurities, and p-type crystals have some uncharged arsenic ('positive holes'). If electrons are driven across a junction from n to p they will combine with arsenic to give As^-. Such electron-hole recombinations are non-radiative for most materials, but with GaAs and GaP, infra-red or red emission can be obtained. With ZnTe preparations yellow-green electroluminescence has been found. GaAs has been used as a semiconductor laser. It is also possible to excite some materials by electron bombardment and in this way laser action has been obtained, for instance, in a ZnO crystal emitting radiation near 380 nm.

Luminescence of Pure Alkali Halides

In contrast to the sulphide phosphors there is no edge emission in the alkali halides. On the other hand broad-band luminescence is observed in undoped crystals at low temperatures. For instance in potassium iodide under X-ray excitation at 4°K there is emission around 371 nm which may be compared with the absorption edge near 220 nm. The explanation lies

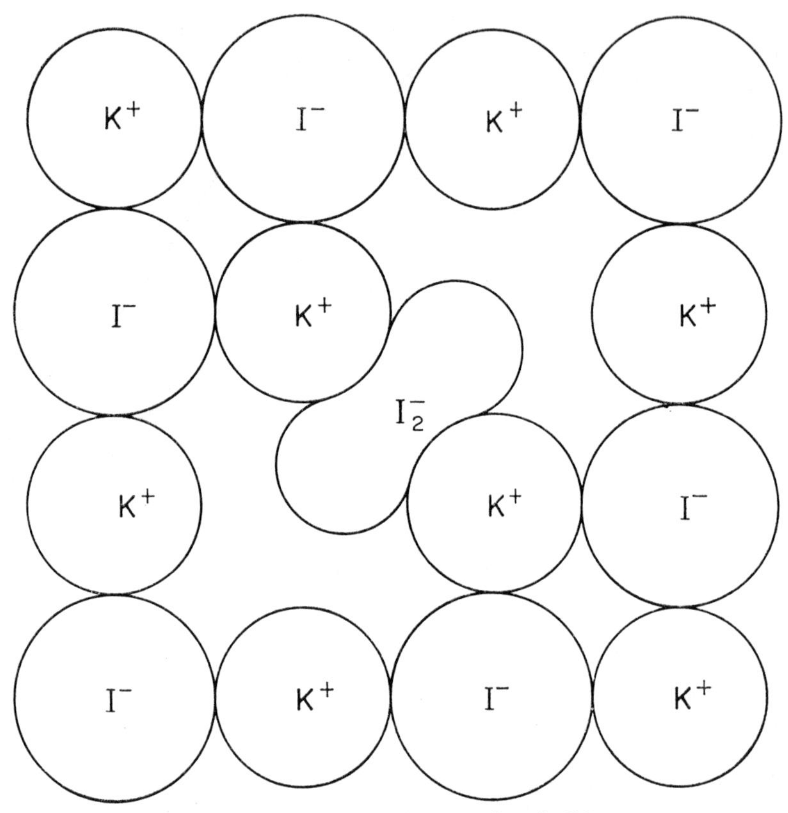

FIG. 5.7 V_k centre in potassium iodide.

in the different nature of the exciton in the alkali halides. Whereas the electron is mobile, the hole is localized. The localized hole is known as a V_k centre and consists of an I_2^- molecule occupying two I^- sites, Fig. 5.7. In other words the missing electron or hole is shared between two negative ions which then relax to form a molecular ion. An exciton is formed when an electron is bound to this hole, and recombination can take place radiatively or non-radiatively. When radiation is emitted there is a large Stokes shift caused by the large number of phonons created when the ions relax back to their normal sites. This mechanism of luminescence has been verified by polarization measurements since, in a crystal at low temperatures, it is possible to form stable V_k centres. These can then be oriented under the action of light in the V_k absorption band. When electrons are released from traps by irradiation in excess electron bands, then the electrons combine with the oriented V_k centres, and polarized light is emitted. Similar results have been found for KCl and KBr and for KI. Thallium-doped sodium iodide is used for scintillation counters, but this aspect will not be pursued here.

Recent measurements by the author have shown that all alkali halides fluoresce when suitably excited, although the NaF fluorescence is very weak. It is believed that many of the emission bands result from the type of transition described above. More than one emission peak may be observed if higher exciton states are involved. It seems natural to ask whether similar processes occur in other crystals, for instance, the oxides. Again, all seem to luminesce, but with greatly varying output. Sometimes the fluorescence may be due to impurities as in the case of fused silica where the characteristic blue luminescence is thought to be due to water molecules bound within the glassy structure. Synthetic sapphire—or corundum—consists of crystalline alumina, Al_2O_3, and is of interest since an undoped crystal has been found to luminesce quite strongly at 165 nm, thus extending the study of luminescence into the vacuum ultra-violet.

The luminescence of diamond is not at present well understood. One diamond has shown recombination electroluminescence at 235 nm. More commonly, many type 1 diamonds, which are those containing nitrogen platelets, fluoresce under ultra-violet excitation, in a line at 415 nm with an associated vibrational band. After electron irradiation and annealing a further line appears at 503 nm.

The use of LiF as a radiation dosimeter depends on the crystal damage caused by the radiations, with the development of electron traps. When the material is heated, it emits thermoluminescence proportional to the amount of radiation received.

The Effect of Temperature on Phosphorescence

Mention has already been made of the 'glow curve' maxima observed during the heating of excited phosphors containing traps. At low temperatures some sulphide phosphors store much of the excitation energy received, and emit only when the electrons are released from their traps, by warming or by specific infra-red radiation. The luminescence of phosphors without traps depends reversibly on temperature; higher temperatures favour conversion of electronic energy into heat. Most phosphors have a temperature range of about 100°C over which the luminescence intensity drops from its low temperature value to zero. Some show a steeper drop than this, and their fluorescence intensity is used to estimate surface temperatures in engineering and medicine.

BIBLIOGRAPHY

No attempt has been made to attribute work to individual authors in view of the very extensive literature of the subject. The following books and review articles will provide plenty of individual references for further study.

DIEKE, G. H. and DUNCAN, A. B. F. *Spectroscopic Properties of Uranium Compounds*, McGraw-Hill, New York (1949).

GARLICK, G. F. J. *Luminescent Materials*, Oxford University Press, Oxford (1949).

PRINGSHEIM P. *Fluorescence and Phosphorescence*, Interscience, New York (1949).

RUNCIMAN, W. A. 'Absorption and fluorescent spectra of ions in crystals', *Rep. Prog. Phys.* **21**, 30–58 (1958).

MCCLURE, D. S. 'Electronic spectra of molecules and ions in crystals. Part II. Spectra of ions in crystals', *Solid State Physics* (edited by F. Seitz and D. Turnbull). Academic Press, New York, Vol. 9, 399–525 (1959).

BIRKS, J. B. *The Theory and Practice of Scintillation Counting*, Pergamon Press, Oxford (1964).

WYBOURNE, B. G. *Spectroscopic Properties of Rare Earths*, Interscience, New York (1965).

Proceedings of the International Conference on Luminescence, Budapest (1966).

Chapter 6

FLUORESCENCE OF ORGANIC COMPOUNDS

J. W. BRIDGES

FLUORESCENCE is the result of the rapid emission of light energy from a molecule which has become excited by light absorption. In order to fluoresce, therefore, a compound must first absorb light. However, relatively few light-absorbing organic materials are strongly fluorescent, either in the solid state or in solution. The first fluorescent organic substances known were vegetable dyes, and until quite recently it was thought that fluorescence was a characteristic property of all dyes. It is now realized that this is not true, for less than 7% of the known dyes are fluorescent. Progress in the understanding and use of fluorescence was severely limited by lack of suitable instrumentation, but during the last decade the development of modern spectrofluorimeters has stimulated considerable new interest in the subject of fluorescence. Thus, in 1965, over seven hundred papers were published on various aspects of fluorimetry. Fluorimetric methods have proved to be invaluable for investigating many aspects of molecular structure, and have greatly enhanced our knowledge of the mechanisms involved in excited state reactions. They have also been used extensively for analytical purposes, particularly by workers in the biological sciences. As an analytical method, fluorimetry has the virtues of high sensitivity and selectivity compared with colourimetry, for whereas few substances can be estimated colourimetrically below 10^{-7} g/ml (because of factors inherent in spectrophotometer design) many compounds may be measured fluorimetrically at concentrations of 10^{-8} to 10^{-10} g/ml. At the limits of sensitivity, the observed fluorescence of a compound may be markedly increased or decreased by any impurities present, particularly if it is in the solid state. For example, 1 part in 1000 of phenazine will completely quench the fluorescence of solid anthracene, while the claimed blue fluorescence of fluorene is due to traces of carbazole produced during manufacture. For this reason, much published data may have to be treated with caution, since frequently either impure materials have been used, or the conditions under which measurement was carried out were not sufficiently rigorous to exclude such factors as oxygen quenching, inner filter effects and photodecomposition. The majority of studies on the fluorescence of organic compounds has been made on solutions, although the properties of specific types of solids, namely those employed as scintillators and optical brighteners, have been intensely investigated. Indeed, the industrial importance of the latter outstrips all other commercial uses of organic fluorescence, for the annual sales of optical brighteners in 1965 amounted to more than £40 million. The fluorescence characteristics of solid organic compounds usually differ from those of inorganic ones in that the fluorescence of organic crystals is primarily determined by the individual constituent molecules, whereas that of inorganic crystals is complicated by the presence of more powerful crystal forces. Some organic compounds fluoresce in the solid state but not in solution (partly, at least, because of lack of molecular rigidity in solution), whereas others show considerably better spectral definition in the solid state. In a few cases the fluorescence of an organic compound does appear to be determined by the nature of the crystalline lattice (e.g., some crystalline hydrocarbons) [1], [2]. It may, in these instances, provide a useful complement to crystallographic studies.

Fluorescence emission usually occurs from the longest absorption wavelength only. In many cases the fluorescence spectrum when plotted on a wave-number scale is a mirror image of the absorption spectrum. The separation between the two spectra (i.e., Stokes shift) is normally about 3000–5000 cm, unless dynamic quenching occurs, in which case separation may be greater than 8000 cm^{-1}.

Requirements for Fluorescence

(*a*) Electronic Considerations

Much of the light energy absorbed by a molecule may be lost by processes other than fluorescence. Indeed, it is rare for an organic compound, particularly in solution, to emit all of its absorbed energy as fluorescence (i.e., to have a quantum efficiency of unity). The lost energy will be degraded by the competing reactions of internal and external conversions. Energy loss by internal conversions arises either because of intersystem crossing from the excited singlet to the triplet state (whence it will be utilized as phosphorescence, or lost as heat) or through predissociation (i.e., disruption of the chemical bonds of the compound). Energy loss by external conversion (quenching) occurs through collision with other molecules, and may result in the radiationless transfer of energy to the colliding molecules or in the production of heat. The fluorescence process will tend to be favoured if a compound has the following properties:

(1) A longest wavelength of absorption with a high extinction coefficient in the ultra-violet or visible region corresponding to a $\pi \rightarrow \pi^*$ excitation. Molecules which absorb high-energy frequencies (i.e., short wavelength) are prone to predissociation rather than fluorescence. On the other hand, species in which the longest absorption wavelength corresponds to an $n \rightarrow \pi^*$ transition (common in molecules containing nitrogen, oxygen or sulphur) are often phosphorescent, but seldom fluorescent [3].

(2) The excited singlet state should be relatively stable to the deactivating processes described above; i.e., it should have a half-life of about 10^{-8} s. If the half-life is more than 10^{-8} s, intersystem crossing will be favoured, while a shorter half-life implies rapid deactivation of the molecule by other processes.

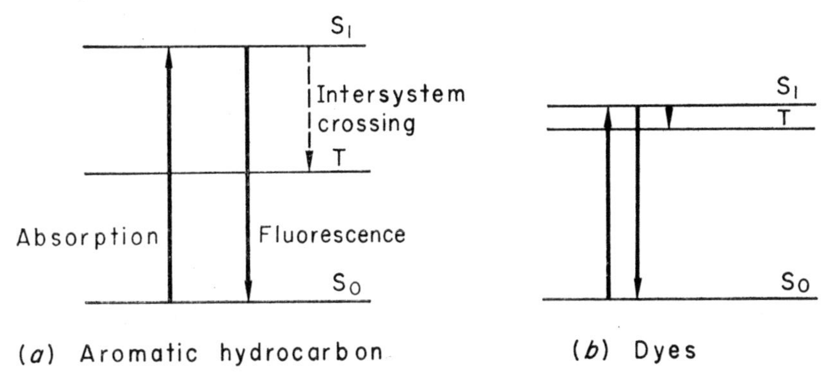

FIG. 6.1 Relationship between the excited singlet, triplet and ground states in (*a*) aromatic hydrocarbons, and (*b*) dyes. (Broken lines represent unlikely transitions, solid lines likely ones.)

(3) The excited singlet S_1 and triplet states T should be well separated as in the case of the aromatic hydrocarbons (see Fig. 6.1), for where the two states are very close together, e.g., in some dyes, intersystem crossing from the excited singlet to the triplet state can occur. Energy decay from the triplet state may produce phosphorescence, but it is unlikely to give fluorescence except in rare instances (see Delayed Fluorescence, Chapter 7). In our present state of knowledge few of these characteristics can be accurately predicted.

(*b*) STRUCTURAL CONSIDERATIONS

Originally it was thought that all fluorescent molecules consisted of two parts, a chromophoric part responsible for light absorption and a fluorophoric portion for fluorescence emission. Many studies were carried out to determine the structural requirements of each part. It is now realized that, with very few exceptions (e.g., $NADH_2$), absorption and fluorescence involve either the same part of the molecule, or, more commonly, the total molecule. Our understanding of the structural requirements for fluorescence is at present rather empirical but certain general rules are apparent. However, from an analytical point of view in particular, these rules can be very useful in predicting which compounds are likely to have a measurable fluorescence. Fluorescent properties are often very sensitive to slight structural modification. The structural requirements for the fluorescence of an organic compound are dependent on three main factors, namely the nature of the carbon skeleton, the geometrical arrangement of the molecule, and the type and position of any substituents.

The Carbon Skeleton

Saturated organic substances are not normally fluorescent in the solid state or in solution, although they may be fluorescent in the vapour phase. In such instances, however, it can usually be demonstrated that the absorbing and fluorescing species are different. The fluorescence is often attributed to radicals resulting from the predissociation of molecular bonds on excitation. In solution, the solvent may either prevent radical formation or it may cause the radicals to be deactivated by external conversions before they can fluoresce. In Table 6.1, the dissociation energies

TABLE 6.1. *Relation between absorption wavelengths, fluorescence and bond energies*

	Absorption maximum†		Least stable bond		
Compound	nm	kcal/mole	bond	cleavage energy‡	Fluorescence
Butadiene	210	~140	C—C	125	–
1,4-Diphenylbutadiene	350	~80	C—C	125	+
Bromobenzene	261	~110	C—Br	66	–
Eosin (tetrabromouranine)	520	~55	C—Br	66	+
Nitrobenzene	260	~110	C—NO_2	55–60	–
5-Dimethyl-4-nitrostilbene	430	~65	C—NO_2	55–60	+
Cyclohexylamine	<200	>150	C—NH_2	75–85	–
Aniline	280	~100	C—NH_2	75–85	+
Cyclohexanol	<200	>150	C—OH	85–100	–
Phenol	270	~105	C—OH	85–100	+

† In ethanol.
‡ At 0°, taken from Cottrell, T. L. *The Strengths of Chemical Bonds* (2nd edn), Butterworths, London, (1958).

of some common bonds are compared with absorption wavelength maxima of some compounds in solution which contain these bonds. It can be seen that compounds which only absorb at higher-energy frequencies than the strengths of their weakest bonds do not usually fluoresce, since under these conditions predissociation is favoured. On the other hand, compounds possessing similar bonds, but capable of absorbing lower-energy frequencies, may be fluorescent. It should be noted that each of these fluorescent compounds contains a conjugated system of double bonds, the presence of which appears to be essential to fluorescence. However, not all compounds with a conjugated system of double bonds are fluorescent, for other requirements must also be satisfied. Thus, 1,3-butadiene possesses a conjugated system but is non-fluorescent, since its low absorption wavelength maximum favours predissociation rather than fluorescence. However, 1,4-diphenylbutadiene which absorbs at a much longer wavelength and hence is unlikely to predissociate, fluoresces strongly. Increasing the extent of conjugation has two main effects, namely, to shift the absorption and therefore the fluorescence wavelengths towards the red end of the spectrum (thus reducing the possibility of predissociation), and to enhance the mobility of the π electrons. Increasing the number and mobility of the π electrons often results in an

TABLE 6.2. *Fluorescence efficiencies of some aromatic hydrocarbons*

Compound	Absorption maximum (nm)	Fluorescence (nm)	Fluorescence efficiency in hexane
Benzene	254	250–300	0·04
Naphthalene	275–314	300–365	0·10
Anthracene	275–380	372–460	0·32
Phenanthrene	250–346	348–407	0·01
Fluorene	246–260	302–370	0·54
Biphenyl	246–260	318	0·23
Triphenylmethane	262	260–310	0·23

increase in fluorescence intensity (see Table 6.2). For example, *trans*-stilbene, although it contains only one more double bond than biphenyl, is considerably more fluorescent. Molecules which contain conjugated double bonds in a cyclic structure are likely to fluoresce more strongly than those contained in a chain arrangement. Thus, naphthalene (Ia) and vitamin A (Ib) possess a system of five conjugated double bonds, but the fluorescence efficiency of naphthalene is at least five times that of vitamin A.

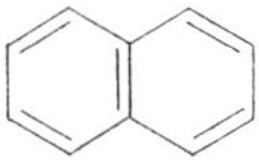

CH_3 H_3C CH_3 CH_2OH CH_3

Ia 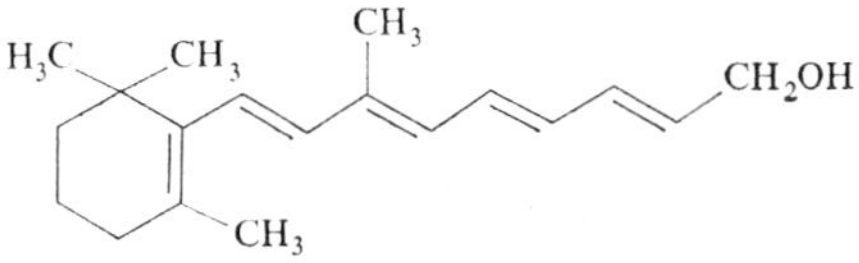 Ib

The effect of increasing the number of benzene rings is generally to increase the fluorescence efficiency. Thus, anthracene has a higher efficiency than naphthalene which in turn is more fluorescent than benzene. Linearly combined rings tend to be more fluorescent than rings in other arrangements, for example, solutions of the linear hydrocarbon anthracene have a fluorescence efficiency of 0·32 whereas those of the angular phenanthrene (II) have an efficiency of only 0·1 [4]. It seems probable that the indentation in the phenanthrene ring system acts as some sort of barrier to the free flow of π electrons around the molecule, whereas in anthracene they are able to circulate readily around the entire molecule. Similarly, solid naphthacene (III) has a fluorescence efficiency more than four times greater than that of 1,2-dibenzanthracene (IV) [5]. If conjugation in a molecule is disrupted, e.g., by the presence of a saturated bond, then that molecule will exhibit two separate fluorescences corresponding to the two separate conjugated systems. Thus, the fluorescence of 9,10-dihydroanthracene (V) corresponds to that of two separate benzene rings. Occasionally,

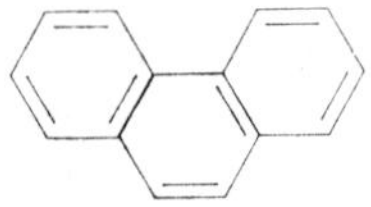

(II) 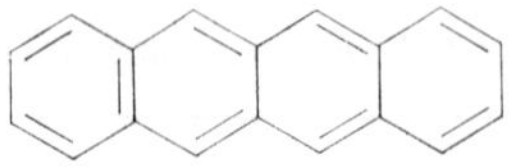 (III)

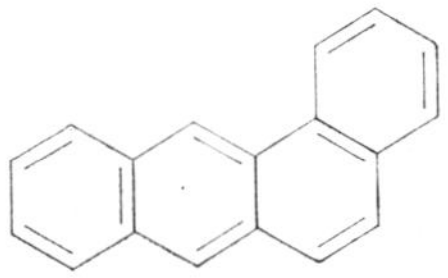

(IV)

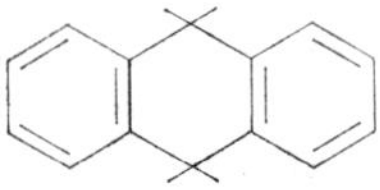

(V) 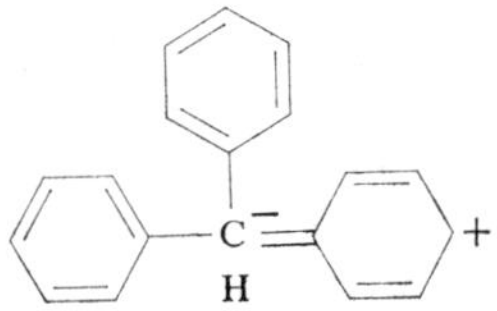 (VI)

however (see Table 6.2), a higher fluorescence efficiency is found than can be explained on the basis of the extent of continuous conjugation, e.g., triphenylmethane, which on the above considerations should have a fluorescence similar to benzene. In such instances it is necessary to invoke the concept of hyperconjugation, in which resonance contributions from structures such as VI are permitted.

The Geometrical Arrangement of the Molecule

In addition to the requirement for a conjugated system of double bonds, certain geometrical considerations also seem to be important. Thus, planarity of the conjugated system appears to be essential for maximum fluorescence. When the planarity of a system is destroyed through steric hindrance, the free mobility of the π electrons will be partially inhibited resulting in a loss of fluorescence. *o*-Diphenylbenzene, for example, although highly conjugated, is not planar, for there is steric hindrance between the phenyl groups, and its fluorescence efficiency is therefore similar to that of benzene. *cis*-Stilbene is less than 1% as fluorescent as its planar *trans*-isomer because it cannot have a planar configuration since molecular overcrowding of its *ortho*-hydrogen atoms occurs. Hexaheliceine (VII) is another instance of an aromatic molecule which is non-planar because of steric hindrance. As a result, it shows a moderate blue fluorescence corresponding to two weakly-linked phenanthrene molecules [6]. The closely related but slightly strained planar compound coronene (VIII) also fluoresces blue.

Molecular rigidity is also necessary for maximum fluorescence. In a non-rigid molecule the absorbed energy can be readily dissipated as heat in much the same way that a loosely-connected vibrating structure tends to produce heat rather than work. This need for molecular rigidity probably explains the much greater fluorescence of cyclic compared with chain systems.

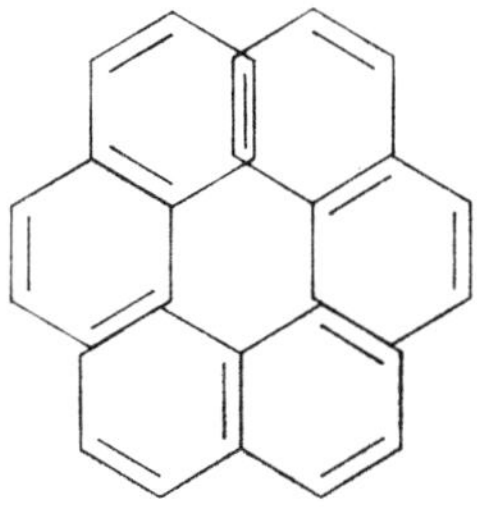

VII 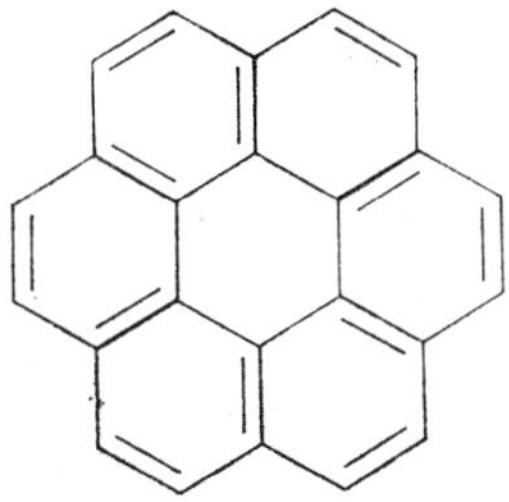 VIII

Comparison of the fluorescence efficiencies of biphenyl and fluorene provides a good illustration of the need for planarity. These compounds possess the same degree of conjugation, but the benzene rings in fluorene (IX) are held rigidly in a planar configuration, whereas those in biphenyl (X) are not. This is reflected in their fluorescence efficiencies, for fluorene has a quantum efficiency of about 0·54 in hexane while that of biphenyl is only 0·23 [4].

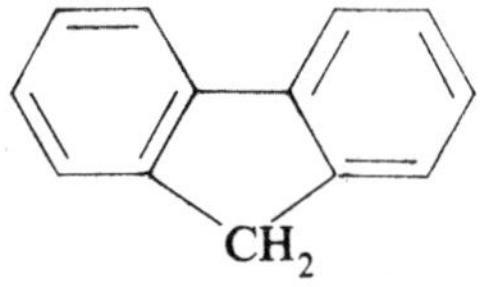

CH_2

IX 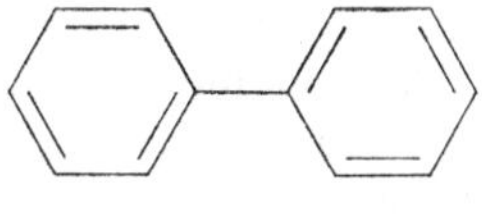 X

Many compounds, e.g., azo dyes, which are apparently non-fluorescent in solution, may become visibly fluorescent when dried on filter paper or other adsorbents. Presumably this is due to the need for molecular rigidity. Although fluorescein (XI) and phenolphthalein (XII) contain the same degree of conjugation, the phenolic rings of fluorescein are held in a tight compact planar structure while those in phenolphthalein are not. As a result, fluorescein is highly fluorescent, while phenolphthalein is virtually non-fluorescent.

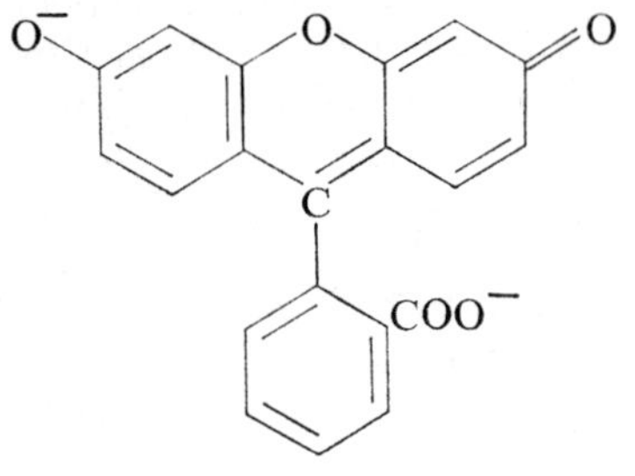

(XI)

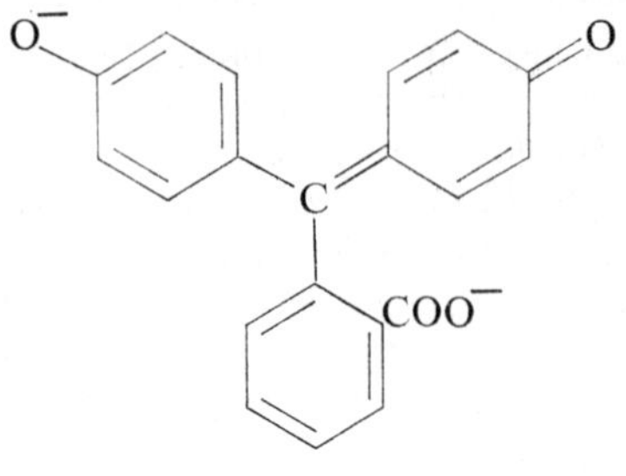

(XII)

The Type and Position of Substituents

Although the addition of a substituent to a saturated ring will not result in a fluorescent molecule, substitution of a conjugated system may have a very profound effect on the fluorescence properties of such a ring. To exert this effect it is generally necessary for the substituent to be either directly attached to the conjugated system or in close steric proximity to it. However, atoms which have a large magnetic field associated with them, (e.g., bromine and iodine) will reduce fluorescence wherever they occur as substituents. It is necessary to distinguish quite clearly between substituents which have a direct effect on π electron mobility and those which, by virtue of their bulk, produce steric hindrance. For example, 2,2′-dibutylbiphenyl is considerably less fluorescent than biphenyl itself because the butyl groups force the rings out of a planar configuration. However, these groups have virtually no effect on the π electron mobility directly. The extent to which the fluorescence properties of a conjugate system are modified by a particular substituent depends largely on the change produced by the latter in π electron mobility. Substituents which enhance π electron mobility will normally increase fluorescence, while those which decrease it will reduce fluorescence. Thus, in general, electron donating (positively mesomeric) substituents tend to enhance fluorescence while electron-withdrawing substituents tend to diminish or abolish it (see Table 6.3). The effect of substitution upon fluorescence can be illustrated with benzene, aniline and benzoic acid. In dilute solution in water, aniline is 40–50 times more fluorescent than benzene, whereas benzoic acid is non-fluorescent. The NH_2 group in aniline tends to activate the benzene ring. At the same time both the absorption and the fluorescence wavelength maxima undergo a shift to a longer wavelength, thus increasing the freedom of the π electrons. The —COOH group of benzoic acid on the other hand partially deactivates the ring by withdrawing the π electrons

TABLE 6.3. *Fluorescence of monosubstituted benzenes* C_6H_5R

Compound	Substituent R	ε max.	Excitation (λ max. nm)	Fluorescence (λ max. nm)	Relative intensity
Ortho-para-directing					
Benzene	H	204	260	291	1
Aniline	NH_2	1430	280	345	46
Monomethylaniline	$NHCH_3$	—	280	360	—
Dimethylaniline	$N(CH_3)_2$	—	283	363	114
Acetanilide	$NHCOCH_3$	—	—	None	0
Fluorobenzene	F	—	262	285	13
Chlorobenzene	Cl	190	265	294	0·02
Bromobenzene	Br	192	—	None	0
Iodobenzene	I	190	—	None	0
Toluene	CH_3	225	265	292	3·8
Ethylbenzene	C_2H_5	240	265	292	1·2
Phenol	OH	1450	272	320	112
Anisole	OCH_3	1480	269	302	92
Phenoxide ion	O^-	2600	289	345	~1
Meta-directing					
Benzoic acid	COOH	970	—	None	0
Benzoate ion	COO^-	560	—	None	0
Nitrobenzene	NO_2	7800	—	None	0
Benzene sulphonic acid	SO_3H	—	—	None	0
Benzene-sulphonamide	SO_2NH_2	740	—	None	0
Benzaldehyde	CHO	11400	—	None	0
Benzenearsonic acid	AsO_3H_2	—	—	None	0
Benzonitrile	CN	1000	273	294	45

from it, thus reducing their mobility. Mono-substituted aromatic compounds containing OH, OCH_3, NH_2, $NHCH_3$, $N(CH_3)_2$, F and C≡N are usually fluorescent, while those containing $NHCOCH_3$, Cl, Br, I, C=O, NO_2, SO_3H and COOH are likely to be weakly or non-fluorescent [8]. Substitution of alkyl groups generally has little effect. The presence of bromine, iodine or heavy metal substituents in either a mono- or a poly-substituted aromatic compound invariably leads to a reduction in fluorescence by encouraging intersystem crossing from the excited singlet to the triplet states. (Iodide ions will also quench fluorescence.) The quenching effect of bromo- and iodo-substituents is well demonstrated with fluorescein solutions. The fluorescence efficiency of fluorescein is 70%, that of its tetrabromo-derivative, eosin, is 15% and that of its tetraiodo-derivative, erythrosine, is 3%. The presence of NO_2 in the molecule always reduces fluorescence, whether the molecule is simple or complex.

The effect of disubstitution and polysubstitution is more complex since the overall effect, on the π electron mobility, produced by the substituents has to be considered. It would appear that the degree of π electron mobility is very critical, for although *p*-hydroxyaniline has about the same fluorescence intensity as aniline, the *p*-substitution of aniline with a weakly electron-withdrawing group (such as COOH or SO_2NH) produces a far more intensely fluorescent molecule (see Table 6.4). Substitution of

TABLE 6.4 *Fluorescence of some disubstituted benzenes compared with related monosubstituted benzenes*

R–C₆H₄–R′

Compound	R	R′	ε max.	Excitation nm	Fluorescence nm	Relative intensity
Aniline	NH_2	H	1,430	280	350	46
Benzenesulphonamide	H	SO_2NH_2	740	—	None	0
Sulphanilamide	NH_2	SO_2NH_2	16,600	260	350	220
Chlorobenzene	H	Cl	190	265	294	0·02
p–Chloroaniline	NH_2	Cl	1,500	290	360	32
Nitrobenzene	H	NO_2	7,800	—	None	0
p–Nitroaniline	NH_2	NO_2	13,500	—	None	0
						Benzene = 1

phenol with such groups also results in an increase in fluorescence intensity. Substitution of aniline with a strongly electron-withdrawing group such as NO_2 abolishes fluorescence. (This is unlikely to be due to predissociation, since the absorption maximum of this compound is at 381 nm.) However, not all nitro compounds are non-fluorescent, for 5-dimethylamino-4′-nitrostilbene is weakly fluorescent (see Table 6.1) presumably because in this molecule there is sufficient conjugation to overcome the deactivating effect of the nitro group [10].

If intramolecular hydrogen bonding occurs between substituents this may profoundly modify fluorescence, although the effect is often unpredictable. Thus, for example, amide or carboxyl anions may be expected to hydrogen bond with phenolic hydroxy groups. In the case of salicylic acid, such bonding results in the enhancement of fluorescence, whereas in proteins it may quench fluorescence [11].

Although a wide variety of heterocyclic compounds is known to be fluorescent, relatively few systematic fluorimetric studies have been made on their structure-fluorescence relationships. Investigations on these compounds are made difficult because their fluorescence characteristics are often dependent on the nature of the solvent used. Generally such compounds tend to be most fluorescent in polar solvents, whilst some are only fluorescent in acid. Under these conditions the lone pair of electrons is bound, and the longest absorption wavelength becomes $\pi \rightarrow \pi^*$ instead of $n \rightarrow \pi^*$.

The fluorescence-structure relationships of heterocyclics, particularly those containing nitrogen, are poorly understood at present and it is therefore difficult to generalize. However, it appears that heterocyclics with a longest absorption wavelength corresponding to an $n \rightarrow \pi^*$ transition are likely to be non-fluorescent, whereas those corresponding to a $\pi \rightarrow \pi^*$ are likely to be fluorescent, provided the requirements already described are fulfilled.

Six-membered ring structures containing a single heteroatom, e.g., pyridine, are non-fluorescent.† If conjugation is increased by the addition of a benzene ring as in the case of quinoline, a compound which is weakly fluorescent in the ultra-violet is obtained. Substitution of a second =N— diminishes fluorescence even further, for pyrimidines, diazines, triazines and their benzo derivatives are non-fluorescent. In these cases the longest absorption wavelength in most solvents, except acid, corresponds to an n–π^* transition. The fluorescence of heterocyclic compounds is particularly sensitive to the effects of substituents, for although pyridine is non-fluorescent, 3-hydroxypyridine in its non-ionized form is almost as fluorescent as phenol. But the non-ionized form of 2-hydroxypyridine is only very weakly fluorescent, whilst 4-hydroxypyridine is non-fluorescent [12]. (The fluorescence of the ions is described later, under pH effects.) Similarly, the fluorescence of the un-ionized form of the hydroxyquinolines is very dependent on the position of substitution (see Table 6.5) for, whereas

TABLE 6.5. *Effect of the position of substitution on the fluorescent intensities of some hydroxyquinolines and indoles*

	Position of OH substitute	λ_a nm	λ_f nm	Relative intensity for mol. forms
QUINOLINE	None	312	400	1·0
	2	332	370	40
	3	333	353	245
	4	330	348	6·0
	5	276	402	8·0
	6	246	370	207
	7	{332, 400}	505	{29, 27}
	8	312	398	8·0
INDOLE	None	285	355	1·0
	2	non-fluorescent		—
	3	292	388	20
	4	non-fluorescent		—
	5	304	342	16
	6	303	362	3·0

the fluorescence intensity of 4, 5 or 8-hydroxyquinoline is only eight times greater than that of quinoline itself, 3- or 6-hydroxyquinolines show a

† In this section, unless otherwise stated, the fluorescence properties referred to are those in alcohol or water.

greater than two-hundredfold increase. Further examples are found among the coumarins and acridines, for while 7-hydroxycoumarin and 9-aminoacridine are intensely fluorescent, 3-hydroxycoumarin is only very weakly fluorescent, and 4-aminoacridine is non-fluorescent.

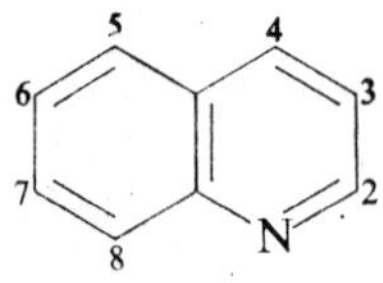

(XIII) 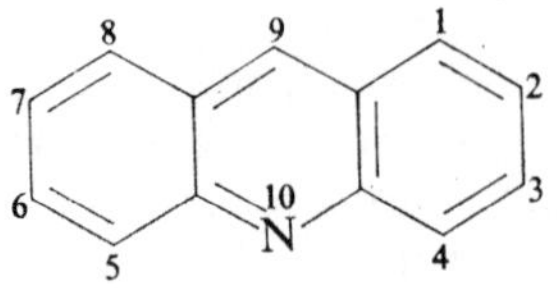 (XIV) 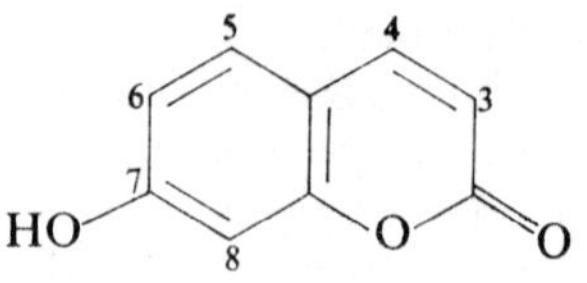 (XV)

Five-membered rings containing a single heteroatom (e.g., O, N) are, like the six-member systems, non-fluorescent. However, when they are fused with a benzene ring as in the case of benzofuran and indole, the fluorescence efficiency becomes comparable with that of naphthalene. In this sense the —NH— and —O— groups in the ring are equivalent to a —CH=CH— grouping. The replacement of a single carbon atom in a ring by a heavier atom tends to produce a bathochromic shift in activation and fluorescence wavelength. This may be associated with a reduction in fluorescence efficiency, since it is likely to encourage intersystem crossing. If intersystem crossing predominates, the molecules are more likely to phosphoresce than fluoresce. Thus, benzothiophen is phosphorescent but not fluorescent. However, some thiophen analogues are fluorescent, e.g., *trans*-thioindigo (XVI) shows a yellow-green fluorescence. The *cis*-isomer is non-fluorescent, presumably because it is non-planar [13].

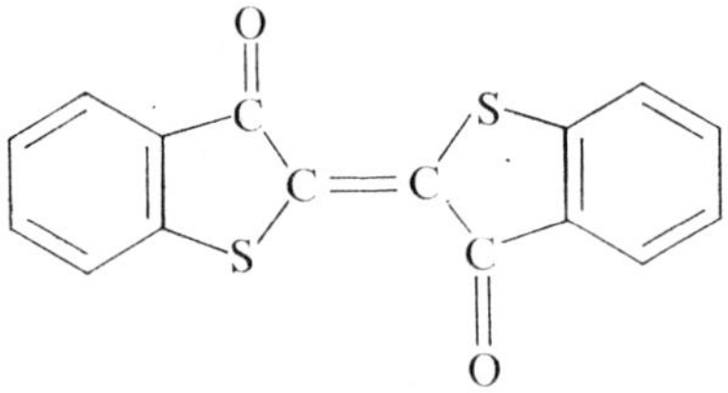

XVI

An increase in conjugation is not an essential requirement for the fluorescence of these five-membered rings, because a number of them, when they contain suitable substituents, are fluorescent. Thus, amino-cyanofuroic ester (XVII) is fluorescent because of the presence of the electron-donating NH_2 and CN groups, while methylaminocitraconic methylimide (XVIII) is fluorescent because of its OH and $NHCH_3$ groups [14]. The position of substitution may be critical in 5-membered hetero-cycles (see Table 6.5), for 2- and 4-hydroxyindole are non-fluorescent, whereas 3- and 5-hydroxyindole are more than fifteen times as fluorescent

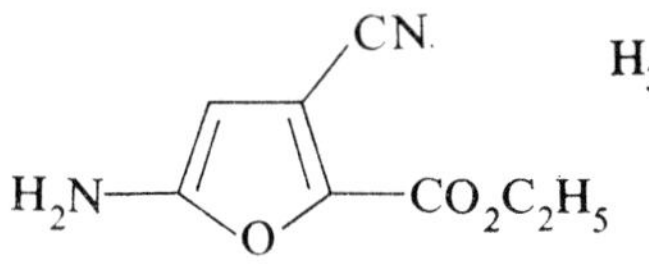

XVII 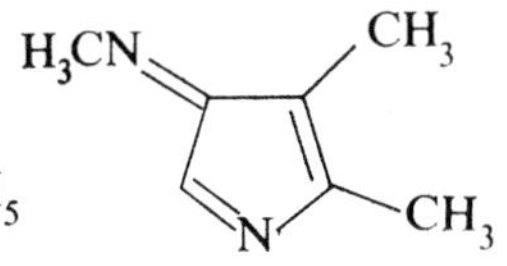 XVIII

as indole. Addition of a double bonded nitrogen into either indole or benzofuran may, depending on the position of substitution, either enhance or reduce the fluorescence. For example, if =N— is substituted in the 2- or 3- positions a more fluorescent compound is produced, while substitution in other positions produces a less fluorescent species. Many oxazole derivatives are highly fluorescent. Indeed, 2,2-diphenyloxazole (PPO, XIX) and 2,2′-*p*-phenylenebis-(5-phenyloxazole) (POPOP, XX) have a fluorescence efficiency, in hexane, of near unity, and emit a bright blue fluorescence [15].

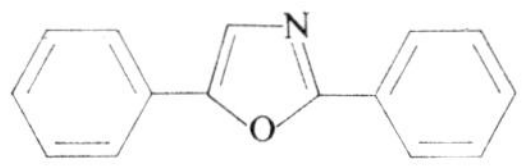

XIX 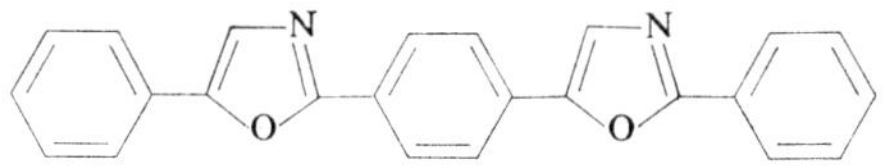 XX

These compounds are both very extensively used as liquid scintillators in radioassays. A number of the more complex heterocyclic ring systems show a visible fluorescence. For example, benzothiazole, benzoxazole and benziminazole derivatives (e.g., XXI) have been employed as fluorescence brightening agents [16].

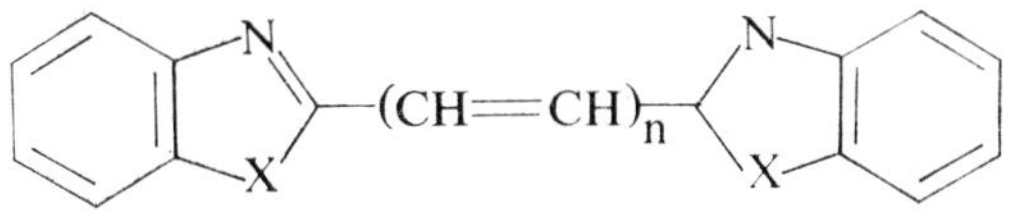

where X = S,O or NH and n = 0 or 1

XXI

The benzoxazole derivatives have the highest fluorescent intensity of the three. Benzothiazole is the least fluorescent, but has the longest fluorescence wavelength. Other fluorescent dyes incorporate phenazine (e.g., XXII, where X=NH) phenthiazine (X=S) or phenoxazine rings, e.g., amethyst violet, methylene blue and capri blue [7]. Again the S analogue methylene blue is the least fluorescent of the three.

Lumiflavin, which is the photochemical conversion product of riboflavin, is highly fluorescent. It contains both a diazine and a pyrimidine ring (XXIII).

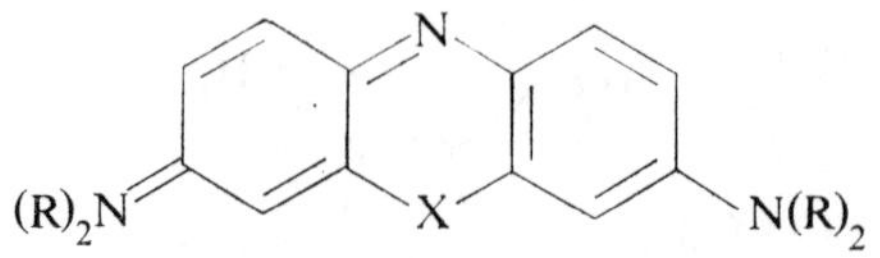

X=NH, S or O
R=alkyl group

XXII

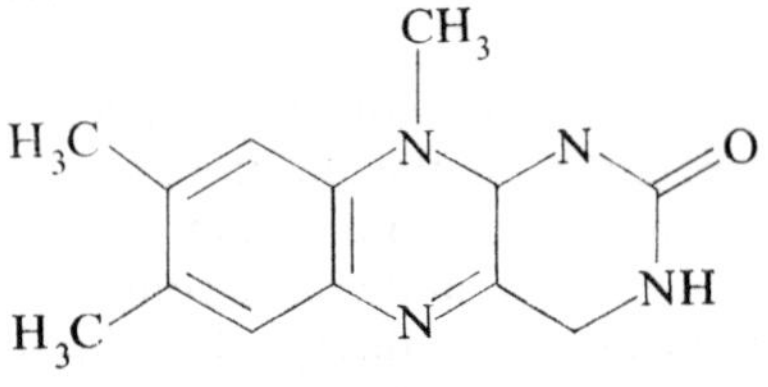

XXIII

Few studies have been made of ring systems other than those containing oxygen, nitrogen or sulphur. Investigations of some selenium compounds have shown that selenium exerts a quenching effect on fluorescence for if the 2-carbon atom of benzimidazole is replaced by selenium to give piazo-selenol (XXIV) a sharp reduction in fluorescence occurs. In fact the weak

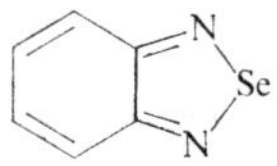

XXIV

fluorescence attributed to this compound may be due to impurities or photodecomposition products. Addition of another benzene ring, however, increases the conjugation sufficiently for 3,4-benzopiazoselenol to be fluorescent. This fluorescence has been used as the basis of a method for the estimation of selenium [17].

Chemically-induced Fluorescence

Organic compounds which are fluorescent *per se* are said to possess 'native' fluorescence. In many cases, simple chemical reactions make it possible to convert non-fluorescent compounds into fluorescent ones, in order to facilitate their estimation. Fluorescence of this type is described as chemically-induced. For example, most aromatic nitro compounds are non-fluorescent, but they may be estimated fluorimetrically if they are converted to their corresponding amino derivatives which are usually highly fluorescent. Chemically-induced fluorescence may also be used to enhance a weak native fluorescence, or to improve the specificity of an estimation method. Thus, both adrenaline and noradrenaline can be oxidized at pH 6·5 to 3,5,6-trihydroxyindoles (adrenolutine and noradrenolutine) which are highly fluorescent in alkali. However, if the oxidation is carried out at pH 3·5, adrenaline can be estimated alone, for at this pH it is oxidized to a trihydroxyindole, whereas noradrenaline remains unchanged [18].

Other Factors Affecting the Observed Fluorescence Intensity

In addition to the structural considerations described above, various environmental factors are also very important in determining the observed fluorescence of a compound. These factors can be considered under the following headings: (*a*) instrumental factors, (*b*) influence of other molecules, (*c*) pH, (*d*) temperature and (*e*) the stability of the compound in the exciting light.

Instrumental Factors

Fluorimeters which automatically correct the absorption and fluorescence for instrumental variables such as lamp output and photomultiplier response are uncommon at present. Most instruments produce spectra which are often grossly distorted versions of the true spectrum, particularly in the ultra-violet region. In the majority of published work no instrumental correction factors have been applied, and it is important to remember this when comparing and evaluating results.

Influence of Other Molecules

The fluorescence properties of a molecule may be profoundly affected by the presence of other molecules. Many compounds fluoresce better in the vapour phase, where the influence of other molecules is at a minimum, than in solution. Interactions with other molecules may produce a shift in the absorption and fluorescence wavelengths and/or a change in the observed fluorescence intensity of a compound by any of the following means (see Fig. 6.2):

1. By absorbing the exciting light, thus preventing it from exciting the fluorescent substance (A), or by absorbing the emitted fluorescent light, thereby stopping it from reaching the photodetector (inner filter effects).
2. By interacting with the fluorescent substance (or by self-interaction) in the ground state to form an aggregate of some type (static quenching) which becomes the light-absorbing species.
3. By interacting with the fluorescent substance in the excited state to form a new molecular species from which the fluorescent light will be emitted (dynamic or diffusional quenching). The new molecular species may be a dimer (for self-interaction) or a charged species resulting from proton or electron transfer. Dynamic quenching is dependent on the rate of diffusion in solution, therefore it is temperature and viscosity dependent. Dynamic quenching has proved to be invaluable for studying the properties of the excited state and reaction rates.

4. By contributing spurious fluorescence or by reacting chemically with the fluorescent substance.

From a practical point of view it is important to understand these processes. The quenching of a fluorescence material may be due either to molecules of the same species (concentration quenching) or to those of a different species (solvent or impurities). True quenching (i.e., static and

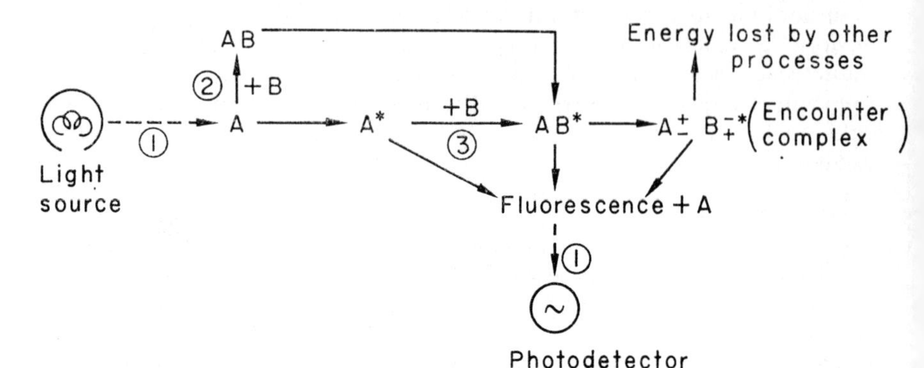

FIG. 6.2 Diagrammatic representation of the ways in which other molecules may 'quench' the observed fluorescence intensity of a compound (A). (1) reprerents inner filter effects, (2) static and (3) dynamic quenching.

dynamic) can be distinguished from inner filter effects because changes in the concentration of a 'true quencher' will alter the quantum efficiency of a fluorescent material, but changes in the concentration of an 'inner filter quencher' will not. Most true quenchers obey the Stern–Volmer equation:

$$\frac{Q_0}{Q} = 1 + K_q C$$

where Q_0 = efficiency of fluorescence without quencher
Q = efficiency at a concentration of C (mole/litre)
K_q = Stern–Volmer quenching constant

The rate constant k for encounters between solute molecules in a liquid is given *approximately* by the diffusion formula: $k = 8\boldsymbol{RT}/2000\eta$ where η = liquid viscosity in poises. For liquids of viscosities ~0·01 (e.g., water or benzene) k is about 10^{10} litre mole^{-1} s^{-1}. By equating the bimolecular rate of quenching (constant k_q) against the rate of emission it can easily be shown that $K_q = k_q \tau$, where τ is the *actual* mean life of the fluorescence, i.e. (quantum efficiency) × (true radiational life). Since K_q often has values near 100 and τ is usually about 10^{-8} s, k_q is often near k, and this shows that quenching then occurs at every molecular encounter. The constants for quenching by oxygen molecules are higher than expected, by a factor of about 3, because for small molecules the diffusion equation is inaccurate. Weak quenchers fall outside the range of the diffusion equation altogether, and their rate constants are independent of viscosity like any ordinary bimolecular reaction.

INNER FILTER EFFECTS

Inner filter effects generally result from using a light-absorbing solvent, or from carrying out fluorescence measurements at high concentrations. The intensity of fluorescence of a compound is directly proportional to its concentration only in highly dilute solutions. (See Fig. 6.3.) In

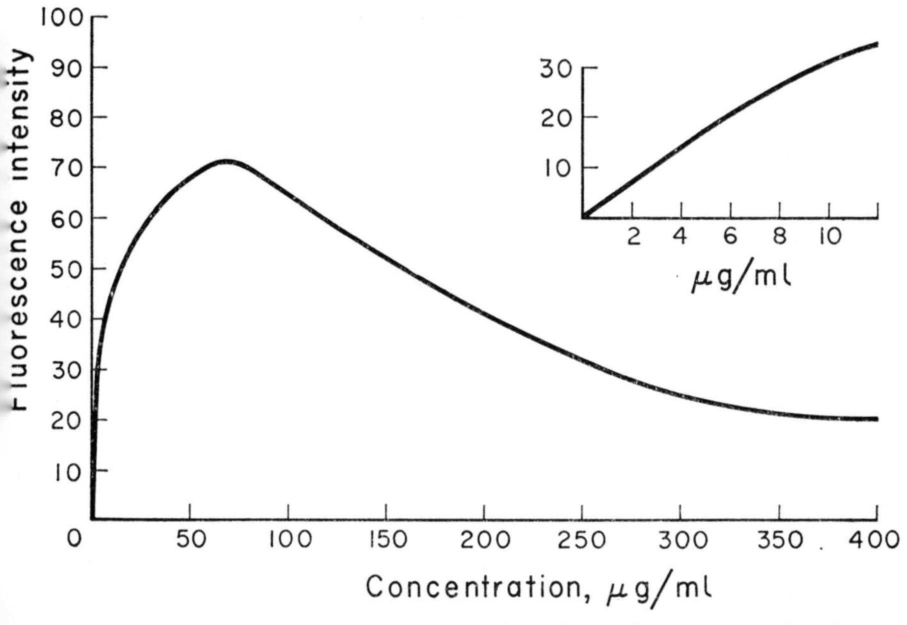

FIG. 6.3 Plot of measurable fluorescence intensity against concentration for phenol solutions.

many cases this only applies at concentrations below 5–10 μg/ml. (The exact relationship between the concentration of a compound and its observed fluorescence is largely dependent on instrumental design. See Chapter 2.) In instrumental set-ups in which the fluorescence is measured at right-angles to the exciting light, the emitted fluorescence has to pass through the solution to the detector, and during this passage some of it will be reabsorbed by other molecules of the compound under examination. The higher the concentration of the compound, the greater the proportion of emitted fluorescence reabsorbed. Therefore linearity between fluorescence intensity and concentration can only be expected at high dilutions, where the number of molecules present is small enough to make the extent of reabsorption unimportant compared with the amount of fluorescence emitted. This effect is most likely to occur in compounds whose activation and fluorescence wavelengths are close enough together to overlap considerably. Such inner filter effects may not only reduce the observed fluorescence intensity but also result in an apparent shift in the activation wavelength. Thus, quinine bisulphate (in $0{\cdot}1\text{N}\,H_2SO_4$) at 1 μg/ml normally absorbs maximally at 247 nm, but at 100 μg/ml the observed excitation spectrum actually passes through a minimum at this wavelength and the remainder of the excitation spectrum is also distorted [19]. Care must be taken, when selecting a solvent, to ensure that its light-absorbing characteristics do not overlap with absorption or fluorescence spectra of the compound under study. If such overlap does occur, the solvent may well act as an inner filter. Benzene and acetone, for example, are unsuitable solvents for studying ultra-violet fluorescence, although they can be used for work in the visible spectral regions. Traces of light-absorbing contaminants may exert a similar effect. Thus, chromic acid remaining from the incomplete washing of glassware may partially quench ultra-violet fluorescence.

Static and Dynamic Quenching

Static and dynamic quenching may be due to like molecules forming dimers or excimers, or to chemically different molecules such as solvent molecules and contaminants.

Solvents

If a change in solvent produces alterations in both the activation and fluorescence wavelengths, static quenching is indicated, that is, the solvent must be able to combine in some way with the solute in the ground state (see Fig. 6.2). However, if a solvent change produces a shift in fluorescence wavelength only, then interaction between the solute and the solvent occurring in the excited state is indicated, i.e., dynamic quenching. Hydrogen bonding, the dielectric constant of the solvent, or its viscosity, is frequently invoked to explain these solvent-fluorescence relationships. However, the picture is highly complex, for these effects may either reinforce or cancel one another, thus the solvent-fluorescence relationships are largely unpredictable. When a molecule is electronically excited its dipole moment is changed and it is not in immediate equilibrium with its solvent environment. If the molecule relaxes, i.e., equilibrates with its surroundings, in a time shorter than its radiational lifetime, some energy will be dissipated and the fluorescence emission will lie well to the red of the absorption band. Such shifts will depend both on solvent viscosity and polarity [10]. Thus, indole (see Table 6.6) shows a pronounced bathochromic shift in fluorescence wavelength with increasing dielectric constant of the solvent, while the activation wavelength remains constant. 1,2-Dimethylindole, which is incapable of hydrogen bonding, shows similar wavelength changes to indole, therefore these effects must be attributable to the dielectric properties of the solvent [20]. In the case of 3,4,5,8-tetra-azanthracene (XXV), the activation and, to a greater extent,

TABLE 6.6 *Solvent effects on the relative intensity (R.I) of the fluorescence of some heterocycles*

Solvent dielectric constant 20°C	Cyclohexane 2·023 λ_a–λ_f	R.I.	Benzene 2·284 λ_a–λ_f	R.I.	Dioxane 2·209 λ_a–λ_f	R.I.	Chloroform 4·806 λ_a–λ_f	R.I.	Ethanol 24·30 λ_a–λ_f	R.I.	Water 80·37 λ_a–λ_f	R.I.
Indole	285–297	1·5	285–310	0·15	285–330	1·45	—	—	285–330	1·53	285–350	
Acridine	365–400	<0·01	—	—	—	—	—	—	355–415 380–44	0·13	355–430 455 382–478	1·0
1,4,5,8–Tetra–azo-anthracene	393–405	2·2	404–432	1·8	402–438	2·3	405–462	1·6	422–475	1·2	415–505	1·0
Chlorophyll a.	441–676 ~665–730	<0·01	441–676 658–730	<0·01	441–676 658–730	<0·01	441–676 658–730	<0·01	441–676 ~665–730	1·0	441–676 ~665–730	1·0

the fluorescence wavelength both undergo a red shift with increasing solvent dielectric constant. Thus, this compound must be able to combine with the solvent in both the ground and excited states. The fluorescence intensity of 3,4,5,8-tetra-azanthracene is greater in benzene and hexane than in water and ethanol, whilst acridine is virtually non-fluorescent in hexane and benzene but moderately so in water and moist ethanol. In

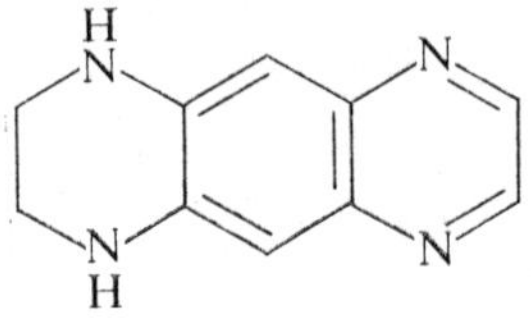

XXV

hexane the lack of fluorescence of acridine, may be ascribed to the fact that the latter's longest absorption wavelength corresponds to an n–π^* transition, whereas in water the lone pair of electrons are hydrogen bonded, and the longest absorption wavelength is due to a $\pi \rightarrow \pi^*$ excitation, which favours fluorescence. The fluorescence of acridine in hexane can be greatly intensified if trichloroacetic acid, which is known to combine with lone pairs of electrons, is added. In fact, acridine in hexane then becomes as fluorescent as acridine in water. This supports the view that the fluorescence of acridine only occurs when its lone pair of electrons is bound. This explanation is not completely satisfactory, however, since the fluorescence properties of some analogues of acridine, e.g., 3,4-, 5,6- and 7,8-benzoquinolines, which might be expected to behave like acridine, are unaffected by solvent changes or by the addition of trichloroacetic acid [21].

The fluorescence of chlorophyll is also very dependent on the nature of the solvent used, for chlorophyll is non-fluorescent in dry hexane and benzene, but highly fluorescent in water. If traces of water or other activators are added to hexane solutions of chlorophyll, the fluorescence intensifies, suggesting that hydrogen bonding is involved [22]. Unfortunately, these hydrogen bonding effects are very largely unpredictable.

A number of triphenylmethane dyes are non-fluorescent in hexane, ethyl alcohol or water but are fluorescent in viscous solvents such as glycerol. Viscosity may enhance fluorescence either by conferring rigidity on the molecule or by suppressing diffusional controlled quenching processes (i.e., dynamic quenching).

Polarization of Fluorescence in Viscous Solution

When a fluorescent material is irradiated with polarized light, those molecules whose axes happen to be orientated in a direction parallel to the light plane will be preferentially excited. If these molecules are held in a rigid medium they will emit polarized fluorescence. However, if the medium is not rigid, the excited molecules may be able to rotate about their axes before the fluorescence is emitted. In this case the fluorescence will be partially polarized or non-polarized. Thus, in a high-viscosity solvent small fluorescent molecules excited with plane polarized light will emit polarized fluorescence, while in low-viscosity solvents the fluorescence will be non-polarized. A small amount of polarized fluorescence may also be observed in high viscosity solvents even if the exciting light is unpolarized. An approximate relationship is:

$$\frac{1}{p}=\frac{1}{p_0}+\left(\frac{1}{p_0}-\frac{1}{3}\right)\frac{RT\tau}{V\eta}$$

where p is the degree of polarization observed at right-angles to the beam of plane-polarized exciting light, η the viscosity of the solvent, τ the mean life of the fluorescence, V the molar volume of the fluorescent molecule, and p_0 the limiting value of the polarization for no molecular rotation [23]. Given τ, therefore, one can determine V, or vice versa. In solutions above 10^{-3}M additional depolarization by energy transfer occurs [50, 51].

Quenching by Other Agents

The fluorescence of crystals is especially prone to quenching even by trace amounts of impurities, because of energy transfer mechanisms. For such quenching to occur, the longest wave energy level of the quencher should be just below that of the excited fluorescent molecule. (See Chapter 8 for a fuller description.) Often the quenching agent is one which on excitation readily reverts to the triplet state, so that when the fluorescent compound is excited its energy is immediately taken up by the quencher, which then rapidly assumes a non-radiative triplet state, thus preventing absorbed energy from being emitted as fluorescence. For example, the presence of small amounts of some organo-tin compounds (plastic stabilizers) may by this mechanism cause a sharp reduction in the fluorescence of some plastic scintillators. Similarly, the fluorescence of anthracene can be eliminated by small amounts of phenazine [24].

Iodo, bromo, nitro compounds and molecular oxygen may produce quenching by encouraging intersystem crossing of the excited singlet state of the molecule to its triplet state. Nitro compounds may also quench by forming π complexes with the fluorescent agent in the excited state. Fluoranthene, for example, is quenched by nitromethane through the formation of a π complex, and this has been used as the basis of a method for estimating fluoranthene quantitatively [24]. Aromatic hydrocarbons and simple phenol anions are peculiarly sensitive to quenching by molecular oxygen, whereas most other compounds are relatively unaffected by it. Oxygen in the ground state exists as a triplet. When it comes in contact with a molecule in the excited singlet state, it may exchange its triplet state with the singlet state of the fluorescent molecule, thus producing quenching; the reaction may proceed via a charge-transfer intermediate:

$$O_2^{\text{triplet}} + A^{*\text{singlet}} \rightarrow (A^+O_2^-)^* \rightarrow {}^{\text{triplet}}A^* + {}^{\text{singlet}}O_2$$

This quenching by oxygen has been used as the basis of an estimation method for determining oxygen concentrations. The method employs a borate-benzoic complex which is quenched in proportion to the oxygen concentration [25]. Oxygen, of course, may also exert an irreversible quenching effect by chemically reacting with the excited state to form a non-fluorescent photo-oxidation product.

Electron Transfer

Electron transfer processes may also be responsible for the degradation of excited state energy, since the excited, but not the corresponding ground state molecule, may be able to donate or accept an electron from the quenching molecule, and in this way the excited state may be destroyed. Thus, ferrous ions can destroy the fluorescence of excited methylene blue by electron donation, while they have no effect on its ground state. Some fluorescent molecules are quenched by electron donating anions such as I^-, Br^-, SCN^- and $S_2O_3^-$, while others are degraded by electron accepting anions such as IO_3^-, NO_3^- and $S_4O_6^-$. The direction of the electron transfer depends on the redox potential of the excited ion, which is likely to differ from that of the unexcited molecule [9]. Organic electron donators or acceptors may also produce quenching. For example, the fluorescence of acridine is quenched by a number of aliphatic amines [26]. Fluorescent compounds may be specifically quenched by one kind of ion but be unaffected by another. Thus, quinine bisulphate is highly fluorescent in $0{\cdot}1\text{N}H_2SO_4$ but is only weakly fluorescent in $0{\cdot}1\text{N}HCl$ because of the specific quenching ability of chloride ions. This specificity of quenching can sometimes be used for identification purposes. Thiosulphate or nitrate ions, for instance, quench the fluorescence of tryptophan and this principle has been used for identifying tryptophan in tissue fluids [27]. Again, benzopyrene and benzofluoranthene have a similar fluorescence, but whereas bromine quenches benzopyrene fluorescence, it has little effect on benzofluoranthene. This has enabled mixtures of these two hydrocarbons to be analysed quantitatively [28]. The term quenchofluorimetric analysis has been coined for this type of assay.

Excimer Formation

Molecules may combine together to form aggregates at high concentrations. A few compounds have been shown to form such aggregates (e.g., dimers) in the *excited* but not in the ground state. Dimerization of this type can usually be detected since it produces a bathocromic shift in fluorescence wavelength without a corresponding change in activation wavelength, when the concentration of the solute is increased. It is most likely to occur in low viscosity solvents, since the rate of formation of dimers is largely dependent on the rate of diffusion of the excited monomer in the solvent. For instance, pyrene in dilute solution (2×10^{-4} mole/litre) in benzene shows a violet fluorescence due to the monomer (λ_F 384 nm) but when the concentration is increased to 2×10^{-3} mole/litre, this violet fluorescence decreases and a new fluorescence peak, corresponding to the dimer, appears at 478 nm [29]. Similar effects have been shown for the fluorescent dye 1,1′-diethyl-2,2′-pyridocyanine iodide [30], chlorophyll [31], benzene, toluene, various anthracene derivatives and PPO.

pH Effects

Many compounds are capable of undergoing ionization. Almost always the ionic form of a compound exhibits different fluorescent characteristics from those of the un-ionized form. The effect of pH upon the fluorescence of a compound is therefore of considerable importance, and a knowledge of the changes in fluorescence resulting from altering the pH of the solvent medium can be valuable in a number of ways. Some compounds (e.g., sulphapyridine) fluoresce over a very limited pH range only. For below an H_0 value of -1 and above pH 4 this compound is non-fluorescent. Other compounds such as phenol, fluoresce over a much wider pH range (see Fig. 6.4). At H_0 of -1 phenol is hardly fluorescent, at pH 0 it has a definite

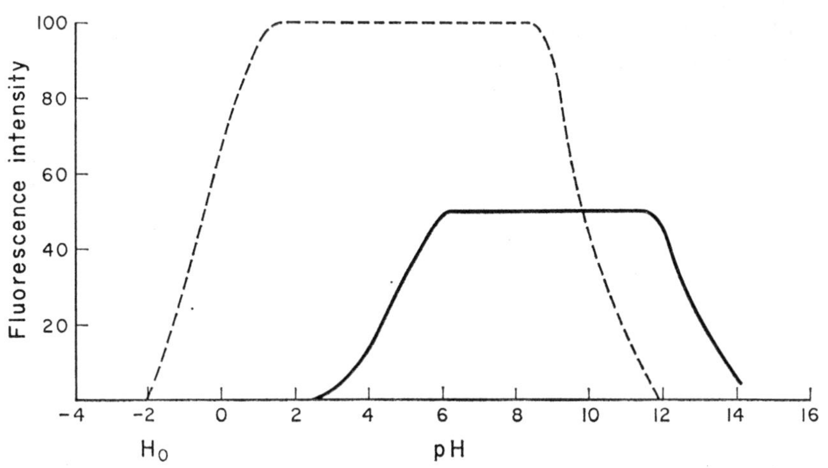

Fig. 6.4 Fluorescence intensities of phenol (– – –) and aniline (——) at various pH values.

fluorescence and at pH 1 it is maximally fluorescent (λ_A 272, λ_F 320 nm). This fluorescence corresponding to un-ionized phenol remains constant up to pH 9·5. Above 9·5 this fluorescence becomes progressively less and a new weak fluorescence is observed (λ_F 289, λ_F 345 nm) corresponding to the phenol anion. The fluorescence of molecular phenol falls to 50% of its maximal value at pH 10 which is the pK_a of phenol. Fluorimetry may in other cases also provide a useful means of determining ground state pK_a values. Aniline is non-fluorescent as the cation (XXVI).

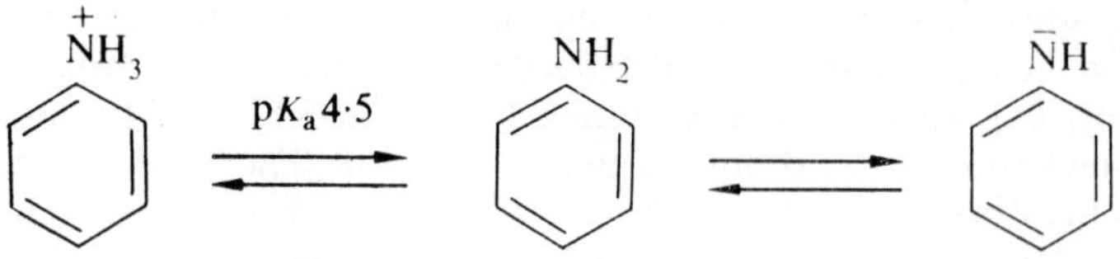

$\overset{+}{N}H_3$ $\underset{}{\xrightarrow[\leftarrow]{pK_a 4{\cdot}5}}$ NH_2 $\rightleftharpoons$ $\bar{N}H$

XXVI XXVII XXVIII

Above pH 2 a small fluorescence is observed (λ_A 280, λ_F 350 nm), which is due to the fluorescence of the molecular form (see Fig. 6.4). This fluorescence reaches 50% of its final value at pH 4·5 (the literature value for the pK_a of aniline) and a maximum value at pH 7·5. The intensity then remains constant until pH 12. Above this value the intensity falls, probably because of the formation of the anion (XXVIII). This anion has not been detected by other classical methods, however; its formation is suggested by the fact that, whereas monomethylaniline, which has a replaceable hydrogen, behaves exactly like aniline, dimethylaniline fluoresces maximally even above pH 14.

pH can also be used to distinguish compounds with similar structures. For example, codeine (XXIX) and morphine (XXX) both have a typical

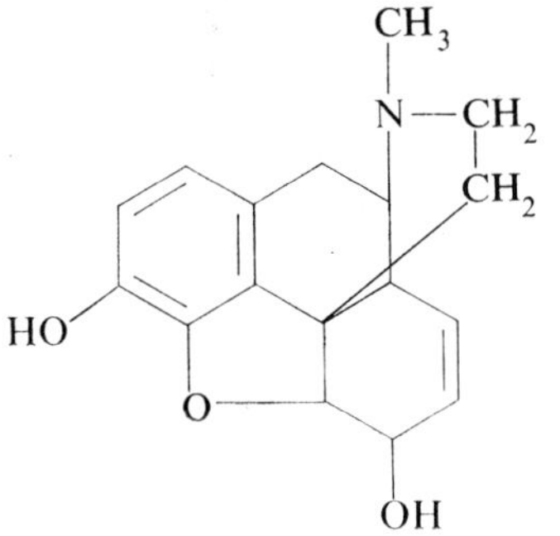

CH_3 N—CH_2 CH_2 HO O OH

XXIX

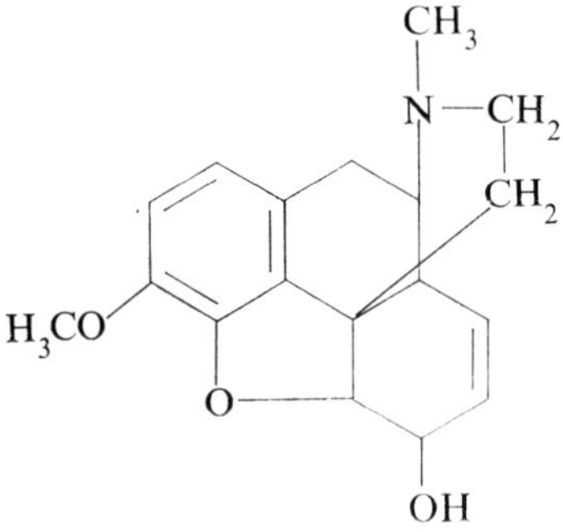

CH_3 N—CH_2 CH_2 H_3CO O OH

XXX

phenol fluorescence at pH values below 10. However, at pH 12 the morphine ionizes to its anion which is almost non-fluorescent, whereas the codeine fluorescence remains unchanged. Tryptamine and 5-hydroxytryptamine both have a similar fluorescence (λ_F 350 nm) at pH 2–10. However, at H_0 −0·5, 5-hydroxytryptamine fluoresces at 540 nm whereas tryptamine is non-fluorescent [18]. Because fluorescence properties are often very pH-dependent a number of visibly fluorescent compounds have been used as pH indicators. Fluorescent pH indicators are particularly useful for titrating coloured or cloudy solutions, for under such conditions the traditional coloured indicators are very difficult to use. A good example of such an indicator is 2-naphthol which, when viewed under ultra-violet light, changes from colourless to blue at pH 6–8. At low pH values 2-naphthol emits the invisible ultra-violet of its molecular form, but above pH 6–8 it emits the blue fluorescence of the naphtholate anion.

Excited State Ionization (proton transfer reactions)

The acidic or basic strength of a compound in the ground state is critically dependent on the electron distribution within it. When the compound is excited this electron distribution may become upset, resulting in the loss of a proton. In such instances the acidic or basic strength of the excited state of the compound will be different from those of the ground state. Thus the charge densities on the C—O—H group of phenol have been calculated to be C, +0·14 and O, −0·25 in the ground state, and C, +0·16 and O, 0·03 in the excited state [32]. In the excited state, therefore, the oxygen will have less affinity for the hydrogen atom than it has in the ground state; i.e., it will be a stronger acid in the excited state. Similar calculations on the charge densities of the C—N—H group of amines have shown that the charge on the N becomes more positive on excitation, and so it will attract protons less; i.e., amines will be weaker bases in excited state than in the ground state [33].

The process of excited state ionization may be represented as follows:

$$\text{Acids:}\quad AH + h\nu_1 \xrightarrow{\text{excitation}} A^{*}H(+H_2O) \xrightarrow{\text{ionization}} A^{*-}(+H_3O^{+})$$

$$A^{*-} \downarrow \text{fluorescence}$$

$$AH \xleftarrow{+H^{+}} A^{-} + h\nu_2$$

$$\text{Bases:}\quad BH^{+} + h\nu_3 \xrightarrow{\text{excitation}} B^{*}H^{+}(+OH^{-}) \xrightarrow{\text{ionization}} B^{*} + H_2O$$

$$B^{*} \downarrow \text{fluorescence}$$

$$BH^{+} \xleftarrow{+H^{+}} B + h\nu_4$$

The process will be favoured if proton acceptors are present. An ionization of this nature in the excited state will produce a shift in the fluorescence wavelength but not in the activation wavelength; a large Stokes shift is therefore usually observed, i.e., 8000 cm^{-1}. Not all fluorescent acids and bases show this excited state ionization, however, for in order that a proton may be lost the ionization equilibrium must be established during the lifetime of the excited state. For example, 2-hydroxybiphenyl clearly shows excited state ionization whereas 4-hydroxybiphenyl does not (see Fig. 6.5) [34]. 4-hydroxybiphenyl (pK_a 9·5) fluoresces at 340 nm (λ_A 288 nm) from H_0 − 1·0 to pH 9·2 corresponding to the un-ionized form and at 401 nm (λ_A 311 nm) from pH 10 to 14 because of the anion. With 2-hydroxy-biphenyl (pK_a 10), however, the fluorescence of the molecular form (λ_F 348 nm) changes to that of the anion (λ_F 415 nm) at pH 1, whereas the excitation wavelength remains that of the molecular form until pH 10 when it changes to that of the anion. Above pH 10 both the activation and

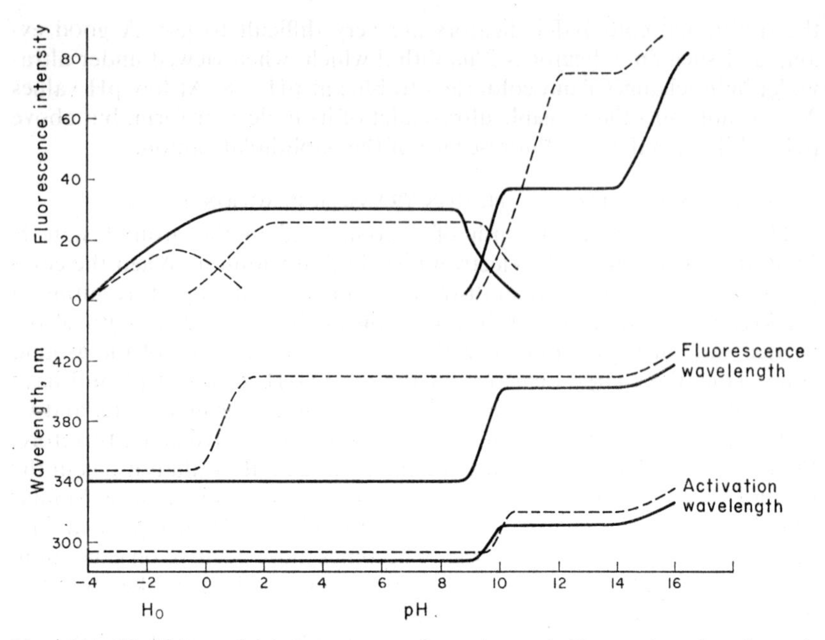

FIG. 6.5 Variation of (*a*) fluorescence intensity and (*b*) wavelengths of maximum activation and fluorescence of 4-hydroxybiphenyl (——) and 2-hydroxybiphenyl (— — —) with pH.

fluorescence wavelengths correspond to those of the anion. From the fluorescence wavelength curve the excited pK_a of 2-hydroxybiphenyl is 1·5 so that in the excited state this compound is a hundred million times stronger as an acid than in the ground state. These two compounds are produced enzymically from biphenyl by liver microsomes, and estimation of these compounds has proved valuable in investigating various aspects of liver biochemistry. A sensitive assay for 2- and 4-hydroxybiphenyl, based on their fluorescence-pH differences, has been developed. *p*-Anisidine (XXXI) also shows excited state ionization. The cation of *p*-anisidine is only very weakly fluorescent whereas the un-ionized form is strongly fluorescent. Although the ground state pK_a is 4·7, the fluorescence

OCH_3 (ring) NH_3^+ $\underset{}{\overset{pK_a\ 4\cdot 7}{\rightleftharpoons}}$ OCH_3 (ring) NH_2

XXXI

corresponding to the molecular form can be seen at pH 0 while the activation wavelength at this pH corresponds to the absorption wavelength of the cation. Actual rates of ionization of excited states have been measured by balancing them against rates of fluoresence emission [52].

It is possible to calculate the excited pK_a for the first excited singlet state of acids and bases from the formula:

$$pK_a - pK_a^* = 21187 \left(\frac{\lambda' - \lambda}{\lambda\lambda'}\right)$$

where λ and λ' are the maximum absorption and fluorescence wavelengths (nm) of the acid and anion (or cation and base) respectively [35]. In general, good agreement is found between the theoretical excited state pK_a and that determined from the fluorescence wavelength or fluorescence intensity curves. For example, in Table 6.7 the theoretically and practically determined excited state pK_as are given for 3-hydroxypyridine and some of its derivatives. In each case the pK_a shift is between 7·4 and 8·4 pH units [12]. A number of other fluorescent compounds have been shown to exhibit excited state ionization, including hydroxy- and amino-pyrenesulphonic acids, some hydroxyquinolines, particularly 3-hydroxyquinoline, 1- and 2-naphthylamines [36a], 1- and 2-naphthols and 2-naphthoic acid [36b].

Temperature

Fluorescence intensity tends to decrease with increasing temperature, because of the increased conversion rate of electronic into vibrational energy. Energies of activation for degradation can be derived, and possibly arise from the easier change-over to a higher, rather than a lower, triplet state. For most practical purposes slight temperature changes are probably unimportant since the fluorescence intensity seldom falls by more than 1% for a 1° rise in temperature. However, a few compounds are known (e.g., *p*-anisidine [37], indole-3-acetic acid and tryptophan [18]), in which the intensity may vary by about 5% for a 1° temperature change. A few instances have been found in which the fluorescence efficiency actually increases as the temperature is raised, as when a non-fluorescent molecular complex dissociates on heating. At low temperatures a considerable improvement in spectral resolution may occur, particularly when the solvent is frozen, and often the fine line structures may be clearly seen. Many studies have been made of the fluorescence spectra of aromatic hydrocarbons in frozen aliphatic hydrocarbon solvent. In many cases the resolution and fluorescence emission wavelength are dependent on the solvent used [21].

The Stability of the Compound in the Exciting Light

In concentrated solutions the photodecomposition of fluorescent compounds under ordinary conditions of measurement is generally small. In dilute solution, however, photodecomposition or photo-oxidation may present a considerable problem. Thus, very dilute solutions of quinine bisulphate (0·01 μg/ml) are unstable in ultra-violet light whilst more concentrated ones (1 μg/ml) are stable. An intense light source, whilst greatly enhancing the potential sensitivity of an instrument, may at the same time increase the photodecomposition rate. From a practical point of view, fluorimetric measurement can usually be made before photodecomposition becomes appreciable. Thus, pyridoxal (1μg/ml) rapidly decomposes in ultra-violet light, but provided measurements are made within 30 seconds of exposing the solution to light, the loss in fluorescence intensity, through photodecomposition, is negligible.

Photodecomposition does not always lead to a decrease in fluorescence. Indeed, in some cases (e.g., coumarin) it may result in the fluorescence being intensified.

Practical Considerations in Fluorimetric Analysis

Particularly when very minute amounts of material are being estimated, scattering of the exciting and fluorescent light, due to Raman, Tyndall and Rayleigh effects, may cause considerable interference with fluorimetric measurements. Light scattering may produce either an increase or a decrease in the apparent fluorescence intensity. Samples extracted from biological sources are particularly likely to contain large amounts of light-scattering materials. Raman and Rayleigh effects are inherent in the solvent and solute used, and cannot be eliminated. Tyndall effects, on the other hand, are due to the presence of large particles (e.g., traces of filter paper), proteins and silicates, and can often be removed by high-speed centrifugation. Rayleigh and Tyndall scattering are characterized by the fact that the emission wavelength is always the same as the activation wavelength. Thus, they are most likely to interfere in the estimation of those fluorescent compounds whose activation and fluorescence wavelengths are close together. Their effects can often be minimized by exciting the compound below its activation maximum, which in turn will cause the emitted scatter peak, but not the fluorescence peak, to move to a shorter wavelength. These forms of scatter, together with spurious fluorescence arising from contaminants, constitute the so-called ‘biological blank’ in many assays. In many cases it is the intensity of the biological blank which limits the sensitivity of a fluorimetric method. Spurious fluorescence can come from such diverse sources as solvent impurities, stopcock grease, paper tissues, rubber stoppers, polythene wash bottles, human skin, and ion exchange columns. For very sensitive measurements these contaminants may need to be removed. Other contaminants present may quench fluorescence. Quenching by phosphate ions is common in estimating biological samples, and it is always advisable to run an internal standard. Phosphate is not, however, a universal quencher. Traces of acetone may quench fluorescence by bonding strongly by van der Waals forces to the excited fluorescent molecules. For example, traces of acetone will quench the fluorescence of indole. Apparent quenching effects may also occur in very dilute solution, through adsorption of the fluorescent substance on glassware or protein precipitates [38].

Some Fluorescent Molecules of Particular Interest

PYRIDINE NUCLEOTIDES

Pyridine nucleotides are essential co-enzymes in many biochemical reactions. Each incorporates two separate conjugated double bond systems, namely, a pyridine and an adenine nucleus. The oxidized forms of the pyridine nucleotides (NAD^+ and $NADP^+$) are non-fluorescent whilst the reduced forms ($NADH_2$ and $NADPH_2$) are fluorescent (XXXII).

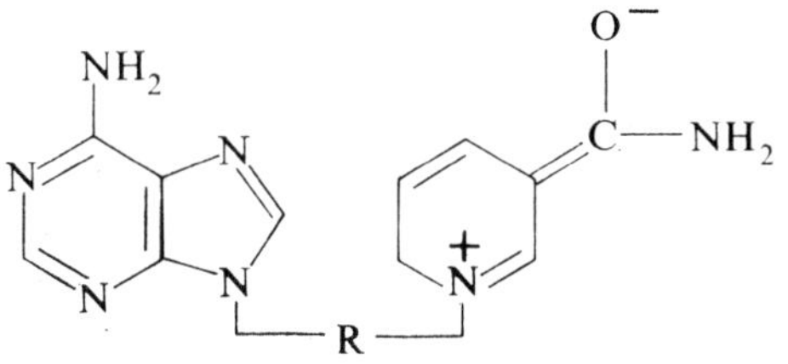

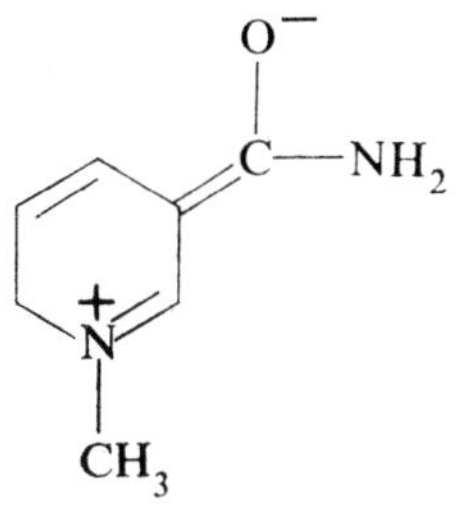

XXXII where R= ribose-phosphate-phosphate- ribose

XXXIII

Neither adenine nor nicotinamide is fluorescent since they both contain groupings which tend to localize the π electrons; however, 4-hydromethyl-nicotinamide (XXXIII) (the reduced form of N-methylnicotinamide) is weakly fluorescent (λ_A 360 nm, λ_F 458 nm). The corrected excitations of $NADH_2$ in aqueous solution are at 260 and 340 nm and the fluorescence is at 457 nm. The 260 nm excitation is due to the adenine moiety. In order to produce fluorescence, this energy, absorbed at this wavelength, is transferred through a highly efficient energy transfer mechanism, to the dihydronicotinamide portion which is excited at 340 nm [39]. This type of intramolecular energy transfer reaction is rare in solution. Fluorescent measurements of $NADH_2$ have been used for the assay of a number of dehydrogenase enzymes including lactate, malate and glutamate dehydrogenases, and to study the oxidation-reduction mechanisms of intact cells and subcellular particles.

PYRIDOXINE (VITAMIN B_6) AND RELATED COMPOUNDS

Pyridoxine, pyridoxal, pyridoxal-5-phosphate and pyridoxamine act as co-enzymes for a number of transaminase and decarboxylase enzymes. These compounds are all derivatives of 3-hydroxypyridine and their fluorescent characteristics closely resemble those of the parent molecule [12]. 3-Hydroxypyridine may exist in any one of four different forms, according to the pH of the medium. Each form has a characteristic fluorescence (XXXIV).

The cation, neutral molecule and dipolar ion are only moderately

cation ⇌ [neutral molecule ⇌ dipolar ion] ⇌ anion

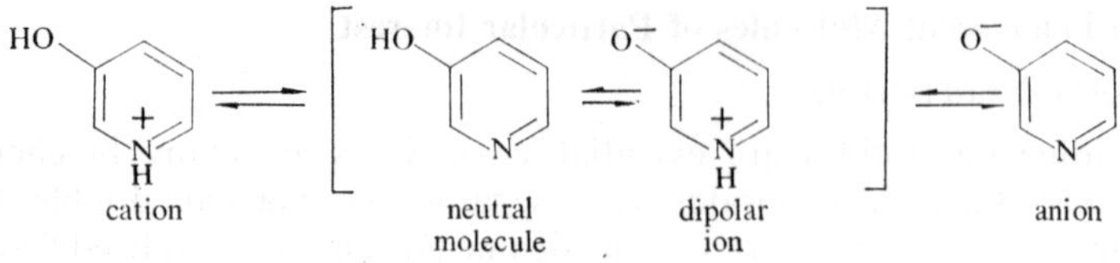

XXXIV

fluorescent, whereas the anion is strongly fluorescent. Thus, the fluorescence intensity of the anion is 50 times that of the cation and 16 times that of the dipolar ion. The cation shows excited state ionization above $H_0 - 1$; thus the fluorescence of the cation (340 nm) is only observed below this pH. The fluorescence of the dipolar ion is observed between pH − 1 and 8·5. The neutral molecule appears to make only a slight contribution to the observed fluorescence. Above pH 8·5 the fluorescence of the anion is seen. Pyridoxine, pyridoxal, pyridoxal phosphate and pyridoxamine also exhibit excited state ionization (see Table 6.7). However, all the forms of these molecules are more fluorescent than the corresponding forms of 3-hydroxypyridine. Pyridoxal and its phosphate, like 3-hydroxypyridine, are maximally fluorescent as anions while the dipolar ions are the most fluorescent forms of pyridoxamine and pyridoxine. Usually these compounds are assayed by oxidation to pyridoxic acid which is then converted to the highly fluorescent lactone. The intense fluorescence of the lactone can be ascribed to resonance structures such as (XXXV).

XXXV

STILBENES

The *trans*-stilbene nucleus is contained in the majority of optical brighteners. Two essential criteria for an optical brightener are a bright blue fluorescence and a good water solubility. In order to achieve a blue fluorescence, electron donating groups (e.g., NH_2 and OH) must be substituted in the stilbene nucleus which will shift its fluorescence emission into the blue region, the intensity of which will be enhanced. The diamino derivative 4,4′-diaminostilbene is a good example (XXXVI). To enhance the water solubility of this compound, sulphonic acid groupings are frequently substituted in the 2,2′-position. Sulphonic acid groups are weakly deactivating in an otherwise unsubstituted aromatic system, but in 4,4′-diaminostilbene-2,2′-disulphonic acid they actually enhance the fluorescence slightly.

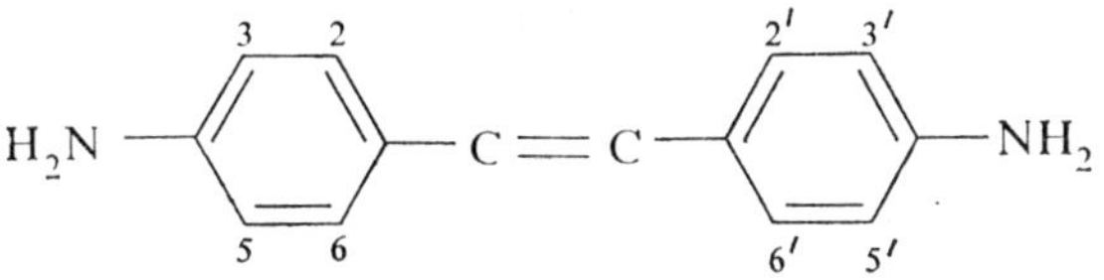

$$H_2N-C_6H_4-C\equiv C-C_6H_4-NH_2$$

XXXVI

CHELATING AGENTS

A number of inorganic metal ions may be estimated by fluorimetric methods (see Table 6.8). Usually, the metal ion is reacted with a chelating agent to form a fluorescent complex. Ideally the chelating agent itself should either be non-fluorescent or have a quite different fluorescence from the metal complex. Few, if any, chelating agents have an absolute specificity for a particular cation; however, the specificity can in many cases be extremely good, provided the pH at which the fluorimetric estimations are made is closely controlled. Large excesses of chelating agent often interfere with estimations. Very few investigations have been made on the structural requirements for the fluorescence of cation-chelate complexes. 8-Hydroxyquinoline-5-sulphonic acid is often used as a chelating agent. At pH 7·1–8·5 it will form a strongly fluorescent complex with Cd^{2+}, Zn^{2+}, Mg^{2+}, Ca^{2+}, Al^{3+}, La^{3+}, Ce^{4+}, Th^{4+}, Sn^{4+}, but not with Li^{+}, Na^{+}, K^{+}, Ag^{+}, Ba^{2+}, Hg^{2+}, Cu^{2+}, Co^{2+}, Ni^{2+}, Sn^{2+}, Pd^{2+}, B^{3+}, As^{3+}, Sb^{3+}, Bi^{3+}, Cr^{3+}, Fe^{3+}, Zr^{4+}, Ta^{5+}, Nb^{5+} or W^{6+} [40]. At this pH, 8-hydroxyquinoline-5-sulphonate exists predominantly in the enol form (XXXVII) and is only very weakly fluorescent (λ_A 339 nm, λ_F 454 nm). In acid, however, 8-hydroxyquinoline is in the form of the cation (XXXVIII) and is strongly fluorescent (λ_A 344, λ_F 454 nm) probably because its lone pair of electrons is bound.

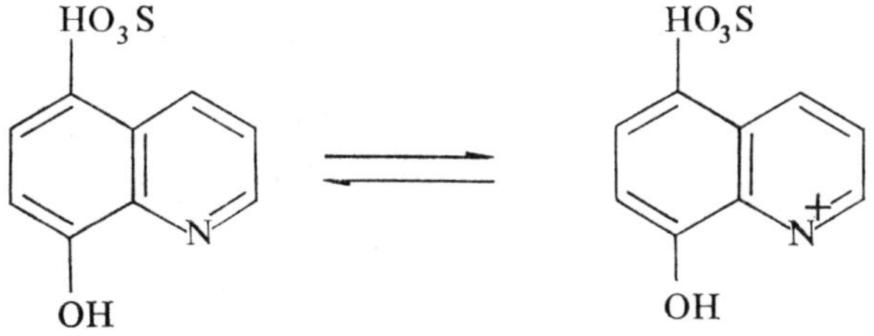

HO_3S HO_3S

N N^+

OH OH

XXXVII XXXVIII

The fluorescence of the metal ion chelate (e.g. XXXIX) at pH 7·1–8·5 could be due to one of two factors, namely, the binding of the lone pair of electrons on the nitrogen by the metal ion or the increased rigidity afforded the molecule by ring closure. Probably both factors make a contribution.

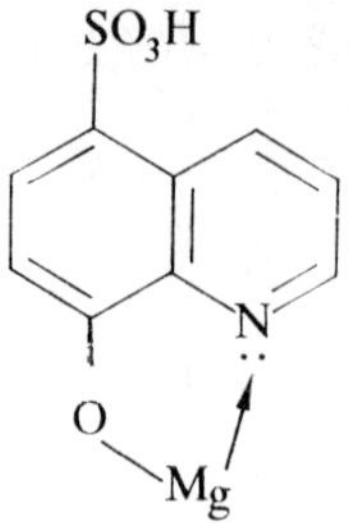

XXXIX

Salicylate may also be used as a fluorescent chelating agent for estimating Mg^{2+}. In the absence of Mg^{2+}, at pH 10 salicylate fluoresces as the monoanion (XL) (λ_A 296, λ_F 420 nm). The fluorescence of the dianion which is more than twice as intense as the monoanion (XLI) (λ_A 316, λ_F 410 nm) is not observed below pH 12.

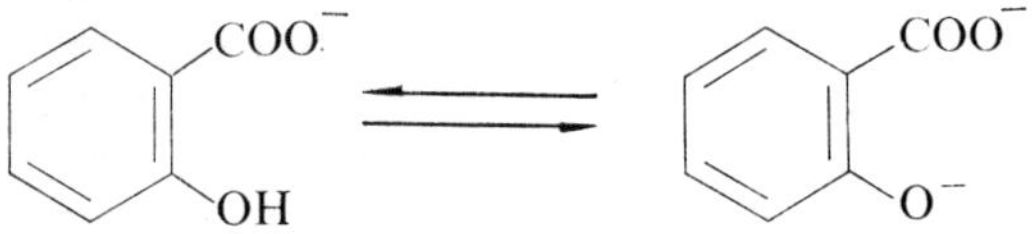

XL XLI

In the presence of Mg^{2+} the fluorescence of the dianion is observed at pH 10, implying that the Mg^{2+} has enabled the phenolic OH group to ionize by forming the complex (XLII). This type of mechanism may account for the fluorescence properties of a number of common chelate-metal complexes.

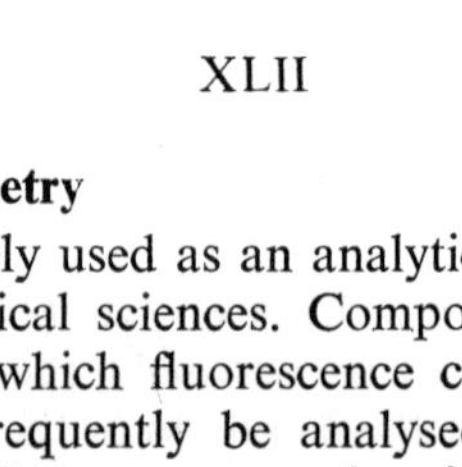

XLII

Analytical Uses of Fluorimetry

Fluorimetry is extensively used as an analytical technique, particularly by workers in the biological sciences. Compounds with a good native fluorescence or those in which fluorescence can be induced by simple chemical reaction may frequently be analysed with a high degree of sensitivity and selectivity. Some compounds exist in such small quantities in organisms that fluorimetry provides the only reasonable means of estimating them. For others, fluorimetry has enabled much smaller samples to be investigated or has greatly reduced the time taken for an analysis to be carried out, and improved its reliability. A wide variety of different materials can be estimated fluorimetrically including carbohydrates, lipids, amino acids, proteins, enzymes, steroids, vitamins, purines, pyrimidines, porphyrins, inorganic ions, drugs, aromatic hydrocarbons, insecticides, herbicides, food additives and dyes.

Five types of fluorimetric assay may be distinguished:

1. 'Native' assay may be used for substances which possess a high 'native' fluorescence.
2. Chemical induction assay may be applied to substances which are non-fluorescent, weakly fluorescent, or whose fluorescence is not sufficiently specific. By simple chemical conversion, such compounds may often be converted to highly fluorescent derivatives.
3. Quenchofluorimetric assays may be used for compounds which act as quenching agents by estimating their quenching effect on a highly fluorescent substance.
4. Enzyme-cofactor assay can be utilized for estimating substances which can act as substrates for dehydrogenase enzymes or other enzymes in which a non-fluorescent cofactor is converted into a fluorescent one, or for estimating the enzymes themselves.
5. Enzyme-substrate assay can be used for determining enzyme activity if the enzyme is capable of converting a non-fluorescent substrate into a fluorescent product.

Of these five types, the native assay and the chemical induction assay are the most widely used at present. Only one example of each type will be considered here; for other methods the reader is referred to Udenfriend's book [18], and to various reviews [8, 41, 42].

Native Assay, e.g., Serotonin

Serotonin like many other indole derivatives shows a strong fluorescence at 350 nm in aqueous solution. However, whereas other indole derivatives are non-fluorescent in 3N acid, serotonin fluoresces at 540 nm and can therefore be estimated without interference from other indolic contaminants [18].

Chemical Induction Assay, e.g., Cortisol

The estimation of cortisol is probably the most important of all clinically-used fluorimetric assays at the present time. Cortisol (XLIII) is non-fluorescent; in fact an examination of its structure shows that it is not sufficiently conjugated to be fluorescent. However, when it is treated with ethanolic sulphuric acid it is converted to highly fluorescent products (λ_A 470, λ_F 530 nm) [43]. The nature of these products is unknown, but presumably they must incorporate a conjugated system of double bonds.

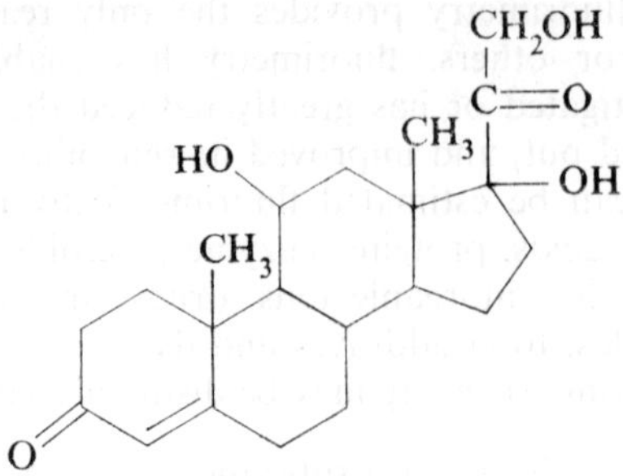

XLIII

Quenchofluorimetric Assay, e.g., Acetone

This technique has barely been exploited, but it appears to be of great potential importance. Thus, the strong fluorescence of β-naphthol in aqueous solution is quenched by acetone, and the extent of quenching is proportional to the concentration of acetone. This principle has been employed in the estimation of acetone in the urine of suspected diabetics [44]. Quenching agents on paper chromatograms may be detected by spraying with a highly fluorescent compound such as quinine.

Enzyme-Cofactor Assay, e.g., Lactate

$NADH_2$ and $NADPH_2$ are highly fluorescent whereas NAD^+ and $NADP^+$ are not. Lactate is converted by the enzyme lactate dehydrogenase to pyruvate, while at the same time NAD is reduced to $NADH_2$.

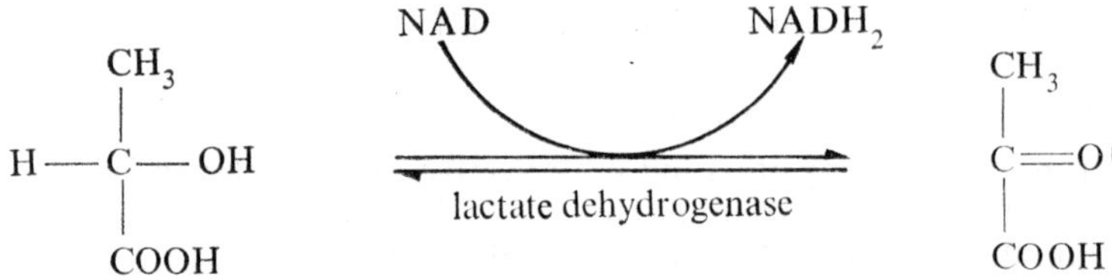

$$\mathrm{H{-}C(CH_3)(COOH){-}OH} \underset{\text{lactate dehydrogenase}}{\xrightleftharpoons{NAD \;\rightarrow\; NADH_2}} \mathrm{CH_3{-}C(=O){-}COOH}$$

Provided that both NAD and lactate dehydrogenase are in excess, and the pH of the medium is suitable, the amount of fluorescent $NADH_2$ produced will be entirely dependent on the concentration of lactate [45].

Enzyme-Substrate Assay, e.g., β-Glucuronidase

Substrates which incorporate a fluorescent compound can be made for a number of enzymes, particularly hydrolytic enzymes. Thus, umbelliferone glucuronide is only weakly fluorescent, but when it is acted upon by the enzyme β-glucuronidase, highly fluorescent umbelliferone is liberated. Provided that the substrate is in excess, and the pH of the incubation medium is suitable, the method provides an extremely sensitive assay for β-glucuronidase [46].

$$\underset{\text{(weakly fluorescent)}}{\text{Umbelliferone glucuronide}} \xrightarrow{\beta\text{-glucuronidase}} \underset{\substack{\text{(highly fluorescent)} \\ + \\ \text{glucuronic acid}}}{\text{Umbelliferone}}$$

Other Uses of Fluorescence

Apart from the analytical uses of fluorimetry described above, fluorescence may be used for a number of other purposes.

OPTICAL BRIGHTENERS

The most important industrial use of fluorescent substances is as optical brighteners in detergents, paper, textiles, plastics and to a limited extent in cosmetics. The quest for whiter and whiter washes has lead to intense research on physical bleaches (as opposed to chemical bleaches such as peroxide and hypochlorite) as a means of removing yellow casts from materials. A material appears to be yellow because it absorbs short wavelength violet and blue light. In order to compensate for this absorption, blue fluorescent optical brighteners are added to the wash so that the material appears white or adopts a blue tinge. For the desired effect to be achieved, it is essential that these brighteners have their absorption maxima in the ultra-violet so that they do not absorb visible light to any appreciable extent. A number of other criteria apart from its fluorescence characteristics also must be considered in choosing an optical brightener, namely, it must be: chemically stable, particularly to chemical bleaches, light stable, non-toxic, readily soluble and easily dispersed, cheap to make and able to bind to the fabric without continual building-up in each wash. Usually each material needs a different brightener.

PAINTS AND DYES

Fluorescent pigments are widely used for warning signs and for advertising posters, but they have not replaced traditional paints to any great extent because they tend to have a relatively poor light stability. Fluorescent pigments are prepared by dyeing a resin matrix such as PVC, sulphonamide resin, etc., with a fluorescent compound such as rhodamine B or fluorescein, and grinding it to uniform particle size of about 7 microns.

FLUORESCENT IMPURITY ESTIMATIONS

Fluorimetry can provide a very sensitive means of detecting trace impurities in a variety of media including solvents, aromatic hydrocarbons, the atmosphere and water. For example, commercially available absolute ethanol may contain unacceptably high levels of aromatic hydrocarbons [47]. Similarly, fluorene and anthracene are frequently found as contaminants in phenanthrene. Carbazole, and polynuclear carbazoles many of which are carcinogenic, may be found in the industrial atmosphere, and fluorimetry provides a very sensitive means of monitoring these [48].

Metal Ion Detection

The fluorimetric analysis of chelate ion complexes has already been described. However, chelates may also be used as endpoint indicators in EDTA titrations. For example, to estimate Ca^{2+} by EDTA titration, a small amount of calcein (chelating agent) is added, producing a green fluorescence. As EDTA is added the Ca^{2+} will be removed from the calcein, and the endpoint is reached when the fluorescence finally disappears.

Tetracycline also avidly chelates with Ca^{2+} (XLIV). When this drug is

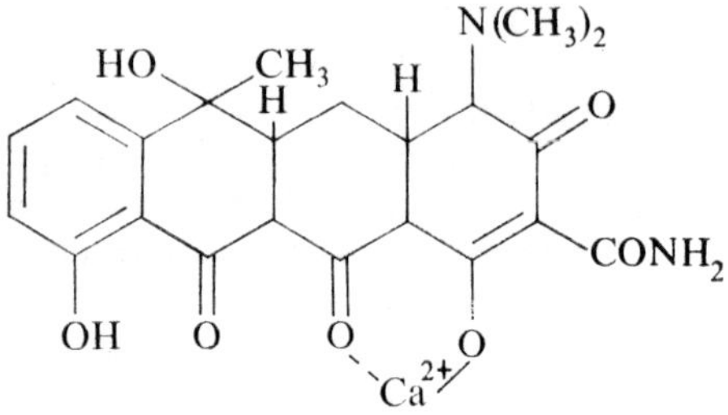

XLIV

administered to man, it will tend to bind to tissues with a high calcium content such as teeth, bone and nails. Indeed, the teeth of such an individual may emit a blue fluorescence when examined with ultra-violet light. This principle has been used to locate pathological lesions of the body which are calcium-rich, including gastric and cutaneous ulcers, neoplastic tissues and various viral and bacterial lesions [49].

Tracer Techniques

Fluorescent indicators are often added to dangerous liquids, e.g., anti-freeze. Fluorescent indicators, particularly fluorescein, have also been employed to trace the course of underground rivers, and to determine the route and rate of removal of materials in effluent disposal studies. Since very small amounts can be detected there is no toxicity problem, and the method may often be more specific, and less of a contamination risk, than radioisotope techniques.

'Invisible' Markers

Ultra-violet fluorescent compounds are used for the 'invisible' marking of laundry, bank-notes, identity cards of various types, and stamps. Because such markers are invisible (except when irradiated with ultra-violet light) they have proved invaluable in criminology. For example, the dusting of bank-notes with solid 'invisibly' fluorescent compounds has been used to prove the transference of money from one individual to another in cases of theft, blackmail and bribery.

TABLE 6.7. *Excited-state dissociation constants of the cations of pyridoxal and related compounds*

Compound	pK^*_a of cation Observed	Calculated	pK_a	$pK_a - pK^*_a$ (obs.)
3-Hydroxypyridine	−3·0	−2·24	5·37	8·37
3-Hydroxy-N-methyl-pyridine	−3·0	−2·39	4·96	7·96
Pyridoxal	−2·2	−2·36	5·05	7·25
Pyridoxal (hemiacetal)	−3·3	−2·53	4·20	7·50
Pyridoxamine	−4·1	−4·25	3·32	7·42

TABLE 6.8. *Some chelating agents used in inorganic fluorimetric analysis*

Element	Reagents	Fluorescence wavelength (nm)	Sensitivity (μg)
Mg	8-hydroxyquinoline-5-sulphonic acid	530	0·2
Cd	8-hydroxyquinoline-5-sulphonic acid	530	0·2
Ca	Calcein	500	0·1
Sn	3-hydroxyflavone	470	0·1
Al	Morin	525–570	0·0005
Be	Morin	510–615	0·004
Se	2, 3-diaminonaphthalene	500	0·002
Zn	Picolinealdehyde-2-quinolylhydrazone	535	0·02
B	Benzoin	480	0·01

REFERENCES

[1] McClure, D. S. *Solid State, Phys. N.Y.* **8**, 1 (1959).
[2] Stevens, B. *Spectrochim. acta*, **18**, 439 (1962).
[3] Kasha, M. *Discuss. Faraday Soc.* **9**, 14 (1950).
[4] Bowen, E. J. *Advances in Photochemistry*, p. 32, Interscience, New York (1963).
[5] Sangster, R. C. and Irvine, J. W. *J. chem. Phys.* **24**, 670 (1956).
[6] Rhodes, W. and El-Sayed, M. A. *J. molec. Spectrosc.* **9**, 42 (1962).
[7] West, W. in 'Chemical Applications of Spectroscopy' (*Techniques of Organic Chemistry*, Vol. 9), p. 734, Interscience, New York (1956).
[8] Williams, R. T. and Bridges, J. W. *J. clin. Path.* **17**, 371 (1964).
[9] Bowen, E. J. and Wokes, F. *Fluorescence of Solutions*, Longmans, Green, London (1953).

[10] LIPPERT, E., LUDER, W. and MOLL, F. *Spectrochim, acta*, **10**, 858 (1959).

[11] COWGILL, R. W. *Biochim. Biophys. Acta.* **112**, 550 (1966).

[12] BRIDGES, J. W., DAVIES, D. S. and WILLIAMS, R. T. *Biochem. J.* **98**, 451 (1966).

[13] ROGERS, D. A., MARGERUM, J. D. and WYMAN, G. M. *J. Am. chem. Soc.* **79**, 2464 (1957).

[14] PRINGSHEIM, P. *Fluorescence and Phosphorescence*, Chapter V, Interscience, New York (1949).

[15] OTT, D. G., HAYES, F. N., HANBURY, E. and KERR, V. N. *J. Am. chem. Soc.* **79**, 5448 (1957).

[16] ADAMS, D. A. W. *J. Soc. Dyers and Colourists*, **75**, 22 (1958).

[17] PARKER, C. A. and HARVEY, L. G. *Analyst*, **87**, 558 (1962).

[18] UDENFRIEND, S. *Fluorescence Assay in Biology and Medicine*, Academic Press, New York (1962).

[19] PARKER, C. A. and REES, W. T. *Analyst*, **87**, 83 (1962).

[20] VAN DUUREN, B. L. *J. org. Chem.* **26**, 2954 (1961).

[21] VAN DUUREN, B. L. *Chem. Revs.* **63**, 325 (1963).

[22] LIVINGSTONE, R., WATSON, W. F. and MCARDLE, J. *J. Am. chem. Soc.* **71**, 1542 (1949).

[23] PERRIN, F. *J. Phys. Radium*, **7**, 390 (1926).

[24] SAWICKI, E., STANLEY, J. W. and ELBERT, W. C. *Talanta*, **11**, 1433 (1964).

[25] PARKER, C. A. and BARNES, W. J. *Analyst*, **82**, 606 (1957).

[26] WELLER, A. in *Progress in Reaction Kinetics*, Vol. 1 (edited by G. Porter), Pergamon Press, Oxford (1961).

[27] DUGGON, D. E. and UDENFRIEND, S. *J. biol. Chem.* **223**, 313 (1956).

[28] HEROS, M. and AMY, L. *Compt. rend. Acad. Sci. Paris*, **255**, 695 (1962).

[29] FÖRSTER, T. and KASPER, K. *Z. Elektrochem.* **59**, 976 (1955).

[30] LEVINSON, G. S., SIMPSON, W. T. and CURTIS, W. *J. Am. chem. Soc.* **79**, 4314 (1957).

[31] STENSBY, P. S. and ROSENBERG, J. L. *J. phys. Chem.* **65**, 906 (1961).

[32] SANDORFY, C. *Can. J. Chem.* **31**, 439 (1953).

[33] COULSON, C. A. and JACOBS, J. *J. chem. Soc.* 1984 (1949).

[34] BRIDGES, J. W., CREAVEN, P. J. and WILLIAMS, R. T. *Biochem. J.* **96**, 872 (1966).

[35] WELLER, A. *Z. Elektrochem.* **54**, 42 (1952).

[36a] HERCULES, D. M. and ROGERS, L. B. *Spectrochim. acta*, **14**, 393 (1959).

[36b] WEHRY, E. L. and ROGERS, L. B. in *Fluorescence and Phosphorescence Analysis* (edited by D. M. Hercules), Chapter 3, Interscience, New York (1965).

[37] BRIDGES, J. W. in *Methods in Polyphenol Chemistry* (edited by J. Pridham), Pergamon Press, Oxford (1964).

[38] BOWEN, E. J. and SAHU, J. J. *J. phys. Chem.* **63**, 4 (1959).

[39] WEBER, G. *Nature (Lond.)*, **180**, 1409 (1957).

[40] RYAN, D. E., PITTS, A. E. and CASSIDY, R. M. *Anal. Chim. acta*, **34**, 491 (1966).
[41] WHITE, C. E. and WEISSLER, A. *Anal. Chem.* **36**, 116R (1964).
[42] WHITE, C. E. *Anal. Chem.* **38**, 155R (1966).
[43] MATTINGLY, D. *J. clin. Path.* **15**, 374 (1962).
[44] HYNIE, I., VECEREK, B. and WAGNER, J. *Cas. Lek. čes.* **99**, 88 (1960).
[45] LOWRY, O. H., Roberts, N. R. and KAPPHAHN, J. I. *J. biol. Chem.* **224**, 1047 (1957).
[46] MEAD, J. A. R., SMITH, J. N. and WILLIAMS, R. T. *Biochem. J.* **61**, 569 (1955).
[47] PARKER, C. A. *Proc. S.A.C. Conference*, Nottingham (1965).
[48] BENDER, D. F., SAWICKI, E. and WILSON, R. M. *Anal. Chem.* **36**, 1011 (1964).
[49] WORSLEY, G. H., MCKENNA, R. D. and BECK, I. T. *Can. med. Ass. J.* **88**, 1272 (1963).
[50] WEBER, G. *Advances in Protein Chemistry*, Vol. 8, p. 415, Academic Press, New York (1953).
[51] HELDT, J. *Acta physica polonica*, **30**, 3 (1966).
[52] WELLER, A. H. *Progress in Reaction Kinetics*, Vol. 1, p. 187, Pergamon Press, Oxford (1961).

Chapter 7

PHOSPHORESCENCE AND DELAYED FLUORESCENCE OF ORGANIC SUBSTANCES

F. WILKINSON and A. R. HORROCKS

THERE is a great deal of confusion concerning the definitions of terms used to discriminate between the different types of 'long-lived' luminescence. Originally phosphorescence was used to describe any 'long-lived' emission, and fluorescence any 'short-lived' emission. In order that this definition be useful, it was necessary to give a more precise interpretation of 'long-lived' and this was usually done by considering any luminescence with a lifetime greater than 10^{-4} s as 'long-lived'. However, this presupposes that all luminescence decay is exponential.

These definitions, which do not distinguish between long-lived luminescences of different spectral distribution, have been abandoned by most chemists dealing with organic substances, in favour of definitions which require a detailed knowledge of the multiplicities of the states involved in the transition. Thus, *phosphorescence* is defined as a radiative transition between states of different multiplicities while *fluorescence* is reserved for light emitted as a result of a transition between states of like multiplicity. For organic molecules, in all media except low-pressure gases, only the lowest excited states of any multiplicity are usually found to emit. Thus in almost every case, phosphorescence is due to emission from the lowest triplet state T_1, and fluorescence is due to emission from the first excited singlet state S_1 (see Fig. 7.1).

These processes may be represented as

$$(a)\ T_1 \rightarrow S_0 + h\nu_p \quad \text{phosphorescence}$$
$$(b)\ S_1 \rightarrow S_0 + h\nu_f \quad \text{fluorescence}$$

Process (*a*) is spin-forbidden and therefore long-lived, while process (*b*) has a mean lifetime in the range 10^{-9}–10^{-6} s, depending on the type of transition. This appears to be in agreement with the original definition. However, under certain conditions the measured luminescence duration is found to be longer than the radiative lifetime for the transition concerned. The increased lifetime of this delayed luminescence is due to the production of the emitting state by some mechanism which is slower than the emission

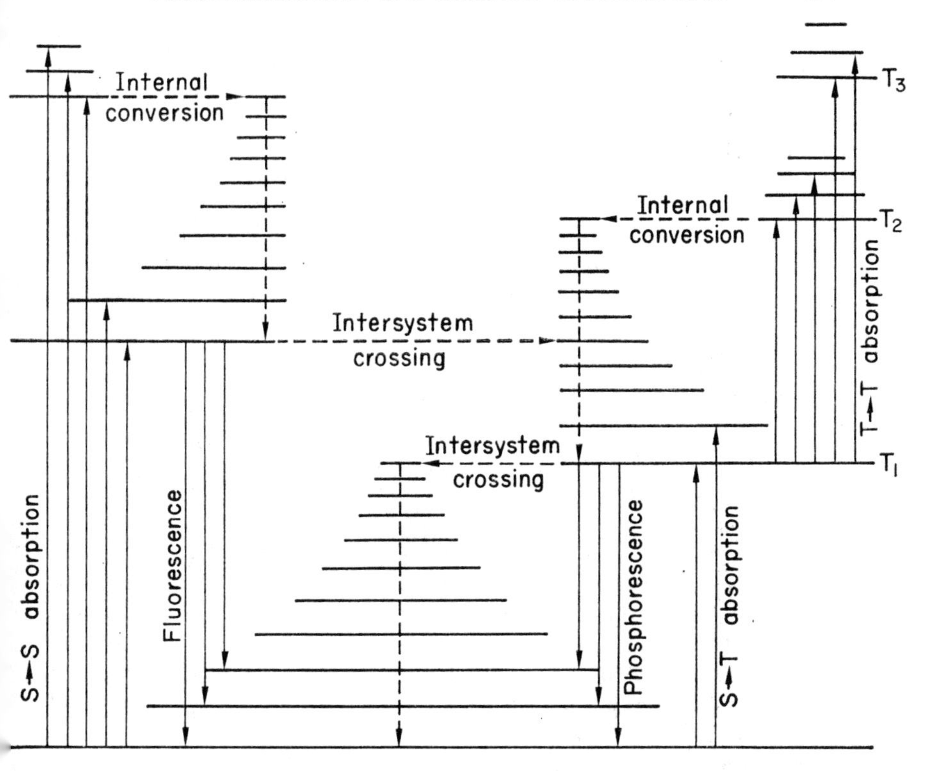

FIG. 7.1 Schematic diagram showing radiative and non-radiative transitions (continuous and broken lines respectively) between singlet (S) and triplet (T) states of a typical organic molecule.

process. Several established mechanisms are capable of delayed production of the first excited singlet states of organic substances which result in emission known as 'delayed' or 'slow' fluorescence. This has the same spectral distribution as 'normal' or 'prompt' fluorescence.

Different types of delayed fluorescence are classified according to the mechanism of production of the first excited singlet state S_1. Unfortunately three recent reviews use different teminology [1]–[3]. Delayed fluorescence due to thermal activation from T_1 to S_1 was originally known as α-phosphorescence. However, Lower and El-Sayed [2] refer to it as 'α-delayed fluorescence', Leach and Migirdicyan [3] suggest 'slow fluorescence' while Parker [1] uses 'E-type delayed fluorescence' (after eosin, which, he has shown, gives delayed fluorescence through this mechanism, under certain conditions). However, there is no good reason why eosin should not be capable of producing delayed fluorescence by other mechanisms. Delayed fluorescence can also be observed as a result of triplet-triplet annihilation. Parker uses 'P-type delayed fluorescence' (after pyrene) to describe delayed fluorescence due to this mechanism while the other authors above just use the term 'delayed fluorescence'. In order to avoid confusion, the mechanism operating should, if possible, be stated. There are many possible mechanisms for the delayed production of excited states of molecules. Other examples, which have been studied in some detail include radical-radical, ion-ion and ion-electron recombination reactions. In fact any chemiluminescent reaction of photoproducts of intermediates might give rise to delayed fluorescence. Leach and Migirdicyan have termed fluorescence following ion-electron recombination 'deferred fluorescence'. This kind of chemiluminescent reaction which follows photodissociation or photoionization has been reviewed recently [3] and will not be discussed here.

Phosphorescence

Identification of Phosphorescence as a Triplet State Phenomenon

The long-lived afterglow of organic molecules dissolved in glassy rigid media has been studied for over a century. However, no satisfactory interpretation of this luminescence was offered until 1935, when Jablonski proposed that phosphorescence was emission from a long-lived metastable state to the ground state of the molecule concerned. He suggested that this state lay between the ground state and the fluorescent state and was populated by a radiationless transition from the latter. Thus he explained the long lifetime of phosphorescence, and its appearance at longer wavelengths than normal fluorescence.

In 1941 Lewis, Lipkin and Magel demonstrated that under intense illumination of fluorescein in boric acid glass, 80% of the absorbing molecules were promoted to the phosphorescent state. Furthermore, under these conditions, these investigators were able to measure the absorption spectrum of this state. They suggested tentatively that the phosphorescent state might be a triplet state, and the new absorption a triplet-triplet transition (Fig. 7.2).

Terenin was of the same opinion in 1943, and a year later the postulate was supported very convincingly by Lewis and Kasha in a paper which gave data on the phosphorescence of some 90 compounds [5]. They showed that each substance has only one phosphorescent state, and proposed that since a triplet state has two unpaired electrons, it should have paramagnetic properties. This was confirmed a year later. More recently the decay of photomagnetism has been shown to follow the phosphorescence decay in a number of cases (see Fig. 7.3).

If the phosphorescent state is a triplet state then the long lifetime of phosphorescence is understandable since the transition is a spin-forbidden one. The reverse transition, namely singlet-triplet absorption, has an equally low probability, and weak absorption bands bearing a mirror-image relationship to the phosphorescence spectra about the 0,0 band were reported for several organic species in 1945.

A natural extension of the photomagnetic studies mentioned above has been the application of Electron Spin Resonance techniques to the study

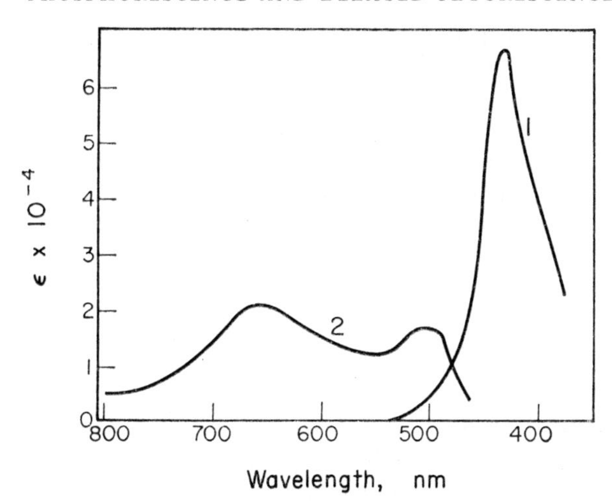

FIG. 7.2 Molar extinction coefficients of fluorescein in boric acid: (1) normal (N) state; (2) phosphorescent (P) state. (Reproduced with permission from LEWIS, LIPKIN and MAGEL *J. Am. chem. Soc.* **63**, 3005 (1941).)

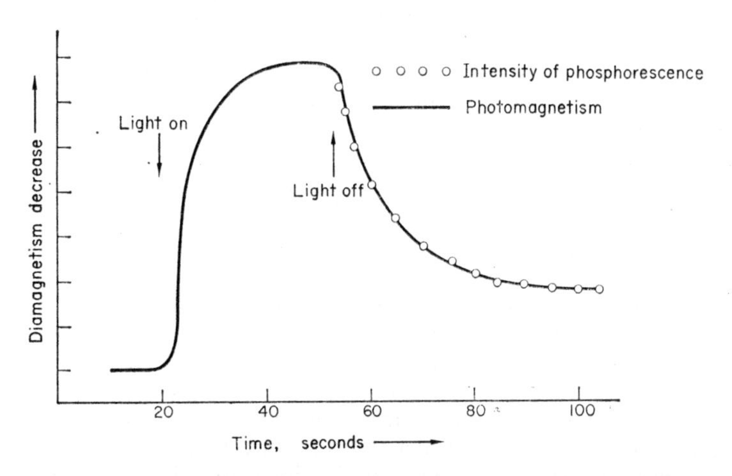

FIG. 7.3 Comparison of photomagnetism and decay of phosphorescence of triphenylene in boric acid 'glass' at room temperature. (Reproduced with permission from EVANS, *Nature*, **176**, 777 (1955).)

of triplet states. The first ESR study was of the phosphorescent state of naphthalene in durene crystals, by Hutchison and Mangum in 1958, and since then many other systems have been studied by means of the technique [6]. ESR measurements have been in striking agreement with theoretical expectations based on the assumption that the phosphorescent and the triplet states are identical.

Population of the Triplet State

Since singlet-triplet absorption involves a spin change and is therefore very weak, population of the triplet state by direct excitation is extremely inefficient. However, the triplet state can be produced in high yields as a result of non-radiative processes following singlet-singlet absorption. Many compounds have a quantum yield of fluorescence independent of wavelength, which indicates that intersystem crossing to the triplet state proceeds from the lowest excited singlet state S_1. Thus a competitive depopulation of S_1 takes place with some molecules fluorescing while others convert to a vibrationally excited triplet state which rapidly loses its excess vibrational energy (see Fig. 7.1). For these compounds the quantum yield of phosphorescence is independent of wavelength, and the phosphorescence excitation spectrum is simply identical to the singlet-singlet absorption spectrum. This is true, for example, in the case of many aromatic hydrocarbons, including naphthalene and some of its halogen derivatives but not for the iodonaphthalenes, which undergo predissociation at high irradiation frequencies.

Intersystem crossing from S_1 is a spin-forbidden non-radiative process, and yet it competes effectively in many molecules with the spin-allowed, non-radiative and radiative $S_1 \rightarrow S_0$ transitions. Although internal conversion between higher singlet states is a very efficient process, internal conversion to the ground state is often absent altogether for fluorescent molecules. Non-radiative transitions from the triplet state to the ground state are also very inefficient. The reasons for this are not completely clear, but an important parameter is the energy difference between the states involved in the non-radiative transition. The larger this difference the less likely is the process to occur. It may be that in many cases intersystem crossing from S_1 to T_1 takes place via intermediate triplet states situated between S_1 and T_1. This would reduce the energy difference and increase the likelihood of intersystem crossing.

Every triplet state formed does not necessarily emit phosphorescence; some undergo intersystem crossing to the ground state. The quantum yield of phosphorescence will therefore be the product of the fraction of molecules excited which crosses to the triplet state, multiplied by that fraction of those triplet state molecules which is found to emit.

Significant population of triplet states by direct $S_0 \rightarrow T_1$ absorption has been achieved by using very high excitation intensities. However, at present direct excitation is the exception rather than the rule.

Spin-Orbit Coupling

In the first approximation, the probability of radiative transitions between states of different multiplicities is zero. However, the assumption made in this approximation is that the orbital and spin angular momenta of an electron in any molecule do not interact, and so the orbital angular momentum of each electron couples to produce a total orbital angular momentum quantum number L and the coupling of the spin angular momentum of each electron gives S, the total spin orbital angular momentum quantum number of the molecule. The multiplicity of any state is $2S+1$, and L and S are combined vectorially to give the total angular momentum quantum number J. This scheme is known as Russell-Saunders (L-S) coupling. It was originally developed for atoms, where it was shown to work well in the case of light atoms, but it becomes progressively worse for heavier atoms, where spin-orbit coupling becomes appreciable. In such cases, other types of coupling (e.g., j-j coupling) are better approximations.

Allowance can be made in the following way for a small amount of spin-orbit coupling in molecules. A nominal triplet state wave function may be written in the form:

$$\Psi_T = \Psi_T^0 + \lambda\Psi_S^0 \text{ where } \lambda = \left| \frac{\int \Psi_S^0 H_{SO} \Psi_T^0 \, d\tau}{E_T - E_S} \right| \tag{7.1}$$

and Ψ_S^0 Ψ_T^0 are wave functions for 'pure' singlet and 'pure' triplet states derived in the absence of spin-orbit interaction. E_S and E_T are the singlet and triplet state energies respectively and H_{SO} is the operator for spin-orbit perturbation. Although Ψ_T is no longer a pure triplet state, it is a nominal triplet state if $\lambda \ll 1$. For the sake of simplicity the equation above is written with only one pure singlet state 'mixing' in with the pure triplet state, but each pure singlet state will have its own mixing coefficient λ, which is inversely proportional to the energy gap. Radiative transitions $S_0 \leftrightarrow T_1$ arise mainly because of the small amount of singlet character possessed by the nominal triplet state. They depend also on the ground state possessing some slight amount of triplet character because of spin-orbit interaction.

The relationship between the radiative lifetime and the integrated absorption spectrum for the weak $S_0 \rightarrow T_1$ transition is given by the equation (see Chapter 1):

$$\frac{1}{\tau_0} = 3 \times 10^{-9} n^2 \nu_0^2 \, . \, \frac{g_l}{g_u} \int \varepsilon \, d\nu$$

where g_l and g_u are the multiplicities, usually taken as 1 and 3 respectively, of the lower and upper states.

PHOSPHORESCENCE SPECTRA

(a) *Experimental Procedure*

The long lifetime of the triplet state makes it extremely vulnerable to bimolecular quenching reactions. It is not surprising therefore that most phosphorescence studies have been made using solid or high viscosity solutions where rates of such quenching reactions are drastically reduced. Originally boric acid glasses were very popular, but this medium suffers from a number of inherent disadvantages. Purity and homogeneity of samples are difficult to obtain, and the high temperatures necessary for preparation often lead to solute decomposition.

Many mixed organic solvents form clear glasses when cooled to low temperatures, two such mixtures being ether, isopentane and ethanol, EPA (5 : 5 : 2 by volume) and methyl cyclohexane with isopentane (5 : 1 by volume) at 77°K. Solutions are made up at room temperature and rapidly cooled in Dewar flasks containing liquid nitrogen. The transparent silica Dewar flasks of the type shown in Fig. 7.4 may be inserted into the spectrophosphorimeter; this causes the exciting and emitted beams to pass through the liquid nitrogen and the sample glass.

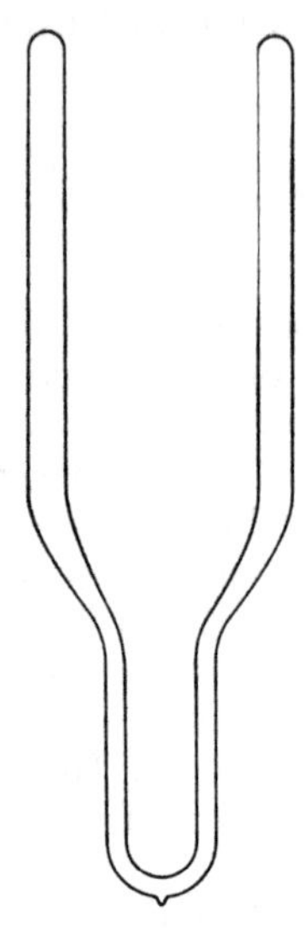

FIG. 7.4 Silica Dewar flask made to fit an Aminco-Bowman spectrophotophosphorimeter. The Dewar flask will accommodate a long narrow cylindrical quartz cell containing the phosphorescent sample.

The utmost care must be taken to attain the highest possible degree of purity during the preparation of these solvents. Untreated solvents, when irradiated alone, usually show quite strong luminescence because of impurities which must be eliminated. These impurities can also lead to quenching of the solute phosphorescence. Purification is usually by chromatography, distillation or recrystallization, or combinations of all three. Solute impurities have often given rise to spurious spectra leading to reports of emission from more than one triplet state in a molecule. Zone-refining, chromatography and sublimation are often successfully employed for solute purification. Care is needed with zone-refining since the high temperatures necessary may lead to some oxidation or general decomposition of the solute during 'purification'. A final precaution necessary during the formation of rigid glasses is the avoidance of microcrystal formation at high solute concentrations.

More recently various plastics have been used as rigid environments for phosphorescence studies. Examples include polymethylmethacrylate, polystyrene, and polycarbonate and cellulose polymers. These are often used in the form of thin films. Other methods of dispersing phosphorescent molecules in rigid environments have involved the solute as an impurity in a host crystal and in rare-gas crystalline 'solvents', where the deposition of molecular beams on quartz windows cooled to 4·2°K has been used.

Phosphorescence has been reported from pure crystals, dilute fluid solutions and the gas phase. In these cases second-order triplet-triplet annihilation, and other decay processes which are first-order with respect to triplet concentration, compete with the phosphorescence. Thus the general lack of phosphorescence in such phases is not due to inherently low yields of triplet state production. Many experiments, including flash photolytic and energy transfer studies, have confirmed the presence of the triplet state in high yields in media where the phosphorescence yield is very low.

Decay by impurity quenching often makes phosphorescence completely undetectable in fluid solution. Oxygen is one such efficient quenching impurity which must be removed by the 'freeze-pump-thaw' degassing procedure. The simpler technique of bubbling an inert gas such as nitrogen through the solution, as used in many fluorescence studies, does not reduce the oxygen content to a sufficiently low level for phosphorescence investigations. Oxygen quenches the gas phase phosphorescence of various aliphatic ketones so efficiently that the phenomenon has been used as a diagnostic test for phosphorescence, since the fluorescence from these molecules is not affected by oxygen. It does, however, quench the fluorescence of aromatic hydrocarbons and some other dyes with high efficiency, although the quenching mechanism is not fully established.

(*b*) *Phosphorescence Spectra of Aromatic Hydrocarbons*

The phosphorescence of aromatic hydrocarbons has been the subject of extensive experimental and theoretical studies. Figure 7.5 shows the total luminescence spectrum of phenanthrene, which exhibits both fluorescence and phosphorescence as do most aromatic hydrocarbons. The phosphorescence spectrum of any compound always lies at longer wavelengths than its fluorescence spectrum since the triplet state is the lowest excited state. As previously mentioned, phosphorescence yields are usually low in all but rigid or viscous media, and so sensitive equipment, similar to that used to obtain the spectra shown in Fig. 7.6, is necessary for measuring the weak phosphorescence from dilute degassed solutions or gaseous samples of aromatic hydrocarbons.

In general, the larger the molecule the lower the energy of the $T_1 \rightarrow S_0$ transition, and the lower is the phosphorescence yield. This is illustrated by benzene which phosphoresces in the ultra-violet region with $\Phi_p = 0{\cdot}2$ whereas naphthalene and anthracene phosphoresce, in the visible and near infra-red regions respectively, with much lower yields. Values of the energy of the lowest triplet state of a number of hydrocarbons are given in

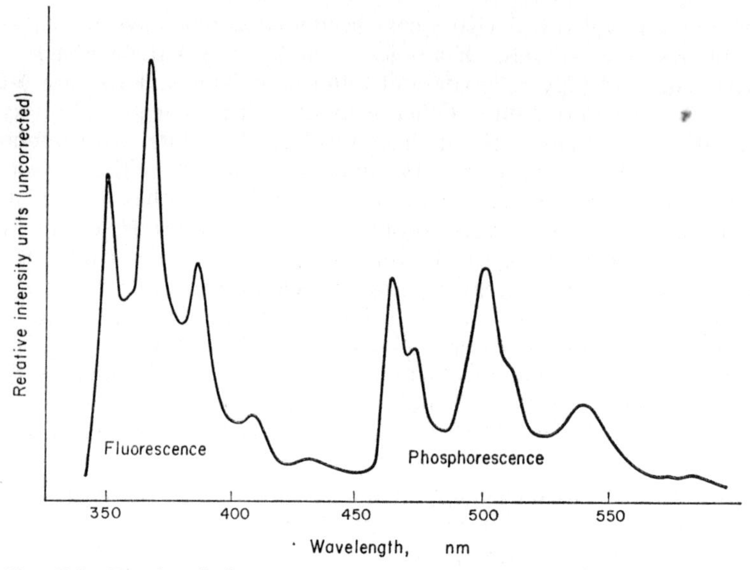

FIG. 7.5 Total emission spectrum of phenanthrene (10^{-4}M) in ethanol at 77°K.

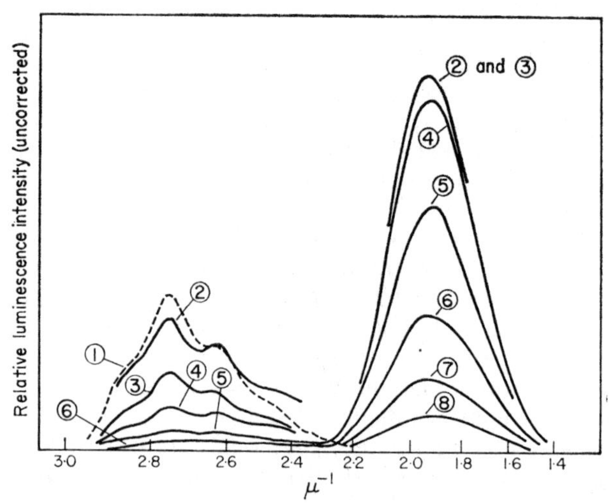

FIG. 7.6 Delayed emission spectra of 10^{-3}M phenanthrene in ethanol. The left-hand sections of the spectra (delayed fluorescence) all were recorded at a sensitivity approximately 1000 times greater than that used for curve (1) (normal fluorescence). The right-hand sections of the curves (phosphorescence) were recorded at the following sensitivities: (2) −107° at ×60; (3) −80° at ×300; (4) −70° at ×600; (5) −48° at ×1000; (6) −32° at ×1000; (7) −14° at ×1000; (8) +13° at ×1000. (Reproduced with permission from PARKER *J. phys. Chem.* **66**, 2506 (1962).)

Table 7.1. The vibrational structure and the bandwidth of the phosphorescence spectra of these compounds are similar to those found for singlet-singlet transitions. At 77°K phosphorescence originates almost exclusively

TABLE 7.1. *Lowest triplet levels, phosphorescence yields and observed lifetimes of a number of aromatic hydrocarbons and their derivatives in solution*

Compounds	$\bar{\nu}_P$ (cm^{-1})	Observed lifetime τ_P (s)	Φ_P	Φ_P/Φ_F	Reference
Benzene	29,400	7·0	0·23	0·98	(*d*)
Toluene	28,800	8·8			(*a*)
Aniline	26,800	4·7			(*a*)
Phenol	28,600	2·9			(*a*)
Benzoic acid	27,200	2·5			(*a*)
Benzoate ion	27,700	0·7			(*a*)
Anisole	28,200	3·0			(*a*)
Naphthalene	21,300	2·6			(*a*)
	21,250	2·3	0·03	0·09	(*b*)
			0·085	0·18	(*d*)
1–Methylnaphthalene	20,850	2·5			(*a*)
	21,000	2·1	0·023	0·053	(*b*)
1–Aminonaphthalene	19,000	1·5			(*a*)
1–Naphthol	20,500	1·9			(*a*)
2–Naphthol	21,100	1·3			(*a*)
2–Naphthoic acid	20,900	2·5			(*a*)
1–Nitronaphthalene	19,200	0·049			(*a*)
				1000	(*b*)
1,5–Dinitronaphthalene	19,900	0·11			(*a*)
1–Fluoronaphthalene	20,970	1·5			(*a*)
1–Chloronaphthalene	20,700	0·30			(*a*)
		0·29	0·16	5·2	(*b*)
1–Bromonaphthalene	20,700	0·018			(*a*)
	20,650	0·018	0·14	164	(*b*)
1–Iodonaphthalene	19,000	0·0025			(*a*)
	20,500	0·0020	0·20	1000	(*b*)
2–Chloronaphthalene	21,000	0·47			(*a*)
2–Bromonaphthalene	21,100	0·021			(*a*)
2–Iodonaphthalene	21,040	0·0025			(*a*)
Triphenylene	23,800	15·9			(*a*)
				1·2	(*c*)
			0·53		(*d*)
			0·42	2·8	(*e*)
Coronene	19,100	9·4			(*a*)
				1·0	(*c*)
Fluorene	23,750	4·9			(*a*)
				0·7	(*c*)
2–Nitrofluorene	20,600	0·13			(*a*)
Phenanthrene	21,700	3·3	0·135	1·1	(*b*)
				0·3	(*c*)
			0·09		(*e*)
Chrysene	19,800	2·5			(*a*)
				0·03	(*c*)
1,2–Benzanthracene	16,500	0·3			(*a*)
				0·001	(*c*)
1,2,5,6–Dibenzanthracene	18,300	1·5			(*a*)
				0·01	(*c*)
Pyrene	16,800	0·2			(*a*)
				0·001	(*c*)
Anthracene	14,700				(*a*)
				0·0001	(*c*)
Biphenyl	23,000	3·1		0·8	(*b*)
p–Nitrobiphenyl	20,500	0·08			(*a*)

(*a*) McClure, D. *J. chem. Phys.* **17**, 905 (1949). Solvent: EPA glass at 77° K.
(*b*) Ermolaev, V. *Soviet Physics* (*Uspekhi*), **6**, 333 (1963). Solvent: Alcohol-ether at 77°K.
(*c*) Kasha, M. *see* (*a*) *above*.
(*d*) Gilmore, E., Gibson, G. and McClure, D. *J. chem. Phys.* **20**, 829 (1952); **23**, 399 (1955). Solvent: EPA at 77°K.
(*e*) Kellogg, R. E. and Bennett, R. *J. chem. Phys.* **41**, 3042 (1964). Solvent: EPA at 77°K.

from the lowest vibrational level of T_1 (see Fig. 7.1). Therefore the vibrational splitting corresponds to ground state vibrations, and the relative intensities of the vibronic bands are governed by the Franck–Condon principle. The vibrational splitting for the reverse transition of singlet-triplet absorption depends on the vibrational levels of T_1. These spectra ($S_0 \rightarrow T_1$ absorption and $T_1 \rightarrow S_0$ emission) show a mirror-image relationship about their 0,0 bands, and this illustrates that the geometrical structure of T_1 is very similar to that of the ground state. In the case of benzene, Sklar has calculated that in its triplet state the molecule is still a planar regular hexagon but the C—C bond lengths are 0·06 Å longer than are those in ground state benzene.

For aromatic hydrocarbons the difference in energy between the lowest excited singlet and triplet states, both of which are π,π^* states, is relatively large. This energy gap generally decreases with increasing size of the molecule. In the series benzene, naphthalene and anthracene, however, the values are in reverse order. This is because, although the lowest excited singlet is often ^1L_b (Platt notation) the larger linear polyacenes have ^1L_a as the lowest excited singlet state (see Figs. 7.7 and 7.8). The lowest

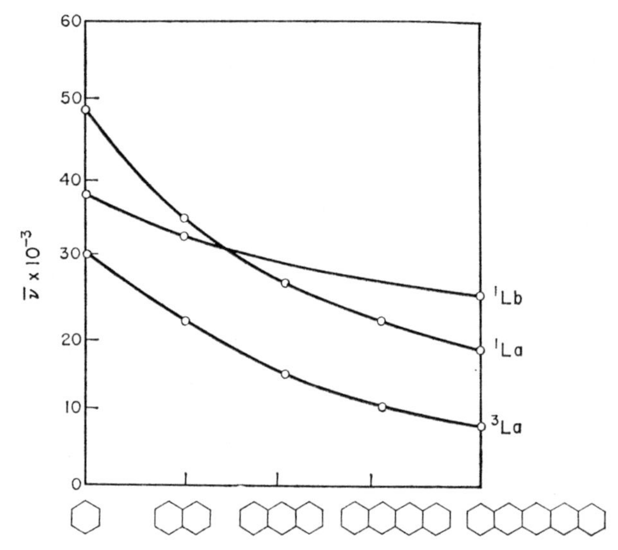

FIG. 7.7 Lower energy levels of the linear polyacenes showing the cross-over of the ^1L_a and ^1L_b levels. (Adapted from PLATT *J. chem. Phys.* **17**, 470 (1949).)

excited singlet state of both benzene and naphthalene is of ^1L_b type, whereas that of anthracene is ^1L_a. The lowest triplet state assignments are not known with absolute certainty, but they are usually taken to be ^3L_a. The ^1L_a–^3L_a gap is large, and decreases regularly with increase in the size of the molecule (cf. Fig. 7.7). Exact energies of the ^3L_b states are not known, and many theoretical papers have included estimates which place

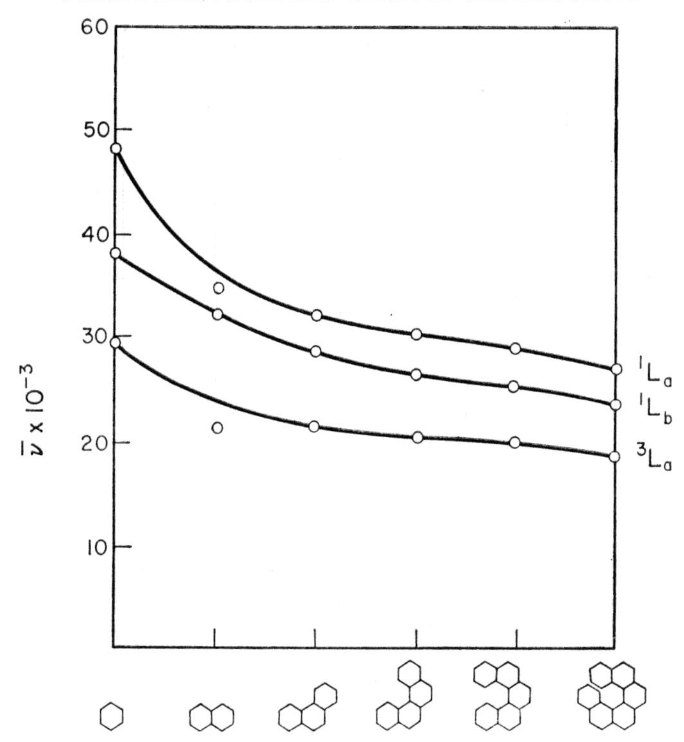

FIG. 7.8 Lower energy levels of the cyclopolyacenes. (Adapted from MCGLYNN *Photochem. Photobiol.* **3**, 269 (1964).)

^3L_b higher than ^1L_b. This seems rather unlikely, and Pariser, who uses a different notation [4], gives ^1L_b and ^3L_b as degenerate levels in the case of polyacenes. There is no doubt that the lowest singlet–lowest triplet energy gap depends on the exact nature of the state. The common statement that all π,π^* states show large singlet-triplet splitting is, to say the least, somewhat misleading.

Substituents usually have more effect on the lifetimes and yields of phosphorescence than on the position of the spectrum, as may be seen for example, for the substituted naphthalene molecules given in Table 7.1. Only those substituents which markedly affect the energy of the π-orbitals are likely to shift the spectra much. The shifts observed for benzene and naphthalene substituents are shown in Fig. 7.9, and the phosphorescence spectra of various substituted naphthalenes in Fig. 7.10.

(*c*) *Phosphorescence Spectra of Compounds Containing Heteroatoms*

Many compounds such as ketones, aldehydes and N-heterocyclic compounds which contain heteroatoms with non-bonding electrons have a $^1(n,\pi^*)$ state as their lowest singlet state. This state is produced by promoting a non-bonding electron on the heteroatom into the lowest anti-bonding

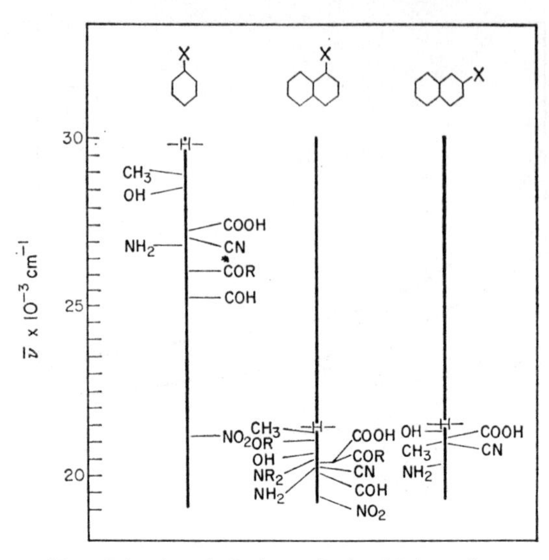

FIG. 7.9 A tabulation of the highest frequency phosphorescence maxima of benzene, and benzene with one auxochrome substituent; also similar data for naphthalene with α- and β-substituents. Left, basic auxochromes; right, acid auxochromes. (Reproduced with permission from LEWIS and KASHA *J. Am. chem. Soc.*, **66**, 2100 (1944).)

π-orbital (π^*) of the molecule. The n orbital in the carbonyl group is essentially a 2*p* orbital on the oxygen atom whereas in an azo-aromatic compound the n orbital is a hybridized sp^2 orbital on the nitrogen atom. Both these orbitals lie in the molecular plane and are symmetric with respect to it. The π-orbitals are of course antisymmetric with respect to the molecular plane, and this explains why these lone pairs are non-bonding, since they cannot conjugate with the π orbital electrons. Such poor orbital overlap between the n and π^* orbitals is reflected in the low oscillator strengths observed for n–π^* singlet transitions, even when these are allowed by symmetry conditions.

The energy difference between ^{1}(n,π^*) and ^{3}(n,π^*) states usually lies in the range 2000–5000 cm^{-1} which is a good deal smaller than many $^1(\pi,\pi^*)$ and $^3(\pi,\pi^*)$ state splittings (see previous section). Six possible arrangements of the energy levels in unsaturated compounds containing heteroatoms exist, and these are shown in Fig. 7.11. It follows that a molecule whose lowest excited singlet is of (n,π^*) type does not necessarily have a ^{3}(n,π^*) state as its lowest triplet state. Those which do, include many aliphatic and aromatic carbonyl compounds, aromatic nitro-compounds and a number of nitrogen heterocyclics, especially the diazines. Such molecules, having lowest singlet and triplet states both of (n,π^*) character, typically exhibit low fluorescence yields and high phosphor-

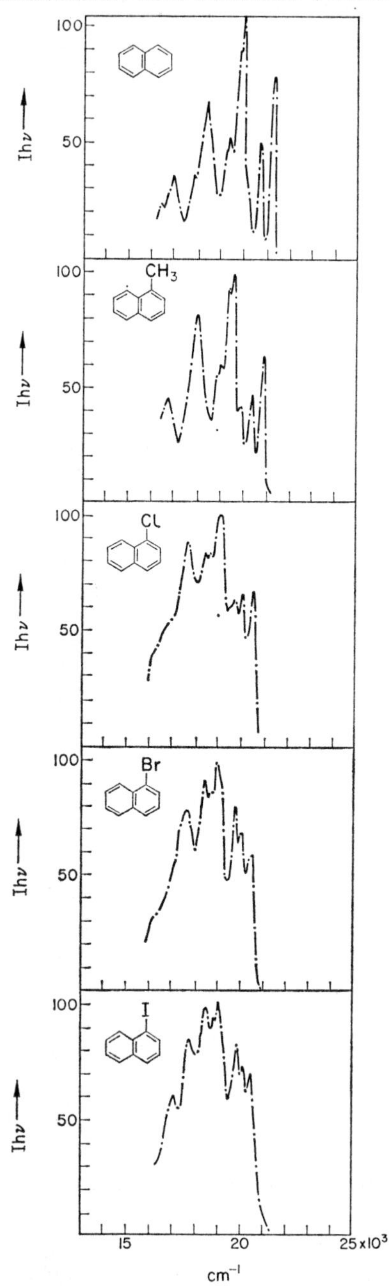

FIG. 7.10 Phosphorescence spectra of napthalene and its 1-halogen derivatives. (Reproduced with permission from ERMOLAEV and TERENIN *J. Chim. phys.* **55**, 698 (1958).)

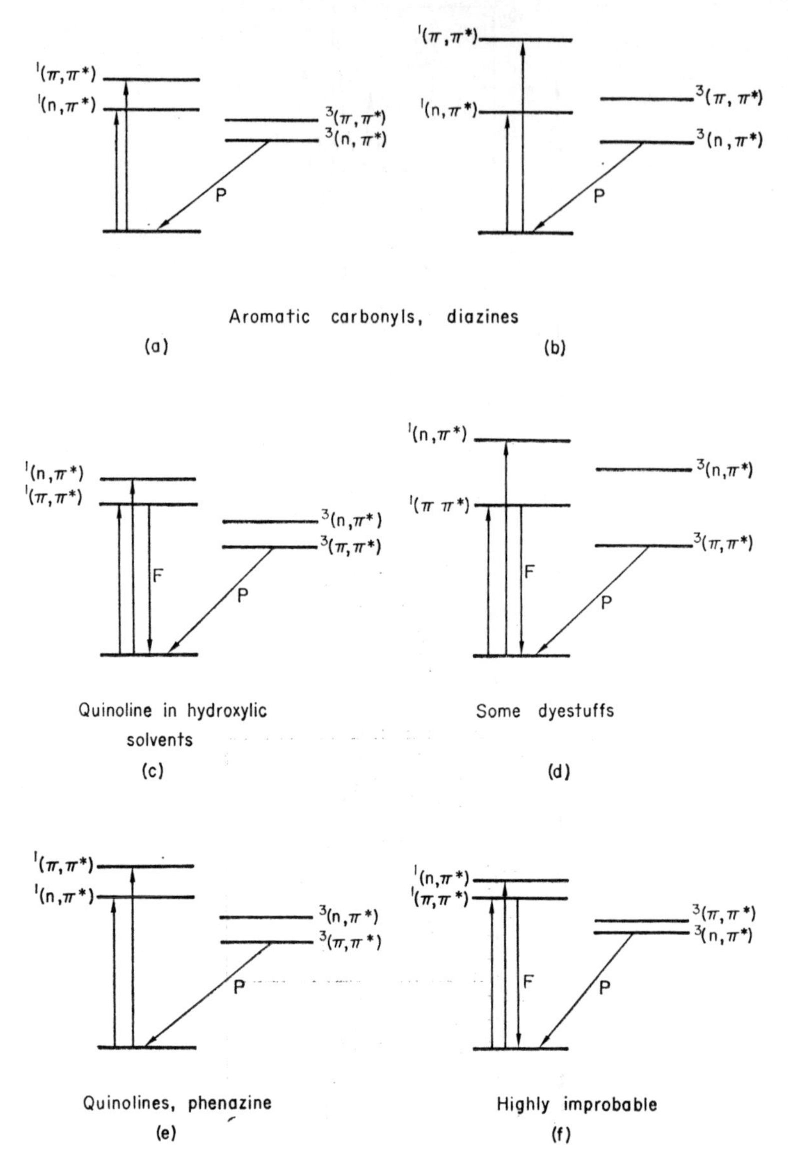

FIG. 7.11 Possible arrangements of the lower $^1(\pi,\pi)$, $^1(n,\pi)$ and $^3(\pi,\pi)$, $^3(n,\pi)$ energy levels.

escence yields. Their phosphorescence spectra are displaced to lower frequencies by less than 5000 cm^{-1} from the $^1(n\text{-}\pi^*)$ transition.

It is often difficult to decide whether the lower singlet or triplet states in a molecule are (n,π^*) or (π,π^*) in nature. The following generalizations, made by Kasha, help with the assignment of singlet states: $^1(n–\pi^*)$ transitions are absent for hydrocarbon analogues, they disappear in acid media, they undergo a blue shift in polar solvents, they have low intensity of absorption representing a long radiative lifetime, and they have a unique polarization. Characterization of triplet states has usually been deduced from phosphorescence lifetime measurements (τ_p values of $^3(n,\pi^*)$ states are much shorter than those of $^3(\pi,\pi^*)$ states). The magnitude of the $S_1 - T_1$ energy difference, and examination of the vibronic structure of the phosphorescence for characteristic vibrations, e.g., C=O vibrations of 1730 cm^{-1}, are also used to help in distinguishing between $^3(n,\pi^*)$ and $^3(\pi,\pi^*)$ states. Singlet-triplet absorption spectra have not until recently been of great use in such assignments because of their very low intensities and frequent overlap with the long-wavelength tail of singlet-singlet absorption. Following the theoretical prediction of El-Sayed that τ_p values of $^3(n,\pi^*)$ states are not susceptible to heavy atom effects whereas $^3(\pi,\pi^*)$ states are, Kearns and his co-workers [7] have developed a method to characterize triplet states which depends on direct excitation of the triplet states followed by measurement of the resulting phosphorescent emission. Using aromatic aldehydes and ketones, whose lowest triplet states are well known to be of (n,π^*) character by virtue of their short τ_p values, small $S_1 - T_1$ splitting and the prominent C=O vibration present in the phosphorescence spectra, these investigators found no enhancement of $S_0 \rightarrow T_1$ absorption on subjecting the molecule to a heavy atom environment. In aromatic hydrocarbons $S_0 \rightarrow T_1{}^3(\pi,\pi^*)$ absorption intensities are well known to be enhanced by heavy-atom effects. Subsequently it would seem that susceptibility to heavy-atom effects may be used to characterize triplet states. Kearns has applied his technique to several substituted acetophenones and has shown, by virtue of the enhancement of their lowest $S_0 \rightarrow T_1$ absorption band in the presence of heavy atoms, that the lowest triplets are $^3(\pi,\pi^*)$ in nature. Acetophenone itself has a $^3(n,\pi^*)$ state as its lowest triplet state.

Aromatic Ketones, Aldehydes and Nitro-compounds

Table 7.2 lists a number of aromatic carbonyl compounds with their corresponding lowest triplet levels and phosphorescent state lifetimes τ_p. The splitting between the first excited singlet and the lowest triplet states is of the order of 5000 cm^{-1} for compounds having the lowest singlet $^1(n,\pi^*)$ and triplet $^3(n,\pi^*)$. For these compounds in rigid glass solutions, no fluorescence spectra are seen, and consequently the total luminescence spectra consist of phosphorescence occurring only from their $^3(n,\pi^*)$ states. Total emission curves for benzophenone, acetophenone and biacetyl in EPA at 77°K are shown in Fig. 7.12. Several carbonyl compounds phosphoresce in phases other than the solid phase. Examples are given in Figs. 7.13 and 7.14. The phosphorescence spectra of benzophenone in the solid and gas phases are very different, whereas there is little difference between gas, liquid or solid phase emission spectra of biacetyl.

Phosphorescence spectra of most aromatic carbonyl compounds show a

TABLE 7.2. *Lowest triplet levels, phosphorescence yields and observed lifetimes of a number of carbonyl compounds*

Compounds	$\bar{\nu}_p$ cm^{-1}	Observed lifetime τ_p (s)	Φ_P	Φ_P/Φ_F	Reference
Benzaldehyde	24,950	$1 \cdot 5 \times 10^{-3}$	0·49	>1000	(*a*)
Benzophenone	24,250	$4 \cdot 7 \times 10^{-3}$	0·74	>1000	(*a*)
			0·70		(*b*)
Acetophenone	25,750	$2 \cdot 3 \times 10^{-3}$	0·62	>1000	(*a*)
			0·63		(*b*)
Ethyl phenyl ketone	26,150	$3 \cdot 8 \times 10^{-3}$		>1000	(*a*)
p-Chlorobenzaldehyde	24,750			>1000	(*a*)
o-Chlorobenzaldehyde	24,350			>1000	(*a*)
m-Iodobenzaldehyde	24,750	$6 \cdot 5 \times 10^{-4}$	0·64	>1000	(*a*)
Anthrone	25,150	$1 \cdot 5 \times 10^{-3}$		>1000	(*a*)
Xanthone	24,800	$2 \cdot 0 \times 10^{-2}$		>1000	(*a*)
Anthraquinone	21,950			>1000	(*a*)
Acetone			0·043		(*b*)
Biacetyl	19,700				
Phenyl-4-benzophenone	21,225	0·3			(*a*)
4-Methoxy phenyl diphenyl ketone	21,350	0·28			(*a*)
2-Naphthaldehyde	20,800	0·35			(*a*)
Methyl 2-naphthalene ketone	20,775	0·97			(*a*)
Phenyl-1-naphthalene ketone	20,100	0·74			(*a*)

(*a*) ERMOLAEV, V. and TERENIN, A. *J. chem. Phys.* **55**, 698 (1958).
ERMOLAEV, V. *Soviet Physics* (*Uspekhi*), **6**, 333 (1963). Solvent: Alcohol-ether at 77°K.
(*b*) GILMORE, E., GIBSON, G. and MCCLURE, D. *J. chem. Phys.* **20**, 829 (1952); **23**, 399 (1955). Solvent: EPA at 77°K.

common feature as regards their vibrational structure. This is the appearance of the aromatic C=O vibrational frequency of about 1730 cm^{-1}. Furthermore τ_p values are very similar in order of magnitude, and quantum yields are close to unity in the rigid solvents.

The non-bonding electrons on the carbonyl oxygen are not very susceptible to substituent effects, whereas the π bonding and anti-bonding orbitals are. Thus the position of the 0,0 phosphorescence bands will have a large dependence on molecular structure. In aromatic hydrocarbons, if both ground and excited π orbitals are affected by substituents then the overall spectral change on addition of a substituent to an aromatic nucleus may be very small. (See Fig. 7.10.)

Aromatic nitro-compounds behave very much like aromatic carbonyls as regards their phosphorescent properties, in that they have τ_p values of a similar order of magnitude and phosphorescence occurs from a $^3(n,\pi^*)$ state. In Fig. 7.15 the spectra of three aromatic nitro-compounds

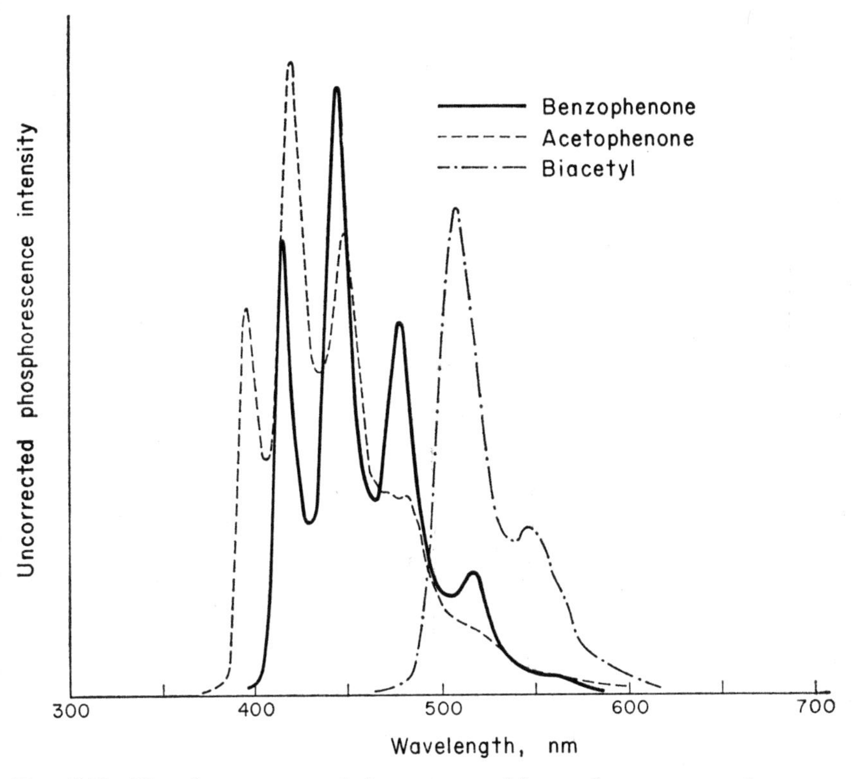

FIG. 7.12 Phosphorescence emission spectra of benzophenone, acetophenone and biacetyl in EPA glass at 77°K.

are illustrated. In most cases the characteristic nitro Raman frequency of 1450 cm^{-1} appears.

NITROGEN HETEROCYCLICS

Pyrazine is a typical diazine in that its energy levels conform to the aromatic carbonyl and nitro-compound scheme and it has a high value of Φ_p. Very recent work has claimed that fluorescence has been observed from pyrazine [8]. Its phosphorescence spectrum is shown in Fig. 7.16 along with that of 2,5-dimethylpyrazine. In the pyrazine spectrum the average spacing between the first six vibrational bands is $\Delta\bar{\nu}=615$ cm^{-1}, and this corresponds to the 609 cm^{-1} angular distortion frequency observed for the ground state of the molecule. This vibrational frequency is also evident in the phosphorescence spectrum of 2,5-dimethylpyrazine. Singlet-triplet splittings for the diazines are in the expected region of 5000 cm^{-1}.

For other N-heterocyclic molecules the lowest triplet state is often of $^3(\pi,\pi^*)$ type, an example of which is phenazine. This has a $^1(n,\pi^*)$ lowest singlet state and shows no fluorescence, but emits an anthracene-like

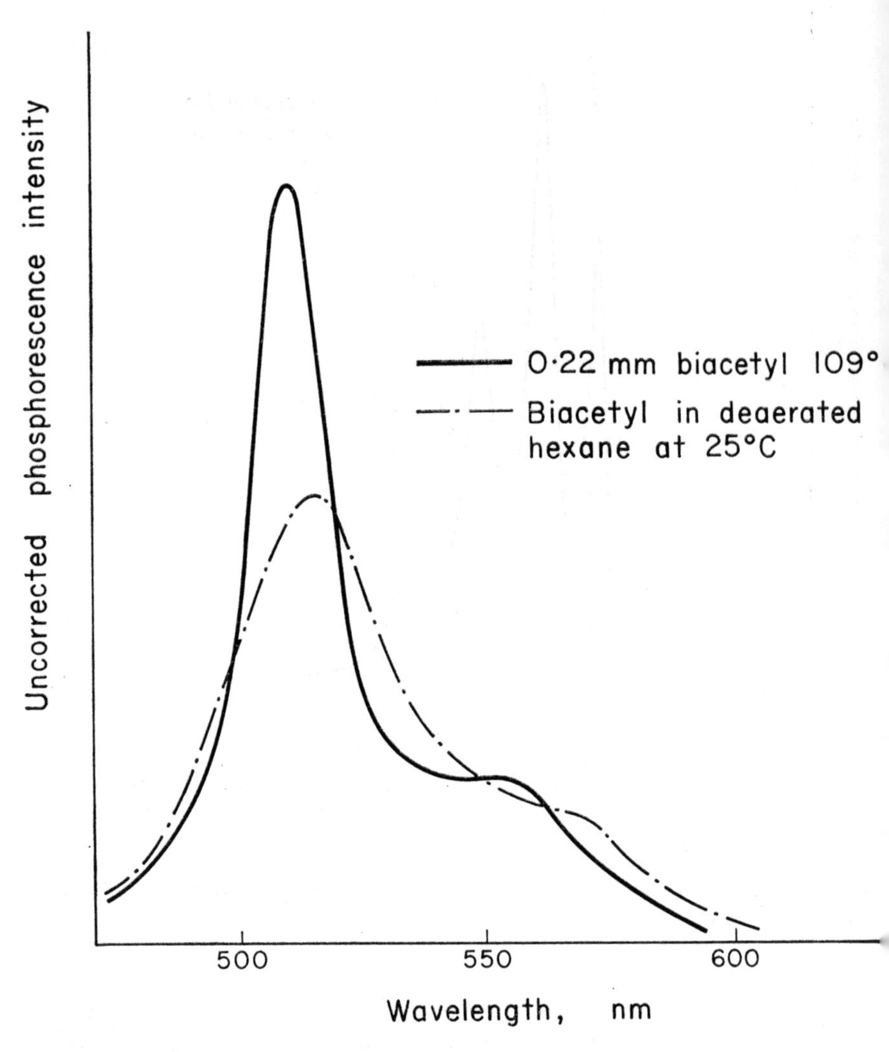

Fig. 7.13 Phosphorescence emission spectra of biacetyl in deaerated hexane solution (0·003M) at 25°C and in the vapour (0·22 mm) at 109°C.

phosphorescence indicating the lowest triplet state to be $^3(\pi,\pi^*)$ in character. Quinoline is another example of a heterocyclic compound showing the phosphorescence characteristic of its parent hydrocarbon. Several naphthyl and diphenyl ketones also have phosphorescence spectra characteristic of their parent hydrocarbons. These coupled with their respective long phosphorescence lifetimes, indicate the lowest triplet states to be $^3(\pi,\pi^*)$ in nature.

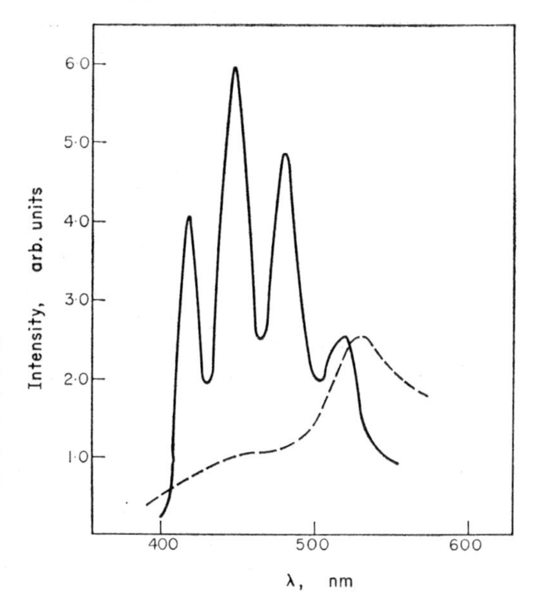

FIG. 7.14 Phosphorescence of benzophenone: ——, in EPA at liquid nitrogen temperature; — — — —, vapour at 170°. (Reproduced with permission from DUBOIS *J. Am. chem. Soc.* **84**, 4041 (1962).)

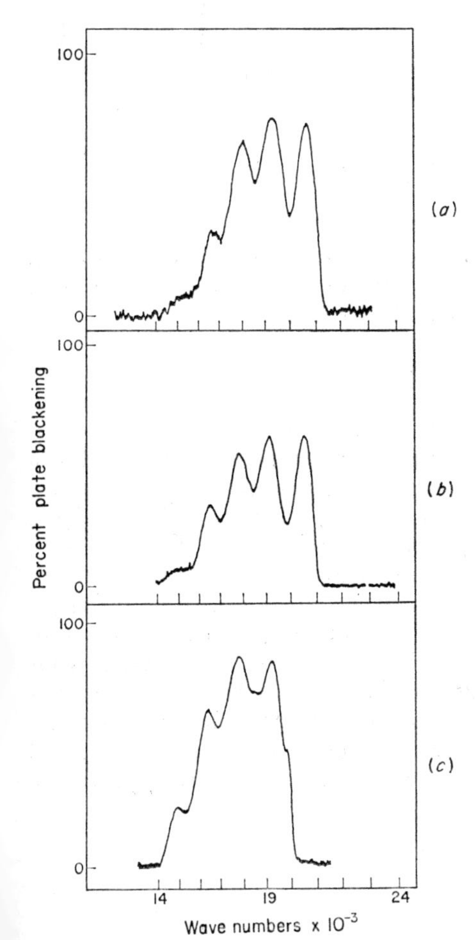

FIG. 7.15 Total emission spectra of some aromatic nitro-compounds in EPA rigid glass solution at 77°K. (*a*) 4-nitrobiphenyl; (*b*) 2-nitrofluorene; (*c*) 1,5-dinitronaphthalene. A unique phosphorescence is observed, from a triplet state of n–π^* promotion type. (Reproduced with permission from KASHA *Radiation Research Supp.* **2**, 243 (1960).)

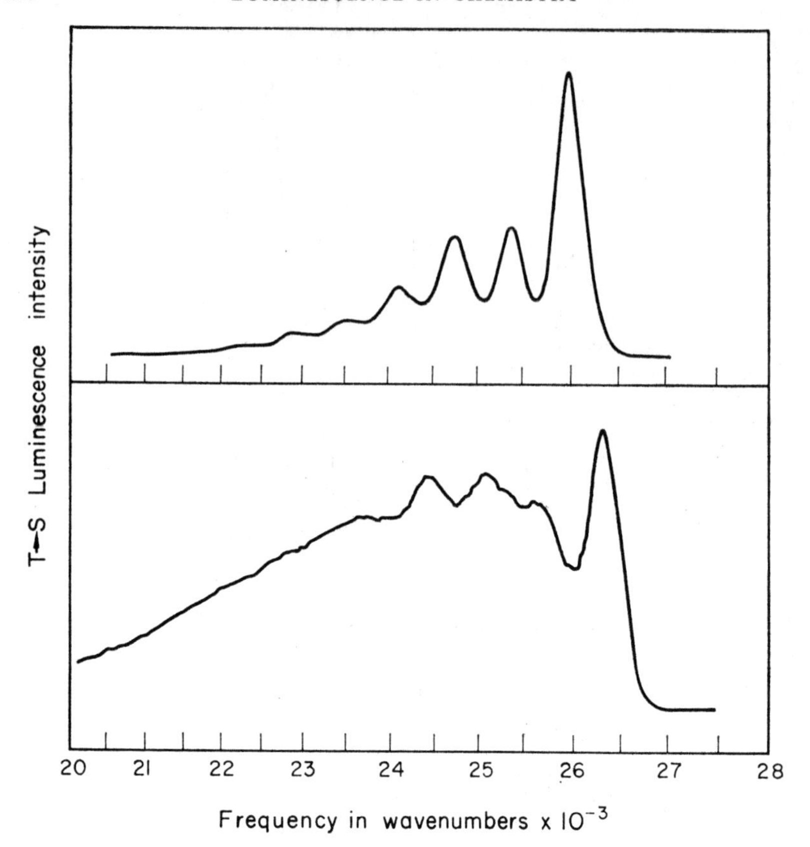

FIG. 7.16 T→S luminescence curves of pyrazine (upper) and 2,5-dimethylpyrazine (lower) at 77°K in EPA glass. (Reproduced with permission from GOODMAN and KASHA *J. molec. Spectrosc.* **2**, 58 (1958).)

PHOSPHORESCENCE YIELDS AND LIFETIMES

Early work on the decay of phosphorescence showed it to be exponential, which suggested that the emission did not arise from a recombination of ions or fragments in the system. More comprehensive studies have shown that phosphorescence of a single pure solute almost always decays exponentially in the liquid state, except at very high excitation energies. This is also true in solid solutions, except where there is a variety of sites in the host matrix for phosphorescing species, e.g., silica gel absorbates, cracked glasses and some polymeric matrices. In most cases, therefore, characteristic mean lifetimes for phosphorescence decay can be obtained, and these are given, together with values of Φ_p and Φ_f, the quantum yields of phosphorescence and fluorescence respectively, in Table 7.1. Measurements of absolute phosphorescence yields are very difficult; only a few values are known with an expected accuracy of better than 10%, and many of the values in Table 7.1 are probably subject to even bigger errors.

As can be seen from Tables 7.1 and 7.2, many compounds have total luminescence yields ($\Phi_f + \Phi_p$) which are much less than unity even at 77°K. Radiationless transitions therefore play a prominent role in determining the values of both Φ_p and τ_p. The following reaction scheme may be used to show the dependence of phosphorescence yields and lifetimes on the processes shown in Fig. 7.3.

		Rate
$S_x \xrightarrow{h\nu} S_x$	Excitation	I_a
$S_3 \rightarrow S_1$	Internal conversion	$k_r[S_x]$
$S_1 \rightarrow S_0 + h\nu_f$	Fluorescence	$k_f[S_1]$
$S_1 \rightarrow S_0$	Internal conversion	$k_s[S_1]$
$S_1 \rightarrow T_1$	Intersystem crossing	$k_i[S_1]$
$T_1 \rightarrow S_0 + h\nu_p$	Phosphorescence	$k_p[T_1]$
$T_1 \rightarrow S_0$	Intersystem crossing	$k_t[T_1]$

where S_x represents any vibrationally excited and higher electronically excited singlet state.

The observed mean lifetime for phosphorescence decay is given by

$$\tau_p = \frac{1}{k_p + k_t} \quad \text{and} \quad \tau_0 = \frac{1}{k_p}$$

where τ_0 is the radiative or natural lifetime for phosphorescence.

$$\Phi_p = \frac{k_p[T_1]}{I_a} \quad \text{and} \quad \Phi_f = \frac{k_f[S_1]}{I_a}$$

Under stationary state conditions

$$\frac{d[S_1]}{dt} = 0 \quad \text{and} \quad \frac{d[T_1]}{dt} = 0$$

Thus

$$[S_1] = \frac{I_a}{k_f + k_s + k_i} = I_a \tau_f$$

where τ_f is the lifetime of the first excited singlet state

and

$$[T_1] = \frac{k_i[S_1]}{(k_p + k_t)} = k_i \tau_p [S_1]$$

Combining these equations we obtain

$$\Phi_p = \left(\frac{k_i}{k_f + k_s + k_i}\right)\left(\frac{k_p}{k_p + k_t}\right)$$

$$= \Phi_i \Theta_p$$

where Φ_i is the quantum yield of intersystem crossing to the triplet state and Θ_p represents the fraction of triplet molecules formed which phosphoresce.

In the previous section the position of the observed phosphorescence was shown to depend on the nature of the lowest energy levels of the molecule concerned. This is true also for lifetimes and yields, although from the equations above it can be seen that radiationless as well as radiative transition probabilities have to be considered.

Radiative Processes

Aromatic hydrocarbons have radiative lifetimes for the phosphorescence process in the range 10–100 s. Similar radiative lifetimes are obtained for $T_1 \rightarrow S_0$ transitions from all molecules constructed of atoms of low atomic number where T_1 is a $^3(\pi,\pi^*)$ state. However, when the lowest triplet state is of $^3(n,\pi^*)$ character τ_0 values are usually in the range 10^{-2}–10^{-3} s. The reverse situation holds for the $S_1 \rightarrow S_0$ transition, $^1(\pi,\pi^*)$ states having a much shorter natural lifetime than $^1(n,\pi^*)$ states. This is because the probability of a radiative transition from a nominal triplet state is almost entirely dependent on the transition probabilities of the admixed single states which cannot be of the same configuration. Thus, spin-orbit perturbation mixes $^3(\pi,\pi^*)$ states with $^1(n,\pi^*)$ and $^1(\sigma,\pi^*)$ which have low transition probabilities while $^3(n,\pi^*)$ states mix predominantly with $^1(\pi,\pi^*)$ states which have high radiative transition probabilities for return to the ground state.

Non-radiative Processes

Internal conversion between excited singlet-states is very efficient, so much so that only in the case of azulene has fluorescence been observed from any but the first excited singlet state S_1. Internal conversion from $S_1 \rightarrow S_0$ is much less efficient, and measurements of the quantum yield of ordinary and sensitized phosphorescence indicates it is absent from many aromatic molecules. In such cases $\Theta_i = 1 - \Phi_f$ and $\Phi_p = k_p \tau_p\ (1 - \Phi_f)$. This allows the calculation of values of $k_p + k_t$ from measurements of Φ_p, Φ_f and τ_p (see Table 7.3).

As mentioned earlier, the values of k_t decrease with increasing size of the T_1–S_0 energy gap. This is illustrated for various hydrocarbons in Fig. 7.17.

Examination of Table 7.2 shows that many carbonyl compounds which have S_1 a $^1(n,\pi^*)$ state, show high phosphorescence yields but no fluorescence. Intersystem crossing from $S_1 \rightarrow T_1$ depends on spin-orbit perturbation which, to the first order, is forbidden between states of the same configuration. Thus the following selection rules have been given by El-Sayed:

$$^{1\text{ or }3}(n,\pi^*) \leftrightarrow {}^{3\text{ or }1}(\pi,\pi^*),\quad {}^1(n,\pi^*) \not\leftrightarrow {}^3(n,\pi^*)$$
$$\text{and}\quad {}^1(\pi,\pi^*) \not\leftrightarrow {}^3(\pi,\pi^*)$$

He suggests that the large yields of intersystem crossing for many molecules with $S_1 = {}^1(n,\pi^*)$ are due to the presence of lower $^3(\pi,\pi^*)$ states as for example in Fig. 7.11(*a*). In some molecules containing two adjacent ring nitrogen atoms, e g., 9, 10 diazophenanthrene, fluorescence from $^1(n,\pi^*)$ states is observed, and in these cases it may be that the interaction between

TABLE 7.3. *Rate constants for radiative and radiationless intercombination transitions in aromatic molecules in alcohol-ether at 77°K*

Compound	k_p (s^{-1}) $^3\Gamma \to {}^1\Gamma$	k_t (s^{-1}) $^3\Gamma \to {}^1\Gamma$	$\tau_{emission} = 1/k_p$ (s)
m-Iodobenzaldehyde	$1{\cdot}0\times10^3$	$5{\cdot}5\times10^2$	$1{\cdot}0\times10^{-3}$
Benzaldehyde	$3{\cdot}4\times10^2$	$3{\cdot}5\times10^2$	$2{\cdot}9\times10^{-3}$
Acetophenone	$2{\cdot}8\times10^2$	$1{\cdot}7\times10^2$	$3{\cdot}6\times10^{-3}$
Benzophenone	$1{\cdot}6\times10^2$	$5{\cdot}0\times10$	$6{\cdot}2\times10^{-3}$
Ethylphenylketone	$\sim1{\cdot}5\times10^2$	$\sim1{\cdot}2\times10^2$	$\sim6{\cdot}7\times10^{-3}$
1-Iodonaphthalene	$1{\cdot}0\times10^2$	$4{\cdot}0\times10^2$	$1{\cdot}0\times10^{-2}$
1-Bromonaphthalene	$7{\cdot}0$	$4{\cdot}3\times10$	$1{\cdot}4\times10^{-1}$
4-Phenyl-4′-methoxybenzophenone	$2{\cdot}3$	$1{\cdot}3$	$4{\cdot}3\times10^{-1}$
4-Phenylbenzophenone	$1{\cdot}6$	$1{\cdot}7$	$6{\cdot}2\times10^{-1}$
4-(*p*-Methoxyphenyl)-benzophenone	$1{\cdot}0$	$1{\cdot}0$	$1{\cdot}0$
Triphenylamine	$8{\cdot}6\times10^{-1}$	$5{\cdot}7\times10^{-1}$	$1{\cdot}2$
1-Nitronaphthalene	$8{\cdot}6\times10^{-1}$	$2{\cdot}0\times10$	$1{\cdot}2$
1-Chloronaphthalene	$5{\cdot}7\times10^{-1}$	$1{\cdot}7$	$1{\cdot}7$
1-Naphthaldehyde	$3{\cdot}8\times10^{-1}$	$1{\cdot}2\times10$	$2{\cdot}6$
Quinoline	$7{\cdot}7\times10^{-2}$	$6{\cdot}6\times10^{-1}$	$1{\cdot}3\times10$
Carbazole	$6{\cdot}9\times10^{-2}$	$6{\cdot}3\times10^{-2}$	$1{\cdot}5\times10$
2-Naphthylmethylketone	$5{\cdot}1\times10^{-2}$	$9{\cdot}8\times10^{-1}$	$2{\cdot}0\times10$
Phenanthrene	$4{\cdot}6\times10^{-2}$	$2{\cdot}6\times10^{-1}$	$2{\cdot}2\times10$
Biphenyl	$3{\cdot}7\times10^{-2}$	$2{\cdot}9\times10^{-1}$	$2{\cdot}7\times10$
Decadeuterobiphenyl	$3{\cdot}7\times10^{-2}$	$5{\cdot}1\times10^{-2}$	$2{\cdot}7\times10$
1-Fluoronaphthalene	$3{\cdot}6\times10^{-2}$	$6{\cdot}3\times10^{-1}$	$2{\cdot}8\times10$
1-Methylnaphthalene	$2{\cdot}0\times10^{-2}$	$4{\cdot}5\times10^{-1}$	$5{\cdot}0\times10$
1-Naphthol	$1{\cdot}6\times10^{-2}$	$5{\cdot}1\times10^{-1}$	$6{\cdot}3\times10$
Naphthalene	$1{\cdot}6\times10^{-2}$	$4{\cdot}2\times10^{-1}$	$6{\cdot}3\times10$
Octadeuteronaphthalene	$1{\cdot}6\times10^{-2}$	$9{\cdot}0\times10^{-2}$	$6{\cdot}3\times10$

(Reproduced from ERMOLAEV *Soviet Physics* (*Uspekhi*), **6**, 333 (1963).)

the non-bonding sets of electrons causes the lowest $^1(n,\pi^*)$ state to drop below the $^3(\pi,\pi^*)$ state thus reducing the intersystem crossing efficiencies in these molecules (see Fig. 7.11(*b*)).

Apart from these selection rules, theoretical predictions suggest that the energy gap and Franck–Condon factors will be important parameters in determining the efficiency of any radiationless process. It is expected therefore, that the rate constant (k_i) for intersystem crossing from S_1 will depend on the energy gap. However, intersystem crossing from $S_1 \rightarrow T_1$ may proceed via the intermediate triplet states. It is impossible to say what the energy gap is for intersystem crossing from S_1, since a detailed information on the energies of higher triplet states is not available either by experiment or from theory. Intersystem crossings involving $T_1 \rightarrow S_0$ transitions are not affected in this way since there are no intermediate electronic states available.

Triplet-triplet absorption spectra in the ultra-violet and visible region

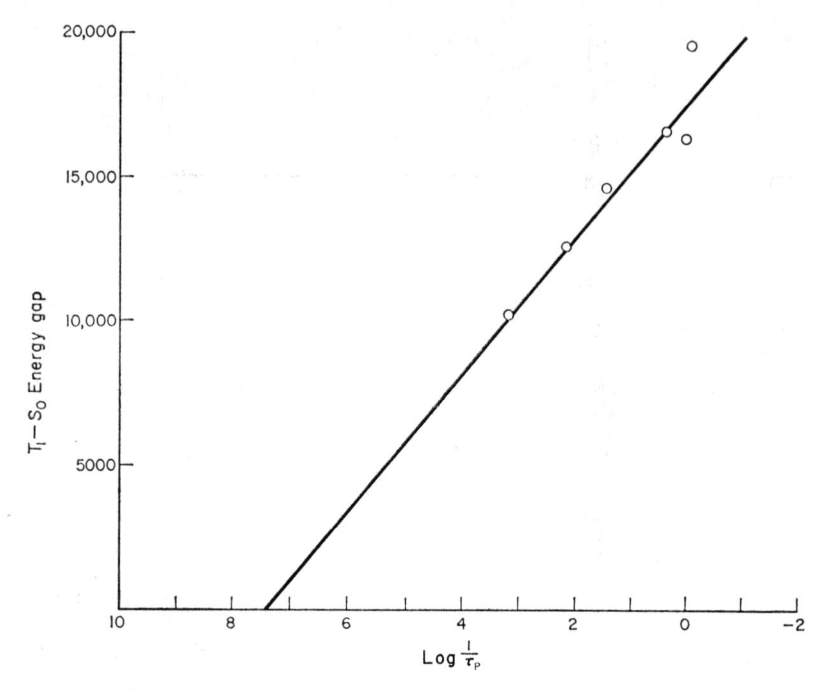

FIG. 7.17 Triplet decay constants for several perprotonated hydrocarbons plotted as a function of $T_1 - S_0$ energy gap. The data were taken for the perprotonated molecules at room temperature in polymethylmethacrylate. (Reproduced with permission from KELLOGG and CONVERS WYETH *J. chem. Phys.* **45**, 3156 (1966).)

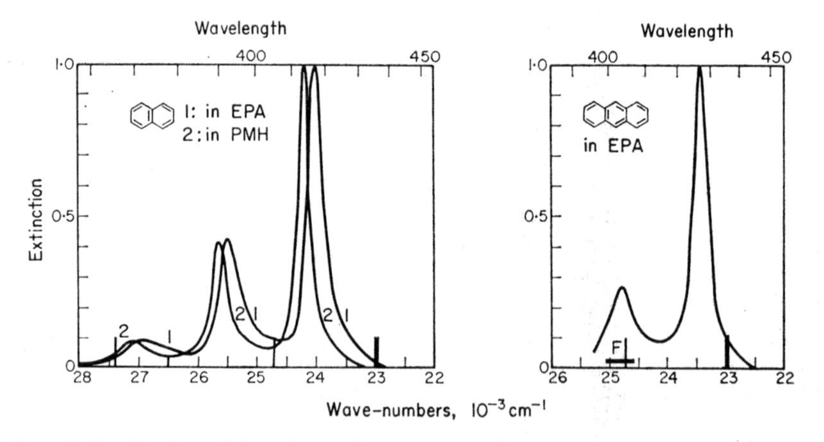

FIG. 7.18 Triplet-triplet absorption spectra of naphthalene and anthracene. (Reproduced with permission from CRAIG and ROSS *J. chem. Soc.* 1589 (1954).)

have been studied in great detail in all three phases, but the spectra obtained give the energies of triplet levels which lie well above S_1. Meso derivatives of anthracene show marked temperature-dependent fluorescence which has recently been shown to be due to intersystem crossing to triplet levels slightly higher than S_1 [9]. Until more reliable estimates of the triplet levels situated near and between S_1 and T_1 are available, either from theory or by experiment, the reasons for the marked variations in values of Φ_f from compound to compound and solvent to solvent are likely to remain obscure.

Effects of Solvent and pH Changes on Phosphorescence

In molecules containing heteroatoms where (n,π^*) and (π,π^*) states are very close together, the relative positions of the states may be altered by a change of solvent. It is well known that on changing from hydrocarbon to hydroxylic solvent $n \rightarrow \pi^*$ transitions show a blue shift whereas $\pi \rightarrow \pi^*$ transitions exhibit a small red shift, as previously stated. Thus the luminescence from quinoline shows a marked solvent-dependence. In a hydrocarbon glass it is non-fluorescent and emits a naphthalene-like $\pi \rightarrow \pi^*$ phosphorescence. Hydroxylic solvents lower the $^1(\pi,\pi^*)$ state below the raised $^1(n,\pi^*)$ state, and naphthalene-like fluorescence occurs with a corresponding reduction in the phosphorescence efficiency from the $^3(\pi,\pi^*)$ state.

Solvent effects on phosphorescence spectra have not been studied in much detail. However, the phosphorescence spectra of benzophenone, anthraquinone and benzil show a blue shift of ~ 300 cm^{-1} as expected for molecules where the lowest triplet states are $^3(n,\pi^*)$, whereas the phosphorescence spectra of naphthalene and phenanthrene show only minor red shifts, since in these cases T_1 is a $^3(\pi,\pi^*)$ state. Very small solvent shifts are found in the phosphorescence spectra of 2-acetonaphthone, carbazole 1-naphthaldehyde, 2-naphthyl phenyl ketone, triphenylamine, etc., which is supporting evidence that in these molecules T_1 is a $^3(\pi,\pi^*)$ state.

Solvent shifts are caused by the differing molecular interactions due to changes in electron distribution upon excitation. In the case of $n \rightarrow \pi^*$ transitions, this effect leads to a reduction in the strength of the hydrogen bonding with polar solvents of the (n,π^*) state relative to the ground state. The change in electron distribution upon excitation also determines the extent to which the chemical properties of the excited state differ from those of the ground state. This sometimes results in pK_a values for excited states being very different from those for the ground state. The acidity constants of excited states have been calculated from the change in luminescence spectra of the protonated and unprotonated species (see Fig. 7.19). An energy level scheme showing both the base and the conjugate acid is shown in Fig. 7.20, from which it follows that

$$\Delta H + \Delta E_A = \Delta H^* + \Delta E_{HA}$$

The standard free energy changes ΔG and ΔG^* for the acid dissociations

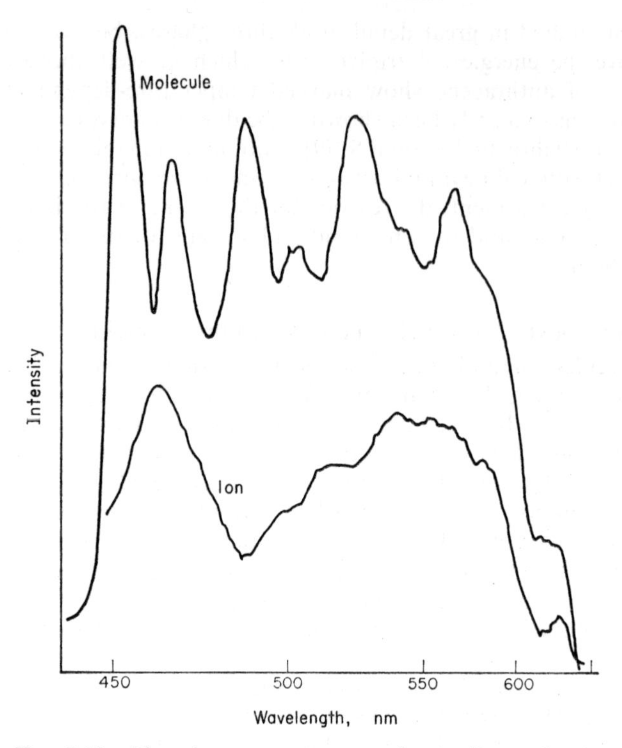

FIG. 7.19 Phosphorescence spectra of quinoline and quinolinium ion. (Reproduced with permission from JACKSON and PORTER *Proc. R. Soc.* **260A**, 13 (1961).)

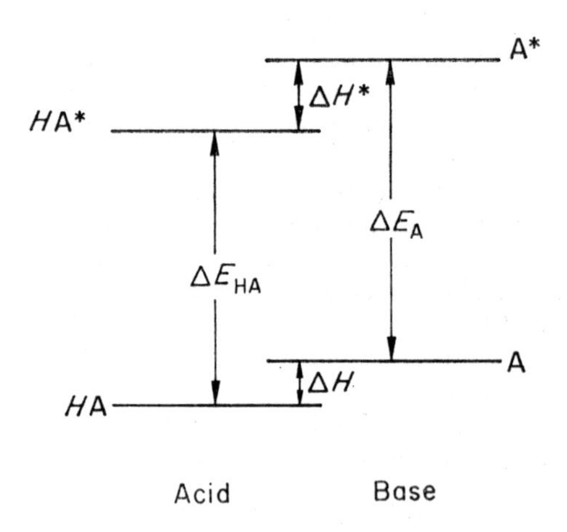

FIG. 7.20 Energy level diagram of an acid and its conjugate base.

are related to K_a and $K_a{}^*$ the acid dissociation constants of the ground state and excited state as shown.

$$\Delta G = \Delta H - T\Delta S = -RT \ln K_a$$
$$\Delta G^* = \Delta H^* - T\Delta S^* = -RT \ln K_a{}^*$$

If the entropy changes of acid dissociation in the ground and excited states are assumed equal:

$$\Delta G - \Delta G^* = \Delta H - \Delta H^* = \Delta E_{HA} - \Delta E_A$$

$$pK_a - pK_a{}^* = \frac{\Delta E_{HA} - \Delta E_A}{RT}$$

$$pK_a{}^* = pK_a - \frac{\Delta E_{HA} - \Delta E_A}{RT}$$

ΔE_A and ΔE_{HA} are usually taken to be the average of the frequencies of the luminescence and absorption maxima for A and HA, respectively. $pK_a{}^*$ values have been obtained for the first excited singlet states of many molecules. Porter and Jackson showed that for the compounds given in Table 7.4 the values obtained from the 0,0 bands of the phosphorescence spectra

TABLE 7.4. p*K of ground and excited states of seven aromatic compounds*

Compound	pK_G (Ground state)	pK_S (First excited singlet state)	pK_T (Flash photolysis)	pK_T (Phosphorescence)
2-Naphthol	9·5	3·1	8·1	7·7
2-Naphthoic acid	4·2	10–12	4·0	4·2
1-Naphthoic acid	3·7	10–12	3·8	4·6
Acridine	5·5	10·6	5·6	—
Quinoline	5·1	—	6·0	5·8
2-Naphthylamine	4·1	−2·0	3·3	3·1
N,N-Dimethyl 1-naphthylamine	4·9	—	2·7	2·9

(Reproduced with permission from JACKSON and PORTER *Proc R. Soc.*, **260A**, 13 (1961).)

are in good agreement with the $pK_a{}^*$ values obtained for triplet states from flash photolytic studies in fluid media. The $pK_a{}^*$ values for the triplet states of the molecules given in Table 7.4 are all similar to those of the ground state, in contrast with the large differences found for the $pK_a{}^*$ values for the first excited singlet states in these molecules. It is unlikely that this is always the case. The thermodynamic arguments made in order to give the final equation for $pK_a{}^*$ require the establishment of equilibrium between the various species. Since it is doubtful that equilibrium is established in rigid glasses, this method would not be expected to give agreement with other methods, especially when $pK_a{}^*$ is very different from pK_a.

Perturbations of S↔T Transitions

Heavy Atom Effects

The τ_p values of the 1-halogenated naphthalenes decreases with increasing atomic number of the halogen substituent (see Tables 7.1 and 7.5).

TABLE 7.5. *Heavy-atom effects on lifetimes of phosphorescence*

Group	Compounds	$\bar{\nu}_p$ cm^{-1}	τ_p (s)
IV	$C(C_6H_5)_4$	28,500	5·4
	$Si(C_6H_5)_4$	28,200	1·24
	$Sn(C_6H_5)_4$	28,400	0·003
	$Pb(C_6H_5)_4$	23,900	0·01
V	$N(C_6H_5)_3$	25,500	0·49
	$P(C_6H_5)_3$	25,200	0·014
	$As(C_6H_5)_3$	25,500	0·0016

(Reproduced with permisssion from McCLURE, D. *J. chem. Phys.* **17**, 905 (1949).)

Other heavy atoms have similar effects. Introduction of a heavy atom into a molecule increases the spin-orbit perturbation and the following five effects are usually observed.

(1) An enhancement of $S_0 \rightarrow T$ absorption.
(2) A decrease in the radiative lifetimes of phosphorescence.
(3) An increase in the intersystem crossing $S_1 \rightarrow T_1$.
(4) A decrease in the fluorescence yield.
(5) An increase in intersystem crossing $T_1 \rightarrow S_0$.

The lifetime of phosphorescence decreases because of factors (2) and (5) and the phosphorescence yield decreases by virtue of (5) but is increased because of process (3), and so it is not easy to predict which effect will dominate. Thus Φ_p increases markedly on going from naphthalene to 1-chloronaphthalene, but drops slightly on going to 1-bromonaphthalene.

The perturbation arising from a substituent in the molecule is termed the internal heavy-atom effect. There is also an external heavy atom effect, as demonstrated by Kasha who observed the enhancement of $S_0 \rightarrow T_1$ absorption of 1-chloronaphthalene when ethyl iodide was added. Extensive studies of the external heavy atom effect were recently undertaken by McGlynn and his co-workers [10]. Using mixtures of 1-halonaphthalenes and alkylhalides (as cracked glasses) they showed that the decrease in τ_p arising from an external iodine atom is greater than that caused by an internal chlorine atom, and roughly equal to an internal bromine. Furthermore they forwarded evidence for the existence of a complex between the alkyl halide as acceptor and the 1-halo-naphthalene as electron donor.

Thus they proposed that external heavy atom perturbation is not described by a collisional process as originally thought, but by a contact process. The latter implies a complex formation of extremely low stabilization energy between the donor-acceptor pair. In a subsequent paper, McGlynn shows that for the 1-chloronaphthalene: ethyl iodide system 1 : 1 stoichiometry is dominant [11].

Graham-Bryce and Corkhill [12] demonstrated that ethyl iodide enhanced the phosphorescence intensity and decreased τ_p of a number of organic species by the external heavy-atom effect. More recent work concerning the external heavy atom effect showed that the radiative lifetime is much more sensitive to this effect than is the non-radiative lifetime of the triplet state. For internal perturbation, however, both radiative and non-radiative lifetimes are very sensitive. (See Table 7.3.)

The measured phosphorescence lifetimes of benzene and perdeuterobenzene at 4·2°K in crystalline rare-gas solvents show a striking decrease on going from argon to xenon as solvent. (See Table 7.6.) The ratio of the

TABLE 7.6. *Effect of rare-gas solvents and deuteration on the phosphorescence lifetime and the ratio of the yield of phosphorescence to fluorescence of benzene*

Solvent	C_6H_6 τ_p (s)	C_6H_6 Φ_P/Φ_F	C_6D_6 τ_p (s)	Reference
Methane	16·0	1	22·0	*(a)*
Argon	16·0	20	26·0	*(a)*
Krypton	1·0	∞	1·0	*(a)*
Xenon	0·07	∞	0·07	*(a)*
EPA	7·0	1	28·0	*(b)*

(a) ROBINSON, G. W. *J. molec. Spectrosc.* **6**, 58 (1961).
(b) LIM, E. C. *J. chem. Phys.* **36**, 3497 (1962).

quantum yield of fluorescence to the quantum yield of intersystem crossing to form the triplet state also decreases in the same order.

DEUTERATION

The phosphorescence lifetimes of many aromatic hydrocarbons have been found to increase considerably when the hydrogen atoms in these molecules are replaced by deuterium, although the phosphorescence spectra are only displaced slightly. The theory of radiationless transitions developed by Robinson and Frosch [13] explains this effect in terms of Franck–Condon factors. The probability of a radiationless transition is proportional to the square of the vibrational overlap integral between the initial and final states. The larger the number of possible degenerate vibronic states of S_0 which can couple with the lowest vibrational level of T_1, the larger this term, but it is reduced for vibronic wave functions of S_0

with higher vibration quantum numbers, because of the oscillatory nature of vibronic wave functions. A pictorial illustration of the reduction expected in vibronic overlap between the lowest state of T_1 and vibronic states of S_0 with high and low values of n'' is shown in Fig. 7.21.

C—H vibrations are the highest frequency vibrations found in aromatic hydrocarbons. Vibronic levels of S_0 which include harmonics and combinations involving C—H vibrations which are degenerate with the lowest

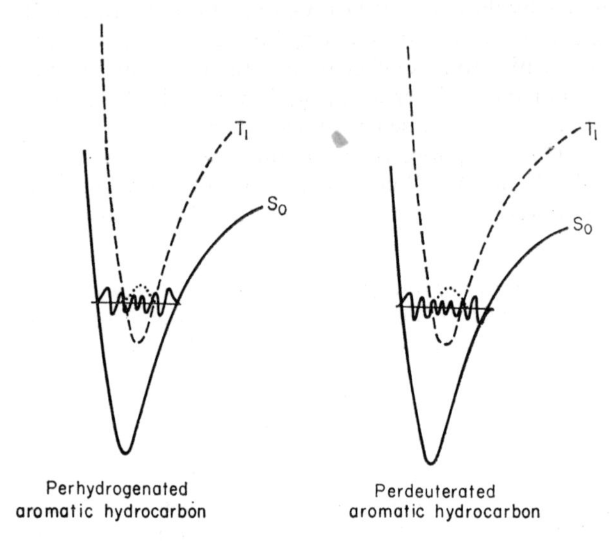

FIG. 7.21 Diagram showing the reduced overlap of wave-functions on deuteration.

vibrational level of T_1, will have the largest overlap integrals, since the oscillation of the wave-function will be less frequent. However, when the C—H bonds are replaced by C—D bonds the degenerate energy levels will have higher vibrational quantum numbers and the radiationless decay may be expected to decrease. This explains qualitatively the increases in phosphorescence lifetime found upon deuteration of aromatic hydrocarbons. The fact that such large increases are possible is additional confirmation that the perprotonated compounds decay predominantly by non-radiative transitions from T_1.

EFFECT OF OXYGEN AND OTHER PARAMAGNETIC SPECIES

The triplet states of organic molecules are quenched very efficiently by oxygen in fluid media at very low concentrations. At high pressures (~ 100 atm) oxygen enchances singlet-triplet transitions in many molecules. Nitric oxide, which is also paramagnetic, produces similar effects. Porter and Wright [14] suggested that quenching by O_2, NO, aromatic triplet states and paramagnetic transition, and lanthanide metal ions could all be due to a catalytic intersystem crossing with overall spin conservation, i.e., to the process

$$A^*(T_1) + Q(M) \rightarrow A(S_0) + Q(M)$$

where M represents any multiplicity other than singlet. This process is spin-allowed if it occurs by an exchange interaction. Two theories have been put forward to explain the enhancement by oxygen of S→T absorption. In the first, the effect is considered to be due to exchange interaction between the paramagnetic perturber and the triplet state. In the second, complex formation is postulated and the enhancement is said to be due to intensity borrowing from the charge-transfer transition of the complex. Whatever the explanation, oxygen enhancement is invaluable in determining the triplet levels of certain molecules which do not phosphoresce (e.g., ethylene, butadiene).

Delayed Fluorescence

Thermally Activated Delayed Fluorescence

The normal mode of population of the lowest triplet state of any molecule is via intersystem crossing from the first excited singlet state. The reverse process is temperature-dependent and leads to long-lived emission from S_1 (see Fig. 7.22). The activation energy for producing delayed

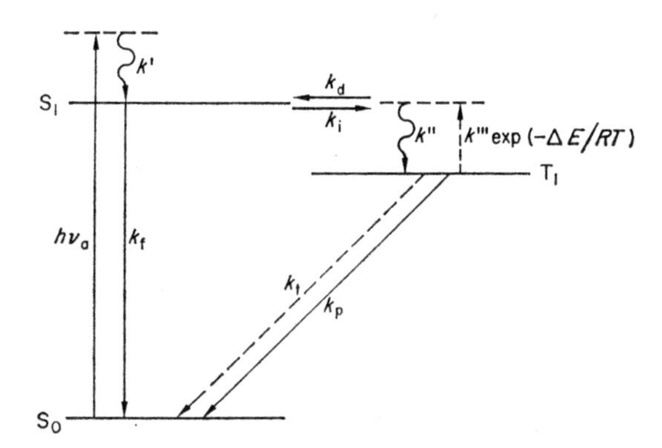

FIG. 7.22 Energy level scheme for thermally-activated delayed fluorescence.

fluorescence according to this mechanism should be equal to the energy gap between S_1 and T_1 (see Table 7.7). The smaller the gap and the higher the temperature, the more likely the process. The lifetime of the delayed fluorescence which arises in this way is governed by the lifetime of the triplet state and is thus equal to the lifetime of the normal phosphorescence.

TABLE 7.7. *Activation energies for the production of delayed fluorescence (ΔE) compared with the energy differences between the lowest singlet and triplet states measured spectroscopically* ($S_1 - T_1$)

Compound	Solvent	ΔE kcal/mole	$\Delta E(S_1 - T_1)$ kcal/mole	Reference
Fluorescein	Boric acid	8 ± 1	9·2	(*a*)
Eosin	Glycerol	10·2	10·5	(*c*)
Eosin	Ethanol	9·7	10·9	(*c*)
Proflavine	Glycreol	8	8	(*d*)
Acriflavine	Glucose	$8{\cdot}0 \pm 0{\cdot}5$	$7{\cdot}5 \pm 0{\cdot}5$	(*e*)
Acriflavine Dimer	Glucose	$5{\cdot}5 \pm 0{\cdot}5$	$5{\cdot}0 \pm 0{\cdot}5$	(*e*)
Acriflavine	Absorbed on silica gel	8	5	(*b*)

(*a*) LEWIS, G. N., LIPKIN, D. and MAGEL, T. T. *J. Am. chem. Soc.* **63**, 3005 (1941).
(*b*) ROSENBERG, J. L. and SHOMBERT, D. J. *J. Am. chem. Soc.* **82**, 3252 (1960).
(*c*) PARKER, C. A. and HATCHARD, C. G. *Trans. Faraday Soc.* **57**, 1894 (1961).
(*d*) PARKER, C. A. and HATCHARD, C. G. *J. phys. Chem.* **66**, 2506 (1962).
(*e*) WEN, WEN-YANG, and HSU, R. *J. phys. Chem.* **66**, 1353 (1963).

DELAYED FLUORESCENCE AS A RESULT OF TRIPLET-TRIPLET ANNIHILATION

Delayed fluorescence due to the following process:

$$A^*(T_1) + A^*(T_1) \rightarrow A^*(S_1) + A(S_0)$$
$$\downarrow$$
$$S_0 + h\nu_f$$

has been observed in gases, liquids and crystalline solids. In crystals, triplet-triplet transfer involves the migration of the energy as triplet excitons. The intensity of this type of delayed fluorescence is proportional to the square of triplet state concentration, which is therefore generally proportional to the square of the absorbed light intensity. The spectral distribution is, in the absence of concentration effects, the same as that for 'normal' or 'prompt' fluorescence (see Fig. 7.23).

Parker and Hatchard [1] have proposed the following mechanism for the production of delayed fluorescence of this type in dilute solution. A similar mechanism operates in the other phases.

	Reaction	Rate
1.	${}^1A + h\nu \rightarrow {}^1A^*$	I_a
2.	${}^1A^* \rightarrow {}^1A$	$k_0[A^*]$
3.	${}^1A^* \rightarrow {}^1A + h\nu_f$	$k_f[A^*]$
4.	${}^1A^* \rightarrow {}^3A^*$	$k_t[A^*]$
5.	${}^3A^* \rightarrow {}^1A$	$k_h[{}^3A^*]$
6.	${}^3A^* + {}^3A^* \rightarrow X$	$k_q[{}^3A^*]^2$
7.	$X \rightarrow {}^1A + {}^1A$	$k_r[X]$
8.	$X \rightarrow {}^1A^* + {}^1A$	$k_s[X]$

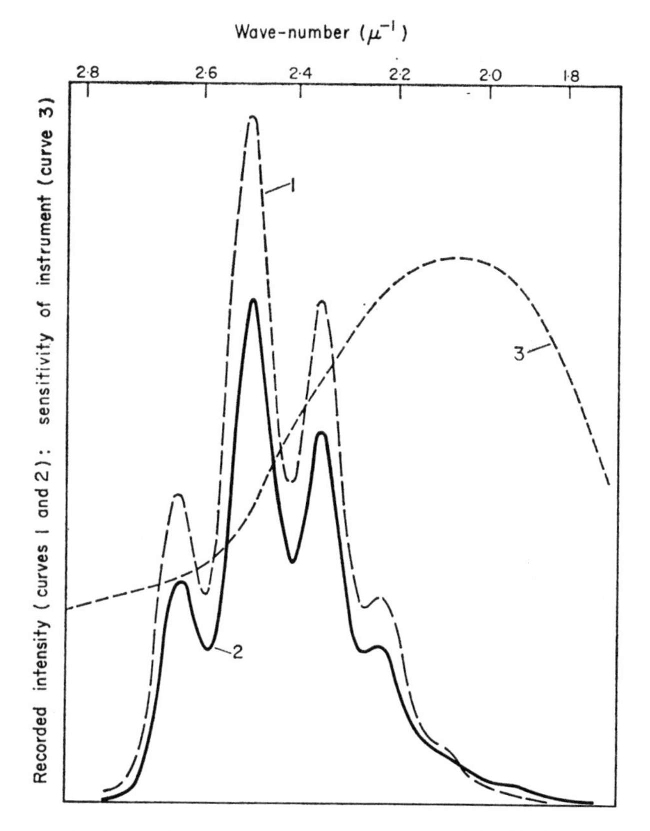

FIG. 7.23 Delayed fluorescence spectrum of 5×10^{-5}M anthracene in ethanol. Half-bandwidth of analysing monochromator was $0{\cdot}05\mu^{-1}$ at $2{\cdot}5\mu^{-1}$. Intensity of exciting light was approximately $1{\cdot}4 \times 10^{-8}$ einstein cm^{-2} s^{-1} at $2{\cdot}73\mu^{-1}$ (366 mμ). (1) Normal fluorescence spectrum (distorted by self-absorption). (2) Delayed emission spectrum at sensitivity 260 times greater than for curve 1. (3) Spectral sensitivity of instrument (units of quanta and frequency). (Reproduced with permission from PARKER and HATCHARD *Proc. R. Soc.* **269A**, 574 (1962).)

At low rates of light absorption the stationary state concentration of triplet states is low, and often $k_q[^3A^*] \ll k_h$. Under these conditions the lifetime of the delayed fluorescence will be first-order and

$$\frac{-\mathrm{d}\ln F_D}{\mathrm{d}t} = k_D = \frac{-\mathrm{d}\ln [^3A^*]^2}{\mathrm{d}t} = 2k_T$$

where F_D is the intensity of delayed fluorescence and k_D and k_T are the first-order rate constants for the decay of delayed fluorescence, and the decay of the triplet state which is equal to the decay of the 'normal' phosphorescence. The relationship $k_D = 2k_T$ has been confirmed in quite a number of cases and over a range of temperatures. (See Fig. 7.24.)

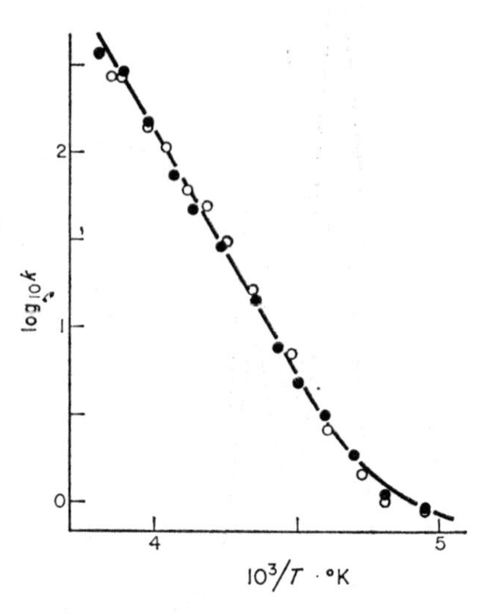

FIG. 7.24 Plot of $\log_{10} k_T$ (solid circles) and $\log_{10} (\frac{1}{2}k_D)$ (open circles) against reciprocal temperature for acenaphthene in liquid paraffin (10^{-3}M). (Reproduced with permission from STEVENS and WALKER *Proc. R. Soc.* **281A**, 420 (1964).)

Triplet-triplet quenching, step [6], takes place in the gas phase and in dilute solutions at every collision. According to Wigner's spin rule two triplet states can yield a singlet, a triplet or a quintet dimeric species. In the mechanism, X may be any or all of these 'excimers' which may rapidly dissociate, undergo internal conversion or intersystem crossing or even emit. For pyrene and many other compounds, the normal or prompt fluorescence consists of both monomer and dimer emission, as does the delayed fluorescence, but the proportions of monomer and dimer emission present are not necessarily identical (see Fig. 7.25).

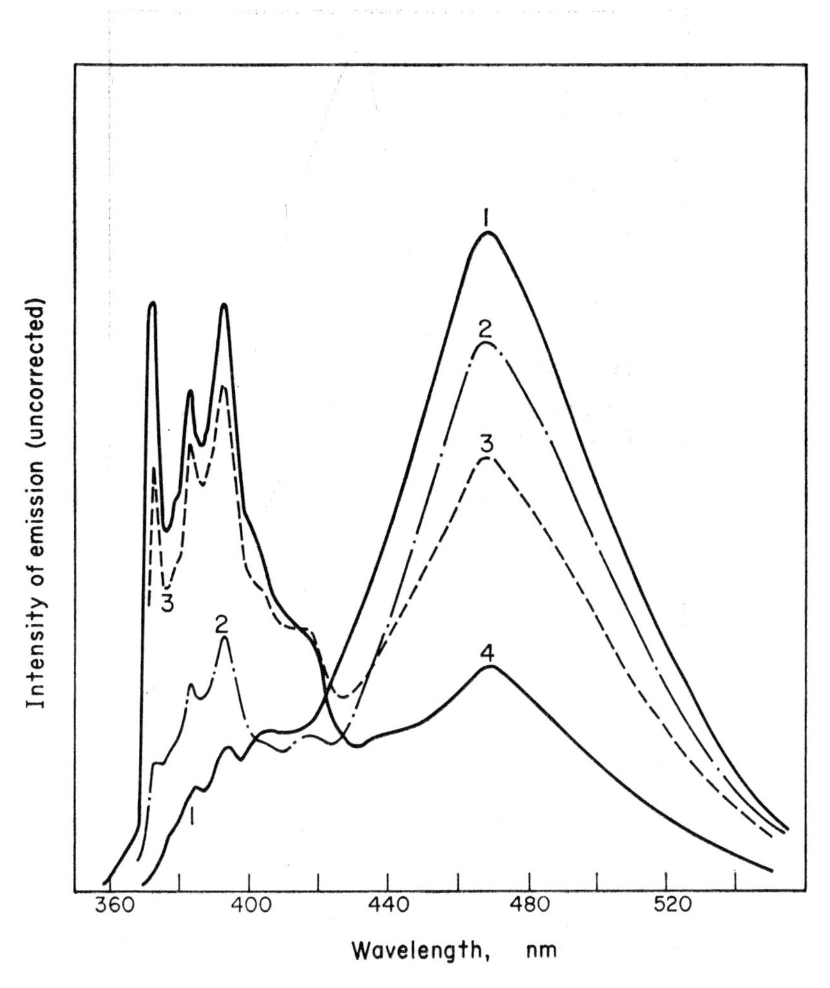

FIG. 7.25(*a*) Delayed fluorescence of pyrene in ethanol. (1) 3×10^{-3}M, (2) 10^{-3}M, (3) 3×10^{-4}M, (4) 2×10^{-6}M. The instrumental sensitivity settings were approximately 1000 times greater than those for the corresponding curves in Fig. 7.25(*b*). The short wavelength ends of the spectra in the more concentrated solutions are distorted by self-absorption. (Reproduced with permission from NOYES, HAMMOND and PITTS *Adv. in Photochems*, **2**, 305 (1964).)

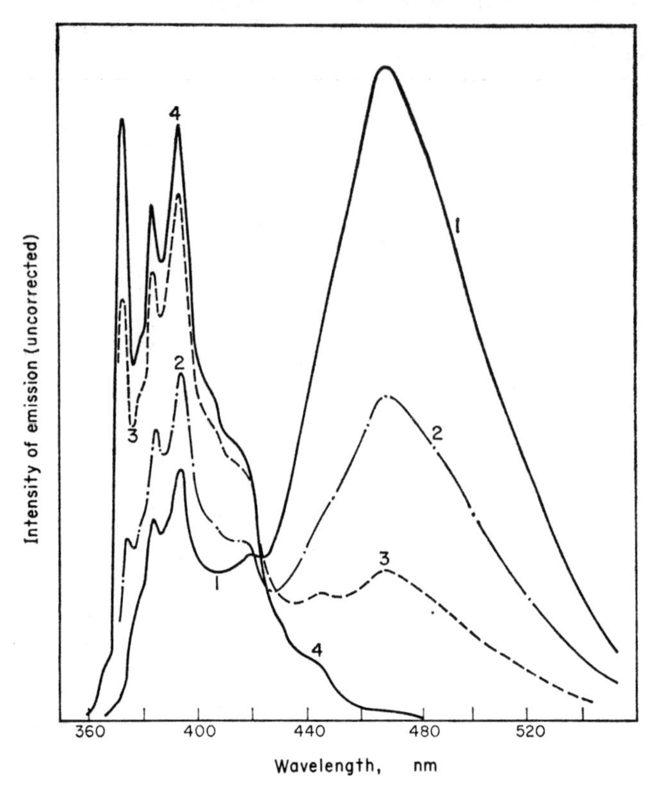

FIG. 7.25(*b*) Normal fluorescence of pyrene in ethanol. (1) 3×10^{-4}M, (2) 10^{-3}M, (3) 3×10^{-4}M, (4) 2×10^{-6}M. The instrumental sensitivity settings for curves 1 and 4 were approximately 0·6 and 3·7 times that for curves 2 and 3. The short wavelength ends of the spectra in the more concentrated solutions are distorted by self-absorption. (Reproduced with permission from NOYES, HAMMOND and PITTS *Adv. in Photochem.* **2**, 305 (1964).)

REFERENCES

[1] PARKER, C. A. *Adv. Photochem.* **2**, 305, Wiley, New York (1964).
[2] LOWER, S. K. and EL-SAYED, M. A. *Chem. Revs.* **66**, 199 (1966).
[3] LEACH, S. and MIGIRDICYAN, E. *Action chimiques et biologiques des radiations*, 199 (1966).
[4] PARISER, R. and PARR, R. G. *J. chem. Phys.* **21**, 466, 767 (1953).
[5] LEWIS, G. N. and KASHA, M. *J. Am. chem. Soc.* **66**, 2, 100 (1944).
[6] HUTCHISON, C. and MANGUM B. *J. chem. Phys.* **29**, 952 (1958).
[7] KEARNS, D. R. and CASE, W. A. *J. Am. chem. Soc.* **88**, 5087 (1966).
[8] LOGAN, L. M. and ROSS, I. G. *J. chem. Phys.* **43**, 2903 (1965).

[9] BENNETT, R. G. and MCCARTIN, P. J. *J. chem. Phys.* **44**, 1969 (1966).
[10] MCGLYNN, S. P., SMITH, F. J. and CILENTO, G. *Photochem. Photobiol.* **3**, 269 (1964).
[11] RAMAKRISHNAN, V., SUNSERI, R. and MCGLYNN, S. P. *J. chem. Phys.* **45**, 1365 (1966).
[12] GRAHAM-BRYCE, I. and CORKHILL, J. *Nature*, **186**, 965 (1960).
[13] ROBINSON, G. and FROSCH, R. *J. chem. Phys.* **37**, 1962 (1962).
[14] PORTER, G. and WRIGHT, M. *Discuss. Faraday Soc.* **27**, 18 (1959).

Chapter 8

INTRAMOLECULAR ELECTRONIC ENERGY TRANSFER BETWEEN ORGANIC MOLECULES

F. WILKINSON

TRANSFER of electronic excitation energy from a donor molecule D to an acceptor molecule A may be represented as

$$D^* + A \rightarrow A^* + D$$

where an asterisk indicates an electronically excited state. This process, which results in quenching of emission from D* and 'sensitized' emission from A*, has been shown to play a significant role in photochemistry, radiation chemistry and in radiation protection. It also is proposed as an important step in photosynthesis.

Electronic energy transfer to acceptor molecules which are already electronically excited can also take place. Mention has already been made in Chapter 7 of triplet-triplet annihilation. This process:

$$D^* \text{ (triplet)} + A^* \text{ (triplet)} \rightarrow \text{(Singlet)} + A^* \text{ (singlet)}$$

has been observed under a variety of conditions, and, together with other examples, it will be discussed later in this chapter. The probability of transfer to an acceptor molecule in an excited singlet state is very small, at all but very high intensities, because of the short lifetimes of such states.

Mechanisms of Electronic Energy Transfer

RADIATIVE TRANSFER

Reabsorption of donor emission by acceptor molecules results in radiative energy transfer. The efficiency of this process depends on the extent to which the emission spectrum of the donor overlaps the absorption spectrum of the acceptor, and on the Beer–Lambert law which states that for monochromatic light

$$I_a = I_0(1 - 10^{-\varepsilon c d})$$

where I_0 and I_a are the intensities of the incident and absorbed light, ε is the extinction coefficient, c the concentration of the absorbing species and d is the path length travelled by the light. Since the excited molecules emit in all directions, the larger the volume of the vessel used, the greater the probability of radiative transfer.

When the emission and absorption spectra of a compound overlap, the emission spectrum observed at right-angles to the exciting source will be distorted because of radiative transfer (Fig. 8.1). The spectrum obtained by

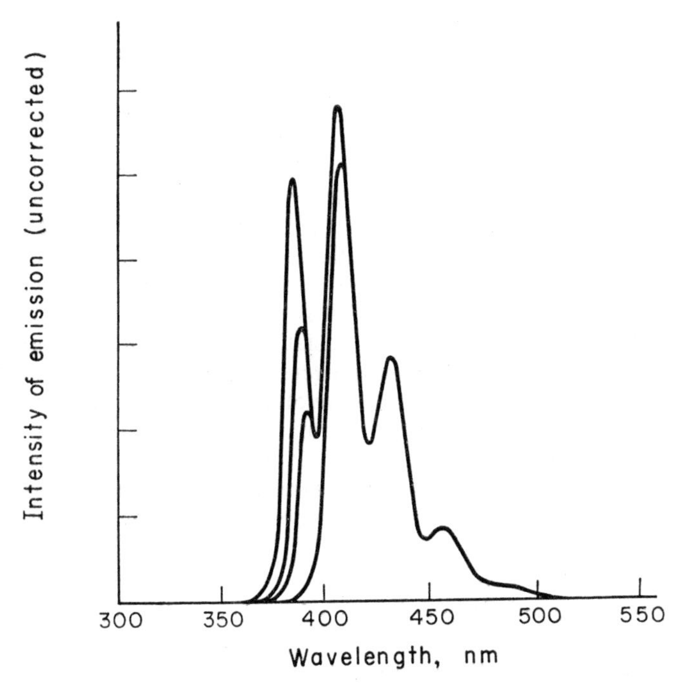

Fig. 8.1 Change in the observed fluorescence spectra of anthracene in benzene solution at 20°C due to reabsorption of the short wavelength band at higher concentrations. (Excitation and observation at right-angles.)

using front-face excitation and viewing will not show this effect, but the intensity observed will be increased. This is because light originally emitted in directions which would not have reached the detector may be re-absorbed, and some emission arising from molecules excited by such an energy transfer process will reach the detector. These effects must be allowed for when experimental investigations into non-radiative processes are being made.

Although minimal reference will be made to radiative transfer in the following pages it is important to remember that life on this planet, and our total source of energy, other than matter itself, are completely dependent on this mechanism, operating at huge distances, allowing radiant energy from the sun to be absorbed in the upper atmosphere or at ground level. It is to be hoped that space scientists will not overlook this process as often as their earthbound brothers have done during their studies of non-radiative energy transfer.

Non-Radiative Transfer

Sensitized fluorescence resulting from a non-radiative process was first observed by Cario and Franck, who irradiated a mixture of mercury and thallium vapours with the 254 nm mercury resonance line. Characteristic emissions from both these atoms were detected, despite the fact that thallium vapour does not absorb at the excitation frequency. Later, Perrin showed that polyatomic molecules are equally capable of exchanging electronic excitation energy. Several theories have been developed to explain the mechanism by which energy absorbed by one molecule is transferred to a second molecule of the same or of a different species. In the case of transfer between different atoms and simple diatomic species there is little chance of exact resonance between the transitions involved in the transfer process, and it is necessary for some of the electronic energy to be converted into translational energy or vice versa. It is found that the efficiency of transfer decreases markedly with increasing energy gap, ΔE, between the states involved. With complex molecules, resonance is possible if vibrations and rotations are included even if there is an electronic energy difference (Fig. 8.2). When there is close resonance between the initial and final states, non-radiative resonance transfer can occur, and may be due to either Coulombic or exchange interaction.

(*a*) *Coulombic Interaction*

The total Coulombic interaction may be taken as the sum of a number of terms including dipole-dipole, dipole-quadrupole and terms involving higher multipoles. The dipole-dipole term, i.e., the term which accounts for interaction between the transition dipoles in the donor and those in the acceptor, is often predominant, especially when the transition is fully allowed.

Förster, in a quantum mechanical treatment of resonance transfer which takes place between two well-separated molecules, has considered only dipole-dipole interaction. Thermal equilibrium in the excited states was considered to be established very rapidly relative to intermolecular transfer, and the transitions in the donor and the acceptor were taken to be continuous as witnessed by their emission and absorption spectra respectively. Resonance transfer is possible in the region of overlap of these spectra. Förster's quantitative treatment leads to the following equation for the transfer rate constant:

$$n(R) = \frac{9000(\ln 10)\kappa^2 \Phi_D}{128\pi^5 n^4 N \tau_D R^6} \int_0^\infty f_D(\bar{\nu})\varepsilon_A(\bar{\nu}) \frac{d\bar{\nu}}{\bar{\nu}^4} \tag{8.1}$$

where $\varepsilon_A(\bar{\nu})$ is the molar decadic extinction coefficient of the acceptor

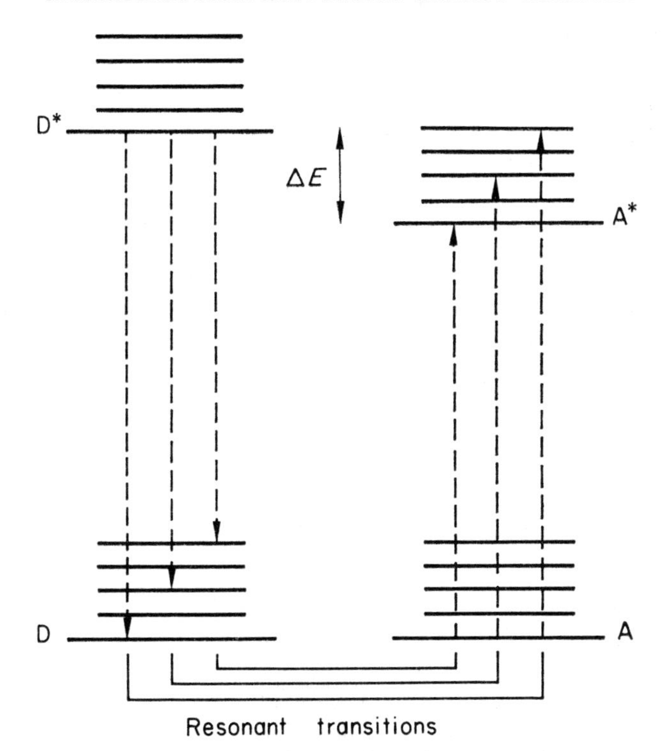

FIG. 8.2 Energy level diagram for a donor D and acceptor A with an electronic energy difference ΔE.

(at wave-number $\bar{\nu}$), $f_D(\bar{\nu})$ is the spectral distribution of the fluorescence of the donor (measured in quanta and normalized to unity on a wave-number scale), N is Avogadro's number, τ_D is the mean lifetime of the excited state, and Φ_D the quantum yield of fluorescence of the donor. n is the refractive index of the solvent, R is the distance between the molecules, and κ is an orientative factor which for a random distribution equals $(2/3)^{\frac{1}{2}}$.

R_0, the distance at which transfer and spontaneous decay of the excited donor are equally probable, is given by:

$$R_0{}^6 = \frac{9000(\ln 10)\kappa^2\Phi_D}{128\pi^5 n^4 N \bar{\nu}^4} \int_0^\infty f_D(\bar{\nu})\varepsilon_A(\bar{\nu})\,d\bar{\nu} \tag{8.2}$$

When the transitions in both the donor and the acceptor are fully allowed and there is good overlap between the emission spectrum of the donor and the absorption spectrum of the acceptor, R_0 values of 5–10 nm are predicted from this equation, and the rate constants for energy transfer due to Coulombic interactions are much greater than those calculated for diffusion-controlled reactions.

When Brownian translational movement is slow, so that each individual transfer may be considered as taking place at a constant distance, and Brownian rotational movement is considered to be fast, which allows the average value of the orientation factor $\kappa^2 = 2/3$ to be used, the following equations are obtained:

$$F_D(t) = F_D(0) \exp \left\{ \frac{-t}{\tau_D} - 2x \left(\frac{t}{\tau_D} \right)^{\frac{1}{2}} \right\} \tag{8.3}$$

$$\frac{\Phi_D{}^0 - \Phi_D}{\Phi_D{}^0} = (\pi)^{\frac{1}{2}} x \exp (x^2)\{1 - \text{erf}\,(x)\} \tag{8.4}$$

where $$\text{erf}\,(x) = 2\int_0^x \exp\,(-x^2)\,dx\,(\pi)^{\frac{1}{2}}$$

$$x = \frac{(\pi)^{\frac{1}{2}}}{2} \frac{C_A}{C_A{}^0} \quad \text{and} \quad C_A{}^0 = \frac{3000}{4\pi N R_0{}^3}$$

$F_D(t)$ and $F_D(0)$ are the donor emission intensities at time t and time zero respectively, Φ_D, $\Phi_D{}^0$ are the luminescence yields in the presence and absence of the acceptor, and C_A is the acceptor concentration. $C_A{}^0$ is called the 'critical transfer concentration' and corresponds on average to one acceptor molecule in a sphere of radius R_0. Equations 8.3 and 8.4 have been found to fit a number of observations on sensitized fluorescence. Galanin has shown that theory predicts a markedly slower decrease in sensitizer lifetime than in sensitizer luminescence yield as the acceptor concentration is increased. This also has been shown to be in agreement with experimental findings (Fig. 8.3).

Attempts have been made to modify eqns 8.3 and 8.4 so that they might be applied to energy transfer taking place at a distance in low viscosity solvents, where allowance for diffusion within the lifetime of the excited state is necessary, but so far experimental tests of these modifications have not been very conclusive.

When solutions of fluorescent substances are irradiated with polarized light, the extent of the polarization of the fluorescence decreases with increasing solute concentration. As early as 1925 Perrin proposed long-range transfer between like molecules, in order to explain this effect. Expressions for the extent of this concentration depolarization can be derived which make use of eqns 8.3 and 8.4.

Strong and Weak Coupling. The application of exciton theory to molecular crystals or aggregates was made by Davydov to explain the effects of aggregation on absorption spectra. Only a very brief outline of energy transfer in crystals will be given here [5, 8]. Where there is strong interaction between excited and unexcited molecules, the excitation energy may be regarded as being delocalized over large regions of the crystal, or alternatively the excitation energy may be thought of as if it were 'hopping' from molecule to molecule. Simpson and Peterson have suggested the term strong coupling for interactions, in molecular crystals, which cause the exciton bandwidth to be much greater than the Franck–Condon bandwidth of the corresponding transition in the individual molecule, the weak coupling exciton case being the reverse condition.

In both these cases very fast transfer rates are predicted, $\sim 10^{15}$ and 10^{12}–10^{13} s^{-1} for strong and weak coupling respectively in pure molecular crystals. The transfer rates are inversely proportional to the intermolecular distance cubed. This contrasts with the dependence on the inverse of the internuclear distance raised to the sixth power predicted for the conditions treated by Förster with well-isolated molecules. These conditions may be regarded as defining a very weak coupling case.

Experimental verification of the predictions of these very fast transfer rates in molecular crystals usually involves the use of crystals containing very low concentrations of impurities and measurement of 'host sensitized luminescence'. The very large transfer distances observed are explained by exciton migration among host molecules and subsequent transfer to the impurities which act as energy acceptors and traps, and emit their characteristic luminescence.

Spin Selection Rules. The spin selection rules for energy transfer by Coulombic interaction are very restrictive. No change of spin in either the donor or the acceptor is allowed, i.e., the multiplicities M_D^* must equal M_D and M_A must equal M_A^*.

Thus processes such as:

$$D^* \text{ (singlet)} + A \text{ (singlet)} \rightarrow D \text{ (singlet)} + A^* \text{ (singlet)}$$

and:

$$D^* \text{ (singlet)} + A \text{ (triplet)} \rightarrow D \text{ (singlet)} + A^* \text{ (triplet)}$$

are fully allowed. When the forbidden nature of the transition in the donor results in a corresponding increase in the lifetime of its excited state, the probability of forbidden energy transfer processes relative to spontaneous decay can still be high, since both processes are forbidden. Thus processes such as

$$D^* \text{ (triplet)} + A \text{ (singlet)} \rightarrow D \text{ (singlet)} + A^* \text{ (singlet)}$$

and

$$D^* \text{ (triplet)} + A \text{ (triplet)} \rightarrow D \text{ (singlet)} + A^* \text{ (triplet)}$$

although they are spin-forbidden by a Coulombic mechanism, have been observed to show transfer over large distances in rigid solvents. The distances are as large as those found for allowed processes, but the transfer rates are much lower.

(*b*) *Exchange Interaction*

When the donor and acceptor molecules are very close together exchange interaction will occur, and if the resonance condition is fulfilled this will result in electronic energy transfer. Dexter has given the following equation for the rate constant for transfer

$$n(R) = \frac{h}{2\pi} Z^2 \int_0^\infty f_D(E) f_A(E) \, dE \tag{8.5}$$

where
$$Z^2 = Y \frac{e^4}{h^2 \delta^2 R_0^2} \exp\left(\frac{-2R}{L}\right)$$

Y is a dimensionless quantity $\leqslant 1$, L is a constant called the effective average Bohr radius, δ is the dielectric constant of the medium, R_0 is a critical transfer distance, $f_D(E)$, $f_A(E)$ represent the donor emission and the acceptor absorption spectrum respectively, normalized so that

$$\int f_D(E) \, dE = 1 \quad \text{and} \quad \int f_A(E) \, dE = 1$$

The transfer rate is independent of the oscillator strength of both the donor and the acceptor. The distance-parameter dependence, however, falls off so much more rapidly than in the case of dipole-dipole transfer, that if it is assumed that the molecules must collide to make possible the energy transfer caused by exchange interaction, this leads to an expectation of a maximum rate equal to a diffusion-controlled reaction. The expected concentration dependence of the lifetime and the emission efficiencies in rigid solvents have been evaluated recently by Inokuti and Hirayama by a method which allows transfer parameters such as γ, defined equal to $2R^0/L$, to be evaluated by a process of curve fitting.

In pure crystals, triplet excitons may result because of exchange interaction. Transition dipoles for most singlet-triplet transitions are so low that Coulombic interaction is prohibitively weak. Early calculations suggested this was also true of exchange interactions, but more recent calculations by Jortner, Rice, Kutz and Choi suggest that the exchange interaction energy in typical molecular crystals of aromatic compounds is $\sim$5–10 cm^{-1}. In the case of the anthracene crystal these authors calculated that the mean lifetime of triplet excitation energy on any molecule is $\sim 10^{-10}$ s^{-1} with a diffusion coefficient for triplet excitation energy of $6{\cdot}2 \times 10^{-6}$ cm^2 s^{-1}. The radiative lifetime for triplet anthracene is $\sim 10^{-2}$ s. If there is no other mode of decay this suggests 10^8 'hops' during the lifetime of the excited state and on the basis of random walk diffusion, the diffusion length would be given by $1 = (2D\tau_0)^{\frac{1}{2}} \approx 3 \times 10^{-4}$ cm. These figures are encouragingly close to some of the experimental values which have been determined recently.

Spin Selection Rules

Transfer due to exchange interaction must take place with overall spin conservation, and is subject to Wigner's spin rule. Allowed transfer processes may be found by applying the following considerations.

If S_a and S_b are the initial spin quantum numbers of the colliding molecules, the resultant spin quantum number of the two species taken together must have one of the values

$$S_a + S_b,\ S_a + S_b - 1,\ S_a + S_b - 2, \ldots\ |S_a - S_b|$$

It follows that the spin quantum numbers of the resulting species can only have values S_c and S_d if at least one of the values

$$S_c + S_d,\ S_c + S_d - 1,\ S_c + S_d - 2,\ \ldots\ |S_c - S_d|$$

is common to the series above.

Thus the processes

$$\begin{array}{llll} \text{D* (triplet)} + & \text{A (singlet)} = & \text{A* (triplet)} + & \text{D (singlet)} \\ S_a = 1 & S_b = 0 & S_c = 1 & S_d = 0 \end{array}$$

$$\begin{array}{llll} \text{D* (triplet)} + & \text{A* (triplet)} = & \text{A* (singlet)} + & \text{D (singlet)} \\ S_a = 1 & S_b = 1 & S_c = 0 & S_d = 0 \end{array}$$

$$\begin{array}{llll} \text{D* (triplet)} + & \text{A* (triplet)} \rightarrow & \text{A* (triplet)} + & \text{D (singlet)} \\ S_a = 1 & S_b = 1 & S_c = 1 & S_d = 0 \end{array}$$

are all spin-allowed by exchange interaction. In the last two processes above, the spin of the system appears to have been reduced and to violate the law of conservation of spin angular momentum; however, it must be appreciated that vector addition is involved.

Experimental Observations

The experimental data quoted in the remainder of this chapter have been chosen to illustrate the various mechanisms already discussed. They should not be considered as a fully comprehensive list. For convenience, energy transfers from singlet and triplet states are presented separately, although this is not a mechanistic distinction.

Energy Transfer from Singlet States

Electronic energy transfer over distances much larger than molecular diameters can take place non-radiatively either as a result of a single-step, long-range process, or because of the excitation energy passing from one molecule to another. For low-viscosity solvents and long-lived excited states, material diffusion followed by energy exchange at normal collision diameters can cause energy to be transported over large distances, although the actual exchange is a short-range phenomenon. When all three mechanisms operate at one time it is very difficult to distinguish the separate contributions.

(*a*) *Single-step Long-range Transfer*

From measurements of the quenching of donor fluorescence and sensitization of the acceptor fluorescence, Bowen and co-workers obtained the first quantitative agreement between experimental results and Förster's theory for dipole-dipole resonance transfer. With 1-chloroanthracene as sensitizer and perylene as acceptor the bimolecular transfer rate constants obtained were $1{\cdot}6 \times 10^{11}$, $2{\cdot}0 \times 10^{11}$, $1{\cdot}4 \times 10^{11}$ and $1{\cdot}5 \times 10^{11}$ litre mole^{-1} s^{-1} for dilute solutions in benzene, chloroform and liquid paraffin, all at room temperature, and in a glassy mixture of tetra- and pentachloroethane at -183°C respectively. These rates show little variation despite the enormous change in viscosity of the media, and this illustrates that material diffusion is unimportant. Even with benzene as solvent, the transfer rate constants are ten times greater than those expected for diffusion controlled reactions. This fact represents transfer between isolated molecules 4 nm apart. In Table 8.1 some further experimental values for R_0 are compared with those calculated from eqn 8.2.

TABLE 8.1. *Long-range transfer distances for singlet-singlet energy transfer*

Donor	Acceptor	R_0 (nm) Calc.	R_0 (nm) Found
Anthracene	Perylene	3·1	5·4[(a)]
Perylene	Rubrenc	3·8	6·5[(a)]
9,10-Dichloroanthracene	Perylene	4·0	6·7[(a)]
Anthracene	Rubrene	2·3	3·9[(a)]
9,10-Dichloroanthracene	Rubrene	3·2	4·9[(a)]
Fluorescein	Fluorescein	5·0	2·6[(b)]
Rhodamine-B	Rhodamine-B	5·5	3·0[(b)]
Chlorophyll-a	Chlorophyll-a	8·0	3·6[(b)]
Chlorophyll-a	Copper Pheophytin-a	3·8 – 4·1	4·0[(c)]
Bis (hydroxyethyl)-2,6-naphthalenedicarboxylate	Sevron yellow G.L.	2·6	2·7[(d)]
Pyrene	Sevron yellow L.	3·9	4·0 – 4·6[(d)]

(*a*) WARE, W. R. *J. Am. chem. Soc.* **83**, 4374 (1961).
(*b*) WEBER, G. *Trans. Faraday Soc.* **50**, 552 (1964).
(*c*) TWEET, A. G., BELLAMY, W. D. and GAINES, G. L. *J. chem. Phys.* **41**, 2068 (1964).
(*d*) BENNETT, R. G. *J. chem. Phys.* **41**, 3037 (1964).

Long-range, single-step transfer is possible between molecules of the same kind when there is appreciable overlap of the absorption and the emission spectra. As explained earlier, if excitation is produced by polarized light, energy transfer to molecules with different orientations will lead to concentration depolarization of the fluorescence. With typical dyes this is observed at concentrations of $\sim 10^{-3}$ mole/litre where the mean distance between molecules is ~ 7 nm. Figure 8.3 shows the effect of increase in concentration on the degree of polarization, and also the fluorescence intensity and the lifetime of the fluorescence of fluorescein in glycerine solution. If the concentration quenching is considered to be due to long range transfer to molecules which have a high probability of being quenched, then all three curves are as expected for transfer caused by dipole-dipole coupling.

The efficiency of transfer has been studied in certain molecules which consist of two independent chromophores separated by one or more saturated bonds. In such cases energy transfer over large distances has been observed to be in agreement with predictions from Förster's theory.

From measurements on the quenching of perylene fluorescence as a function of triplet phenanthrene-d_{10} concentration, Bennett has confirmed

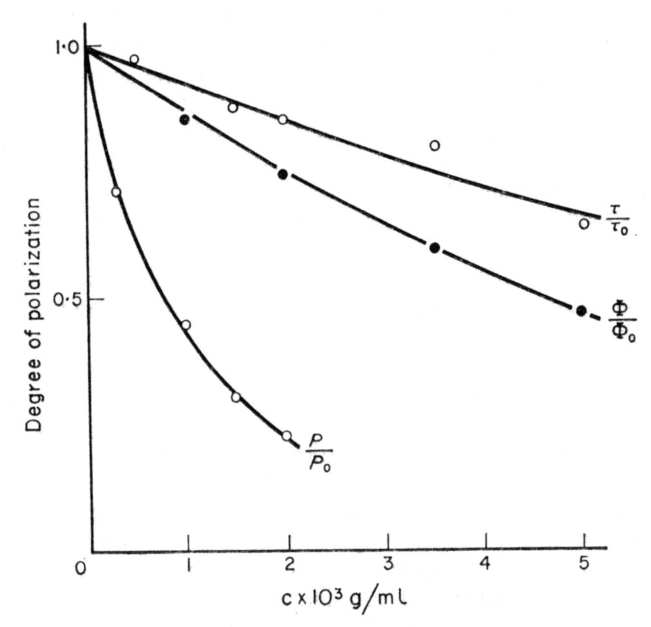

FIG. 8.3 Concentration dependence of the degree of polarization P/P_0, the yields Φ/Φ_0 and the lifetimes τ/τ_0 of the fluorescence of fluorescein in glycerine solution. (Reproduced from GALANIN *J. exp. theor. Phys.* **1**, 317 (1955).)

theoretical predictions of long-range energy transfer from the first excited singlet state of perylene to the triplet phenanthrene-d_{10} in a rigid plastic.

The process

$$\text{Perylene } (S_1) + \text{Phenanthrene-}d_{10}\ (T_1) \rightarrow \text{Perylene } (S_0) + \text{Phenanthrene-}d_{10}(T_x)$$

is fully allowed because of dipole-dipole interaction. Long-range transfer is expected because there is good overlap between the fluorescence spectrum of perylene and the triplet-triplet absorption spectrum of phenanthrene-d_{10}.

(*b*) *Exciton Migration*

Many molecular crystals emit fluorescence. In some cases this resembles the fluorescence of the isolated molecules but shows much sharper vibrational lines, while in other cases, for example pyrene, it is broad and structureless and is similar to the 'excimer' emission first observed in concentrated pyrene solutions by Förster and Kasper. Naphthalene crystals have been found to show a sharp emission band which has no associated vibrational structure. This agrees with theoretical expectations for emission from a free exciton. However, most observations with a variety of crystals suggest that the free excitons generated by absorption generally become localized or trapped near imperfections in, or at the surface of, crystals before emission takes place.

In pure crystals energy migration is usually inferred from the measurement of the absorption spectra and estimates of the interaction energies. It is worth noting that, in typical cases, 'splittings' of as little as 1 cm^{-1} can correspond to $\sim$1000 energy transfer acts during the lifetime of the excited state.

In mixed crystals measurements of host-sensitized fluorescence confirms exciton migration over large distances. A well-known example of sensitized fluorescence is that from anthracene crystals containing very small concentrations of naphthacene. There is no transfer if these crystals are dissolved in fluid or rigid solution, and this confirms the mechanism of exciton migration. The low concentration of guests suggests that the excitation energy explores $\sim 10^6$ lattice sites during the lifetime of the excited state.

Simpson has given a convincing demonstration of exciton migration in polycrystalline anthracene. Layers of different depths were prepared by evaporating known weights of anthracene on to cooled glass plates. A second layer consisting of 1 part in 300 of naphthacene in anthracene was deposited on top of each of the pure anthracene layers. The second layer acted as an exciton detector, since at this concentration the transfer efficiency from anthracene to naphthacene is unity. The thickness of the anthracene layer was varied from 0·01 to 0·6 μ. Excitation was directed at the pure anthracene layer, and the intensity of emission from the naphthacene in the detector layer was measured. Allowance was made for radiative transfer by preparing anthracene and detector layers separated by a small air gap, and measuring the effect directly. The observed 'exciton' diffusion length was 46 nm which would correspond to a root mean square displacement of 112 nm between the points of origin and decay of the 'excitons'.

Exciton migration may also take place in pure liquids. Solvent-solute transfer following solvent-solvent energy migration has often been suggested, especially in scintillator work where the bulk material is excited. Lipsky and co-workers have shown that the rate constant for energy transfer from benzene as solvent to *p*-terphenyl as solute with Co^{60} or 254 nm irradiation is $4{\cdot}2 \times 10^{10}$ litre $mole^{-1}$ s^{-1}. The rate constant for oxygen quenching of the transfer process ($7{\cdot}2 \times 10^{10}$ litre $mole^{-1}$ s^{-1}) is even larger. Both these values are too big to be explained by material diffusion or, for this system, by long-range single-step transfer, and it is therefore necessary to suggest some kind of exciton migration. Voltz and co-workers have suggested that solvent-solvent energy migration may be considered as a term additional to the material diffusion of D* and A, and one which results in a larger apparent diffusion coefficient $D' = D + \Lambda$ where $D = D_A + D_{D^*}$ is the material diffusion coefficient and Λ is the excitation diffusion coefficient.

The overall rate constant for transfer may be expressed in a form analogous to that used for diffusion-controlled reactions which was first derived by Smoluchowski:

$$k_{\mathrm{T}}=\frac{4\pi(r_{\mathrm{D}^*}+r_{\mathrm{A}})N(D+\Lambda)}{1000}\left[1+\frac{r_{\mathrm{D}^*}+r_{\mathrm{A}}}{\{\tau_{\mathrm{D}}(D+\Lambda)\}^{\frac{1}{2}}}\right]$$

k_{T} may be taken to be the sum of the two rate constants k_{D} and k_{Λ}. Various experimental values of k_{Λ} are available which, for solvents such as benzene, toluene, *p*-xylene, etc., are in the range 10^{10}–10^{11} litre mole^{-1} s^{-1}.

Pure liquids often show considerable excimer fluorescence. However, the role of excimers in energy migration in pure liquids is, at present, far from clear.

(*c*) *Transfer at Normal Molecular Diameters*

Although it has been known for over fifty years that singlet-singlet energy transfer, i.e., the process

$$\mathrm{D^*\ (singlet) + A\ (singlet) \rightarrow D\ (singlet) + A^*\ (singlet)}$$

takes place with high efficiency between organic molecules, very few quantitative measurements have been made until recently. One such example was a study of the benzene sensitization of anthracene fluorescence in the vapour phase, with excitation at 265 nm. From the increase in fluorescence of anthracene observed upon the addition of benzene, Stevens concluded that transfer takes place at every encounter at normal collisional distances.

Dubois and co-workers have shown that many compounds sensitize biacetyl fluorescence in aerated solution. An explanation of the results obtained can be based on the following simple mechanism

1. $D + h\nu \xrightarrow{I_{\mathrm{D}}} {}^1\mathrm{D}^*$
2. ${}^1\mathrm{D}^* \xrightarrow{k_2} \mathrm{D} + h\nu_{\mathrm{D}}$
3. ${}^1\mathrm{D}^* \xrightarrow{k_3} \mathrm{D}$ or ${}^3\mathrm{D}^*$
4. ${}^1\mathrm{D}^* + \mathrm{O}_2 \xrightarrow{k_4}$ Quenching
5. $\mathrm{A} + h\nu \xrightarrow{I_{\mathrm{A}}} {}^1\mathrm{A}^*$
6. ${}^1\mathrm{A}^* \xrightarrow{k_6} \mathrm{A} + h\nu_{\mathrm{A}}$
7. ${}^1\mathrm{A}^* \xrightarrow{k_7} \mathrm{A}$ or ${}^3\mathrm{A}^*$
8. ${}^1\mathrm{A}^* + \mathrm{O}_2 \xrightarrow{k_8}$ Quenching
9. ${}^1\mathrm{D}^*\ (\mathrm{S}_1) + {}^1\mathrm{A}(\mathrm{S}_0) \xrightarrow{k_{\mathrm{T}}} {}^1\mathrm{D}\ (\mathrm{S}_0) + {}^1\mathrm{A}^*\ (\mathrm{S}_1)$

This work was carried out at optical densities of less than 0·04, and the following equations can therefore be expected to give accuracy within 5%:

$$\mathrm{I_A} = I_0\varepsilon_{\mathrm{A}}[\mathrm{A}]l \quad \text{and} \quad I_{\mathrm{D}} = I_0\varepsilon_{\mathrm{D}}[\mathrm{D}]l$$

where I_0 is the incident intensity, ε_{D} and ε_{A} are the extinction coefficients at the exciting wavelength, and [D] and [A] the concentration of the donor and acceptor respectively.

Applying the steady state hypothesis to the mechanism:

$$\frac{d[D^*]}{dt} = 0 = I_D - (k_2 + k_3 + k_4[O_2] + k_T[A])[D^*]$$

$$\frac{d[A^*]}{dt} = 0 = I_A - (k_6 + k_7 + k_8[O_2])[A^*] + k_T[D^*][A]$$

The lifetime of the excited singlet state of the donor in the absence of the acceptor τ_D is given by

$$\tau_D = \frac{1}{k_2 + k_3 + k_4[O_2]}$$

If F_A and $F_A{}^0$ are the fluorescence intensities of the acceptor in the presence and absence of the donor and F_D and $F_D{}^0$ are the fluorescence intensities of the donor in the presence and absence of the acceptor, then at constant donor concentration:

$$\frac{F_D{}^0}{F_D} = 1 + k_T\tau_D[A]$$

At constant acceptor concentration

$$\frac{F_A}{F_A{}^0} = \frac{\varepsilon_D[D]k_T\tau_D}{\varepsilon_A(1 + k_T\tau_D[A])} + 1$$

Figures 8.4 and 8.5 show the plots obtained with naphthalene as donor and biacetyl as acceptor. A summary of the results obtained from similar plots is shown in Table 8.2 where K_q and K_s are the constants obtained from the quenching and the sensitization plots respectively. Both these constants are equal to $k_T\tau_D$.

If τ_D is known, k_T can be obtained from these values. The transfer rate constants given in Table 8.2 for various donors are all close to those obtained from the equation

$$k_D = \frac{8RT}{2000\eta} \text{ l.mole}^{-1}\text{ s}^{-1} \tag{8.6}$$

which gives $k_D = 3{\cdot}5 \times 10^{10}$ l.mole^{-1} s^{-1} for hexane at 28°C. Osborne and Porter have shown that eqn 8.6 gives good agreement with experiment for reactions which are diffusion-controlled (η = solvent viscosity). Biacetyl absorption is very weak throughout the near ultra-violet and visible regions, and this makes long-range, single-step transfer unlikely. The actual mechanism operating at collisional distances is difficult to distinguish, but both exchange and coulombic interactions could apply. The k_T values are consistently close to those given by eqn 8.6;

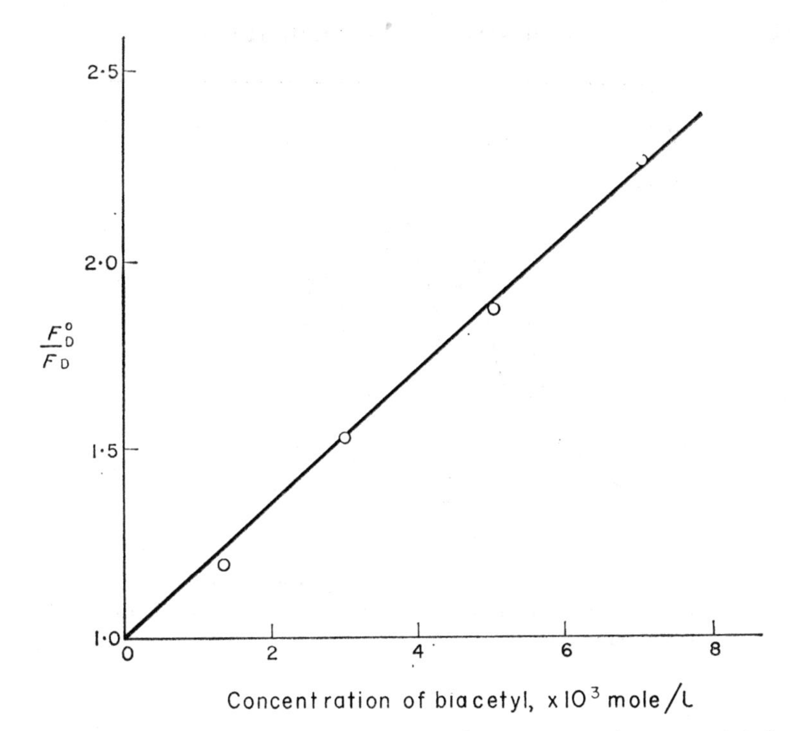

FIG. 8.4 Quenching of the fluorescence of the energy donor naphthalene (10^{-2}M) by the acceptor biacetyl in aerated cyclohexane solution at 25°C (λ_{ex} = 320 nm).

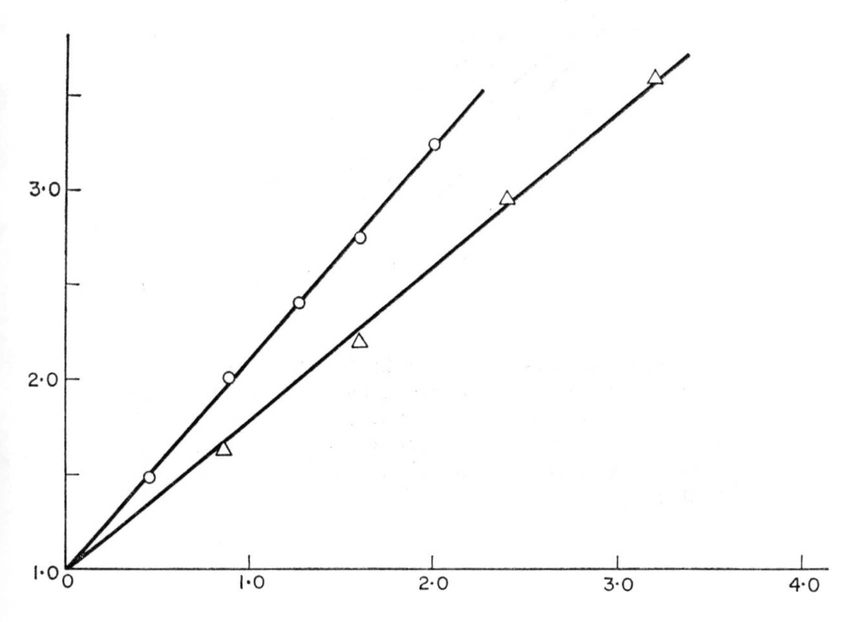

FIG. 8.5 Sensitization of the fluorescence of biacetyl by naphthalene excited at 320 nm in aerated cyclohexane solutions at 25°C. (⊙ = 0·01M biacetyl △ = 0·02M biacetyl.)

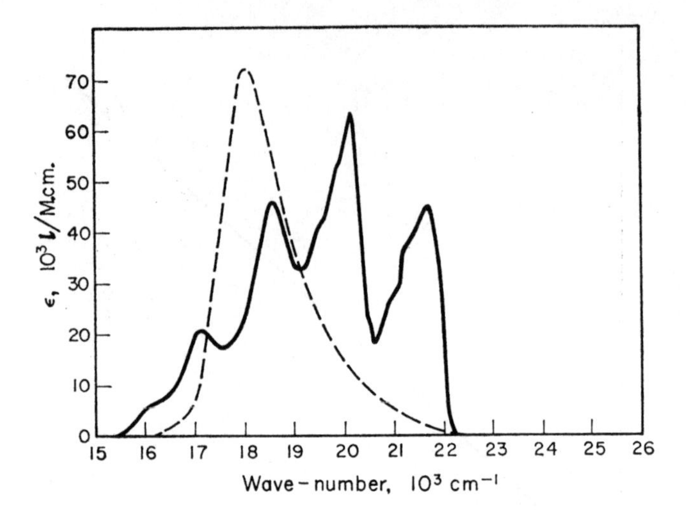

FIG. 8.6(*a*) Solid curve: phosphorescence spectrum of phenanthrene-d_{10}; dashed curve: visible absorption spectrum of rhodamine B.

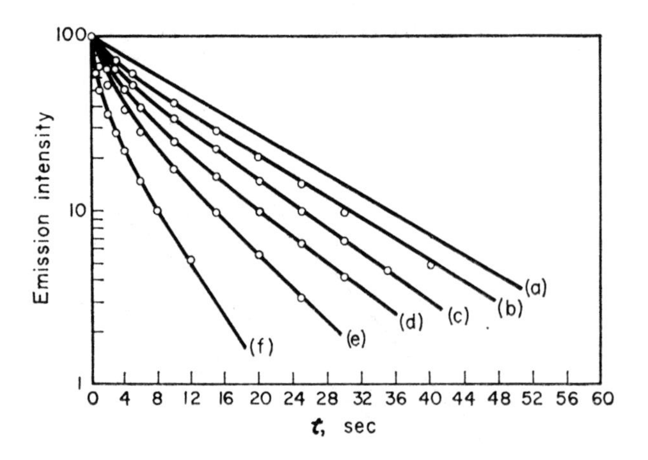

FIG. 8.6(*b*) Time-dependence of the phosphorescence decay of phenanthrene-d_{10} with rhodamine B at the following concentrations ($\times 10^{-3}$ mole/litre): (*a*) 0; (*b*) 0·53; (*c*) 1·0; (*d*) 1·9; (*e*) 3·1; (*f*) 5·0. (Reproduced with permission from BENNETT, SWHWENKER and KELLOGG *J. chem. Phys.* **41**, 3040 (1964).)

thus when τ_D is unknown, k_T may be taken as being equal to k_D and so estimates of τ_D may be made. (See Table 8.2.)

TABLE 8.2. *Rate constants for singlet-singlet energy transfer to biacetyl in aerated solutions at* 28°C

Donor	Solvent	K_q (M^{-1})	K_s (M^{-1})	τ_D[a] (10^9 s)	k_T(l mole^{-1} s^{-1})	Ref.
Benzene	Hexane	190	190	5·7	$3{\cdot}3 \times 10^{10}$	(*b*)
Benzene	Cyclohexane	200	270	23*	—	(*c*)
Toluene	Hexane	214	197	5·8	$3{\cdot}7 \times 10^{10}$	(*b*)
o-Xylene	Hexane	208	—	6·0	$3{\cdot}5 \times 10^{10}$	(*d*)
m-Xylene	Hexane	200	—	6·0	$3{\cdot}3 \times 10^{10}$	(*d*)
p-Xylene	Hexane	210	—	6·1	$3{\cdot}4 \times 10^{10}$	(*d*)
Mesitylene	Hexane	192	—	5·5*	—	(*d*)
Pentamethyl benzene	Hexane	177	—	3·9	$4{\cdot}5 \times 10^{10}$	(*d*)
Hexamethyl benzene	Hexane	80	—	2·0	$4{\cdot}0 \times 10^{10}$	(*d*)
Ethyl benzene	Hexane	208	—	5·7	$3{\cdot}6 \times 10^{10}$	(*d*)
n-Propyl benzene	Hexane	217	—	5·2	$4{\cdot}2 \times 10^{10}$	(*d*)
n-Butyl benzene	Hexane	214	—	6·8	$3{\cdot}2 \times 10^{10}$	(*d*)
Cumene	Hexane	204	—	6·0	$3{\cdot}4 \times 10^{10}$	(*d*)
Pseudo cumene	Hexane	194	—	5·5*	—	(*d*)
Diphenyl	Hexane	152	149	4·3*	—	(*b*)
Naphthalene	Hexane	186	—	8·3	$2{\cdot}2 \times 10^{10}$	(*d*)
Naphthalene	Cyclohexane	180	171	17*	—	(*c*)
Phenanthrene	Cyclohexane	128	131	13*	—	(*e*)
Chrysene	Cyclohexane	243	244	24*	—	(*e*)
Acetone	Hexane	43	45	1·3*	—	(*b*)
Acetone	Cyclohexane	26	14	2·0*	—	(*c*)
Methyl ethyl ketone	Cyclohexane	28	34	3·2*	—	(*e*)
Diethyl ketone	Cyclohexane	47	53	5·0*	—	(*e*)
Cyclopentanone	Cyclohexane	25	19	2·2*	—	(*e*)

* Estimated values.

(*a*) IVANOVA, KUDRIASHOV and SVESHNIKOV *Dokl, Akad, Nauk SSSR*, **138**, 572 (1961).
(*b*) WILKINSON and DUBOIS *J. chem. Phys.* **39**, 377 (1963).
(*c*) STEVENS and DUBOIS *Luminescence of Organic and Inorganic Molecules* (Eds. Kallmann and Spruch), p. 115, Wiley, New York (1962).
(*d*) DUBOIS and VAN HEMERT *J. chem. Phys.*, **40**, 923 (1964).
(*e*) DUBOIS and COX *J. chem. Phys.* **38**, 2536 (1963).

(*d*) *Transfer from Higher Singlet States*

Usually, higher excited states cascade down to the lowest excited singlet state S_1 so rapidly that there is little chance of transfer from such states in dilute solution. A number of claims have been made of observations of transfer from higher singlet states, but the evidence for this is inconclusive. However, recent measurements by Stevens and Dubois, using azulene as either an energy donor or an acceptor, have shown that transfer to and from higher singlet states is definitely possible.

Azulene is an exceptional molecule in that it shows fluorescence from its second excited singlet state, which is ^1L_a in character. Its lower ^1L_b

state is non-fluorescent. In dilute aerated solution naphthalene sensitizes fluorescence from azulene and the ^1L_a state of azulene sensitizes fluorescence from anthracene. At high concentration, azulene quenches anthracene fluorescence, probably as a result of energy transfer, to give the ^1L_b state of azulene.

The processes:

1. Naphthalene* (S_1) + Azulene (S_0) → Naphthalene (S_0) + Azulene* (S_2)
2. Azulene* (S_2) + Anthracene (S_0) → Azulene (S_0) + Anthracene* (S_1)
3. Anthracene* (S_1) + Azulene (S_0) → Anthracene (S_0) + Azulene* (S_1)

take place at every collision. The energies of the states involved are:

Naphthalene $S_1 = 32{,}200$ cm^{-1} (^1L_b); Anthracene $S_1 = 26{,}700$ cm^{-1} (^1L_a); Azulene $S_1 = 14{,}400$ cm^{-1} (^1L_b), $S_2 = 28{,}300$ cm^{-1} (^1L_a).

These experiments illustrate that when the donor has available sufficient energy to promote the acceptor to higher electronic states it can do so (process 1), and also that an energy difference of 12,300 cm^{-1}, as in process 3, does not markedly reduce the efficiency of energy transfer.

Energy Transfer from Triplet States

(*a*) *Long-range Single-step Transfer*

The forbidden nature of the radiative transition from triplet states is often reflected in long phosphorescence lifetimes in rigid solution. If the phosphorescence yield is also high, long-range transfer due to dipole-dipole resonance interaction can sometimes compete effectively with phosphorescence, even at low acceptor concentrations, provided the transition in the acceptor is fully allowed. Thus Ermolaev and Sveshnikova detected long-range, single-step transfer from the triplet state of triphenylamine to the acceptors chrysoidin, chlorophyll a and b and pheophytin a and b in an ethanol glass at 77°K, with production of their sensitized fluorescence. In this case, R_0 can be calculated by using the overlap between the phosphorescence spectrum of the donor and the $S_0 \rightarrow S_1$ absorption spectrum of each acceptor. With chrysoidin as acceptor, R_0 (exp) = 5·5 nm and R_0 (theory) = 4 nm. This process has also been confirmed from the time-dependence of the phosphorescence decay of phenanthrene-d_{10} in the presence of rhodamine B in cellulose acetate films at room temperature (Fig. 8.6). In this case R_0 (exp) = 4·7 nm which confirms that the process:

$$D^* \text{ (triplet)} + A \text{ (singlet)} \rightarrow D \text{ (singlet)} + A^* \text{ (singlet)}$$

is taking place, as expected, by Coulomb interaction.

Kellogg has shown that triplet-triplet annihilation also can take place at large distances in a cellulose acetate film. The process:

$$D^* \text{ (triplet)} + A^* \text{ (triplet)} \rightarrow D \text{ (singlet)} + A^{**} \text{ (triplet)}$$

requires overlap between the phosphorescence spectrum of the donor and the triplet-triplet absorption of the acceptor. For self-transfer with phenanthrene-d_{10} the overlap is good and $R_0 = 4$ nm, which is in good agreement with theoretical expectations.

Triplet state lifetimes in the gas phase and in fluid solution are usually much shorter than in rigid media and the probability of decay by processes other than energy transfer becomes correspondingly greater under these conditions. However, it should be borne in mind that the process of triplet-triplet annihilation is allowed by an exchange mechanism, as we shall see later.

(b) Triplet Exciton Migration

Pure crystals of aromatic hydrocarbons show no phosphorescence even at 77°K, while crystals of aromatic ketones, such as benzophenone, show strong phosphorescence. Both types of crystal show host-sensitized phosphorescence from guests or impurities present in very low concentrations. Delayed fluorescence in anthracene crystals depends on the square of the intensity of the exciting light, indicating the presence of a bimolecular interaction between long-lived excitons. However, since the decay of delayed fluorescence in aromatic crystals is sometimes first-order, triplet-triplet annihilation cannot be the main mechanism of decay under all conditions.

Robinson and co-workers, using isotopic mixed crystals, have made many observations of energy transfer. In the case of benzene-d_6 containing C_6H_5D and C_6H_6 as guests which have triplet levels lower than that of C_6D_6 by 170 and 200 cm^{-1} respectively, only the guest molecules emit. The amounts of emission from the two guests should depend on their relative concentrations, if guest-to-guest transfer does not occur. This is true for the fluorescence observed; however, the host sensitized phosphorescence from C_6H_6 is ten times greater than that from C_6H_5D at equal guest concentrations. It follows that the triplet excitation is transported over larger distances than is singlet excitation energy. Triplet energy transfer between phenanthrene and naphthalene takes place in single crystals of diphenyl at 77°K, and has been studied by magnetic resonance techniques. This transfer process has also been observed at 4°K. Hochstrasser has used benzophenone as a crystalline host with benzanthracene as a guest. Measurement of the ratio of host-sensitized guest fluorescence to phosphorescence intensities shows that the triplet excitation is transported over very large distances in ketone crystals. In fact the triplet exciton transfer rates are $\sim 10^{10}$ s^{-1} which is close to those predicted for exchange interaction. With such high transfer rates, even one part in 100 million of impurity can have a significant effect on the phosphorescence from such crystals.

Various evidence, including that from delayed fluorescence in crystals which is due to triplet-triplet annihilation, yields experimental values for the diffusion length of the triplet exciton in anthracene crystal of $\sim 5 \times 10^{-4}$ cm, which is close to the theoretical estimates of 3×10^{-4} cm. Most of the experimental results suggest that the number of molecules visited by triplet excitation energy during the excited state lifetime is anything from 100–1000 times greater than the number visited by singlet excitons during their much shorter lifetimes.

If the triplet level of the guest molecule is not much lower than that of the host ($\sim kT$), then thermal activation into the host triplet exciton level can take place and allow the excitation energy to migrate large distances from host to host or trap to trap. Excitation transfer in crystals is therefore not dissimilar from that pertaining in fluid solution, except that diffusion involves the excitation energy and not the excited molecule.

(*c*) *Transfer at Normal Molecular Diameters*

(1) *Rigid Solution.* Sensitized phosphorescence due to triplet-triplet energy transfer in rigid media was first reported by Terenin and Ermolaev in 1952. The transfer process:

$$D^* \text{ (triplet)} + A \text{ (singlet)} \rightarrow D \text{ (singlet)} + A^* \text{ (triplet)}$$

is allowed by exchange interaction and forbidden by coulombic interaction. In this case the transitions in both the donor and the acceptor are forbidden, and although increase in the triplet donor lifetime may compensate

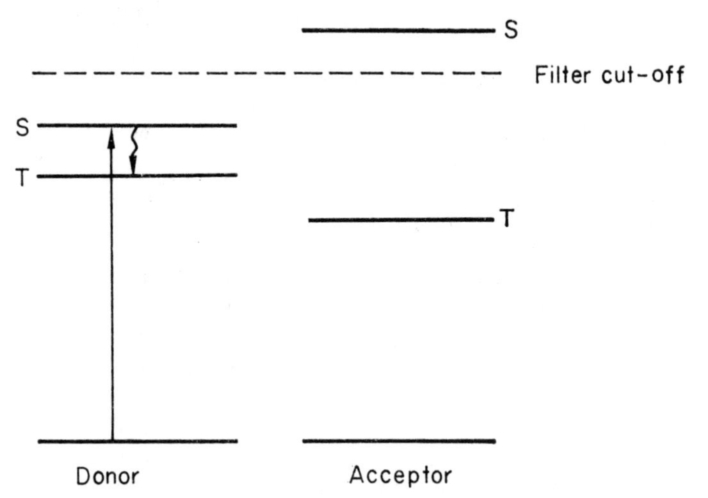

FIG. 8.7 Ideal arrangement of donor and acceptor energy levels for demonstrating triplet-triplet energy transfer.

for the forbidden nature of the donor transition, there are no compensating effects for the forbidden transition in the acceptor.

Unambiguous demonstration of this type of transfer can be obtained by choosing pairs of molecules with energy levels as shown in Fig. 8.7. This excludes the possibility of singlet-singlet energy transfer and makes it possible to excite the donor without directly exciting the acceptor. Aromatic ketones make especially good donors since they have high phosphorescence yields and small energy differences between S_1 and T_1. Aromatic hydrocarbons such as naphthalene have much larger S_1–T_1

gaps and have been used frequently as triplet energy acceptors. (See Table 8.3.) The heights of the energy levels S_1 and T_1 for many of the compounds given in Table 8.3 can be found in Tables 7.1 and 7.2.

Triplet-triplet energy transfer not only leads to sensitized phosphorescence from the acceptor, but also results in strong quenching of the donor phosphorescence and in a slight reduction in its lifetime. For acceptor concentrations in the range 1–10^{-2} mole/litre the relative decrease in the donor phosphorescence can be described by the equation:

$$\frac{\Phi_D{}^0}{\Phi_D} = \exp(\alpha C_A)$$

where $\Phi_D{}^0$ and Φ_D represent the donor phosphorescence yield in the absence and presence of the acceptor at concentration C_A mole/l (Fig. 8.8). Perrin first deduced this equation for quenching within an 'active

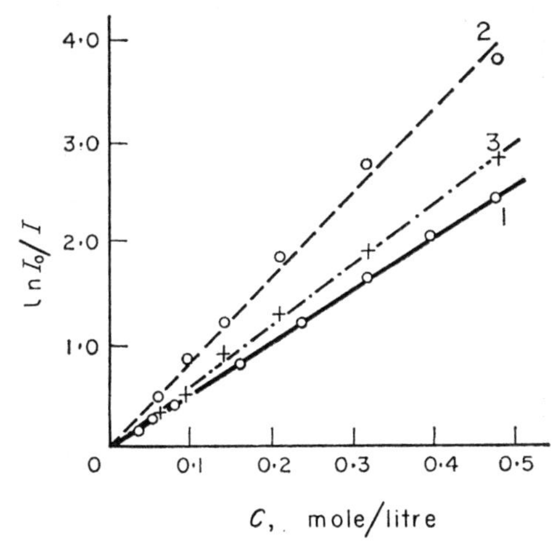

FIG. 8.8 Dependence of the relative donor phosphorescence yields ($\ln I_0/I = \ln \Phi_0/\Phi$) on the acceptor concentration for different donor-acceptor combinations in alcohol ether mixtures at 93°K.

1: benzophenone and 1-bromonaphthalene
2: carbazole and naphthalene
3: phenanthrene and naphthalene.

The excitation was confined to the spectral region where only the donor molecules absorb, viz. $\lambda_{ex} = 365$ nm for 1 and 334 nm for 2 and 3. (Reproduced from ERMOLAEV and TERENIN, Paper presented at the Tenth Conference on Luminescence, Moscow (1961) in *Soviet Academy of Sciences Bulletin, Physical Series*, **26**, 21 (1962).)

sphere', i.e., by assuming that if an acceptor molecule is within the 'active sphere' the probability of quenching is unity, and, if it is not, the probability of quenching is zero.

TABLE 8.3. *Volumes and radii of spheres of action for triplet-triplet energy transfer and yields of sensitized phosphorescence Φ_S in a rigid alcohol ether glass at 90° or 77°K*[(a)]

Donor	Acceptor	Volume (10^{21} cm^3)	R(nm)	Φ_S: $\frac{\Phi_{SPA}}{\Phi_D^0 - \Phi_D}$	Φ_S: $\frac{\Phi_A \text{ phos}[1 - \Phi_D \text{ fluor.}]}{\Phi_D \text{ phos}[1 - \Phi_A \text{ fluor.}]}$
Benzaldehyde	Naphthalene	6·8	1·2	0·13	0·04
Benzaldehyde	1–Chloronaphthalene	7·0	1·2	0·22	0·34
Benzaldehyde	1–Bromonaphthalene	7·2	1·2	0·27	0·29
Benzophenone	Naphthalene	8·6	1·3	0·07	0·06
Benzophenone	1–Methylnaphthalene	9·5	1·3	0·07	0·05
Benzophenone	1–Chloronaphthalene	9·5	1·3	0·12	0·22
Benzophenone	1–Iodonaphthalene	8·6	1·3	0·35	0·27
Benzophenone	Quinoline	7·2	1·2	0·14	0·14
Acetophenone	Naphthalene	6·0	1·1	0·10	0·07
p–Chlorobenzaldehyde	Naphthalene	6·7	1·2	0·14	—
p–Chlorobenzaldehyde	1–Bromonaphthalene	6·2	1·1	0·49	—
o–Chlorobenzaldehyde	Naphthalene	5·4	1·1	0·11	—
m–Iodobenzaldehyde	Naphthalene	5·8	1·1	0·11	0·07
m–Iodobenzaldehyde	1–Bromonaphthalene	5·7	1·1	0·30	0·22
Xanthone	Naphthalene	9·2	1·3	0·11	—
Anthraquinone	Naphthalene	5·9	1·1	0·10	—
Anthraquinone	1–Bromonaphthalene	7·6	1·2	0·27	—
Triphenylamine	Naphthalene	9·3	1·3	0·07	—
Carbazole	Naphthalene	14	1·5	0·08	—
Phenanthrene	Naphthalene	10	1·3	0·30	0·28
Phenanthrene	1–Chloronaphthalene	11	1·4	0·73	1·10
Phenanthrene	1–Bromonaphthalene	11	1·4	0·99	0·94

(*a*) ERMOLAEV *Dokl, Akad, Nauk*, **139**, 348 (1961), and *Soviet Phys. Doklady*, **6**, 600 (1962).

Experimental values for the volumes and the radii of the spheres of influence for various donor and acceptor pairs are listed in Table 8.3. The values of R are only slightly greater than normal collision diameters, and it follows that there is a rapid falling off in the probability of transfer at this distance. The $S_0 \rightarrow T_1$ radiative transition probability increases by a factor of 1000 in the 1-halonaphthalenes, but as can be seen from Table 8.3 this has no effect on the efficiency of transfer. This is consistent with transfer due to an exchange mechanism, since the process is already fully allowed by the spin selection rules for this type of interaction.

Inokuti and Hirayama have applied to Ermolaev's data their suggested analysis for processes believed to be taking place by an exchange mechanism, and have found very good agreement between the measured and calculated values for Φ_D/Φ_D^0 and τ_m/τ_0 as a function of acceptor concentration, as illustrated by Fig. 8.9. The various parameters needed to give this agreement are given in Table 8.4.

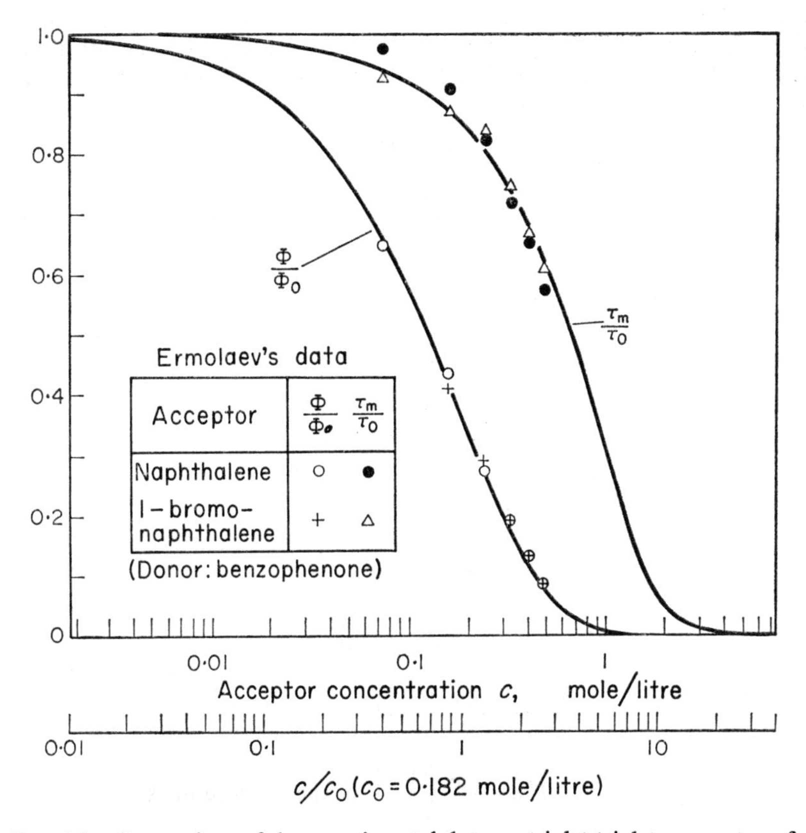

FIG. 8.9 Comparison of the experimental data on triplet-triplet energy transfer in rigid glass with the theory of Hirayama and Inokuti. (Reproduced with permission from HIRAYAMA and INOKUTI *J. chem. Phys.* **43**, 4149 (1965).)

TABLE 8.4. *Result of analysis of Ermolaev's data for triplet-triplet energy transfer from benzophenone*

	Naphthalene	1-Bromo-naphthalene
γ	18·0	19·5
C_0 (mole/litre)	0·186	0·182
R_0 (nm)	1·29	1·30
L (nm)	0·143	0·133

Table 8.3 gives values for Φ_S which Terenin and Ermolaev call the 'quantum yield of sensitized phosphorescence' defined as follows:

$$\Phi_S = \frac{\Phi_{SPA}}{\Phi_D{}^0 - \Phi_D}$$

where Φ_{SPA} is the experimental value of the sensitized phosphorescence yield. This yield is independent of donor and acceptor concentrations, and a value can be calculated on the basis of the mechanism given below:

$$D \xrightarrow{I_D} {}^1D^*$$

$${}^1D^* \xrightarrow{k_{fD}} {}^1D + h\nu_D{}^F$$

$${}^1D^* \xrightarrow{k_{iD}} {}^3D^*$$

$${}^3D^* \xrightarrow{k_{tD}} D$$

$${}^3D^* \xrightarrow{k_{PD}} D + h\nu_D{}^P$$

$${}^3D^* + A \xrightarrow{k_T} {}^1D + {}^3A^*$$

$${}^3A^* \xrightarrow{k_{tA}} {}^1A$$

$${}^3A^* \xrightarrow{k_{PA}} {}^1A + h\nu_A{}^P$$

$$\Phi_{SPA} = \frac{k_{PA}[{}^3A^*]}{I_D}$$

$$\Phi^0{}_D = \frac{k_{PD}[{}^3D^*]}{I_D} \quad \text{in the absence of A}$$

and

$$\Phi_D = \frac{k_{PD}[{}^3D^*]}{I_D} \quad \text{in the presence of A}$$

Assuming photostationary conditions for each excited state it can be shown that:

$$\Phi_S = \frac{\Phi_{S\,PA}}{\Phi_D{}^0 - \Phi_D} = \left(\frac{k_{PA}}{k_{PA} + k_{tA}}\right)\left(\frac{k_{PD} + k_{tD}}{k_{PD}}\right)$$

If it is assumed that the only non-radiative transition from the first excited singlet state of an organic molecule is intersystem crossing then

$$\frac{k_P}{k_P + k_t} = \frac{\Phi_{\text{phosphorescence}}}{1 - \Phi_{\text{fluorescence}}}$$

and

$$\Phi_S = \left[\frac{\Phi_{A\text{ phos.}}}{1 - \Phi_{A\text{ fluor.}}}\right]\left[\frac{1 - \Phi_{D\text{ fluor.}}}{\Phi_{D\text{ phos.}}}\right]$$

Some values calculated in this way are shown in the last column of Table 8.3. The values of Φ_S obtained in this way and from energy transfer experiments are equal within experimental error, which suggests that non-radiative decay from the first excited singlet state of these organic compounds is all due to intersystem crossing which produces triplet states.

Concentration depolarization of phosphorescence which is due to energy transfer between triplet states of identical molecules in rigid solutions is complicated by the possibility of singlet energy transfer. Nevertheless a number of negative results have been reported at concentrations in the range 10^{-2}–10^{-3} mole/litre. A knowledge of the results on sensitized phosphorescence due to transfer between unlike molecules indicates that much higher concentrations are necessary before triplet-triplet energy transfer may be expected to lead to concentration depolarization of phosphorescence.

(2) *Fluid Solution.* In 1958 Backström and Sandros reported that the phosphorescence of biacetyl in benzene solution at room temperature was quenched at almost every collision by aromatic hydrocarbons with lower triplet levels, while hydrocarbons with higher triplet states had almost no effect. They suggested that triplet-triplet energy transfer was responsible for the quenching observed. Porter and Wilkinson, using the technique of flash photolysis, made measurements of the decay and sensitized build-up of the triplet-triplet absorption of various donors and acceptors respectively, and showed that high transfer rate constants were a general phenomenon (see Table 8.5 and Fig. 8.10). They confirmed that quenching of triplet biacetyl by aromatic hydrocarbons results in the promotion of the hydrocarbon to its triplet state (see Fig. 8.11).

From Table 8.5 it can be seen that when the triplet level of the donor is considerably higher than that of the acceptor (greater than 1000 cm^{-1}) the rate constants are close to those expected for diffusion-controlled reactions. The decrease in the measured rate constant for energy transfer as the triplet levels approach each other is due to temperature-dependent transfer taking place in the reverse direction.

Sensitized phosphorescence of biacetyl has been observed using many

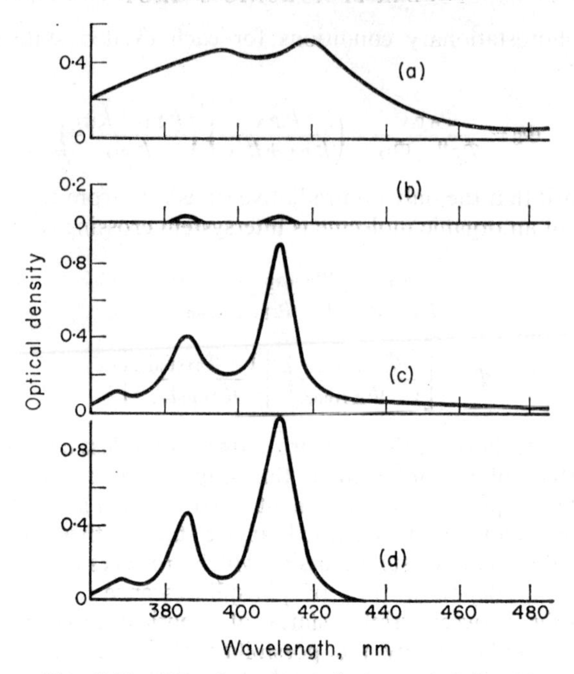

FIG. 8.10 Transient absorption spectra following photolysis in a filtered reaction vessel showing triplet-triplet energy transfer in hexane solution for the donor, triphenylene to the acceptor, naphthalene.

(*a*) $1{\cdot}01 = 10^{-3}$M triphenylene
(*b*) $1{\cdot}15 \times 10^{-4}$M naphthalene
(*c*) Mixture $1{\cdot}01 = 10^{-3}$ triphenylene and $1{\cdot}15 \times 10^{-4}$M naphthalene
(*d*) As (*b*), but the filter was removed.

triplet energy donors, including aromatic ketones, hydrocarbons, etc. These observations all confirm very efficient transfer at normal collision diameters as expected from exchange interaction.

It is uncertain whether triplet biacetyl is produced by energy transfer from triplet benzene in dilute solution. Van Loben Sels and Dubois [10] find that the sensitized ratio of the phosphorescence to fluorescence intensities of biacetyl is greater than that due to self-absorption in pure benzene, but that the effect is no longer detectable when the benzene is diluted with cyclohexane by a factor of seven.

These authors suggest that material diffusion of a triplet 'excimer' of benzene of extremely short lifetime could explain these results. The effect also could be explained by triplet exciton migration in the liquid phase.

(3) *Vapour Phase.* Vapour phase energy transfer, apparently arising from triplet states, has been observed in a number of systems, but little

TABLE 8.5. *Rate constants for triplet-triplet energy transfer as a function of the difference in triplet energy levels of the donor and acceptor ΔE*

Donor	Acceptor	Solvent	ΔE(cm^{-1})	k_T(l. mole^{-1} s^{-1})	Ref.
Biacetyl	3,4 Benzpyrene	Benzene	5000	$8{\cdot}2 \times 10^9$	(*a*)
Biacetyl	Anthracene	Benzene	5000	$8{\cdot}1 \times 10^9$	(*a*)
Biacetyl	1,2-Benzanthracene	Benzene	3200	7×10^9	(*a*)
Biacetyl	Pyrene	Benzene	3000	$7{\cdot}5 \times 10^9$	(*a*)
Triphenylene	Naphthalene	Hexane	2200	$1{\cdot}3 \times 10^9$	(*b*)
Biacetyl	*Trans*-stilbene	Benzene	2000	$4{\cdot}4 \times 10^9$	(*a*)
Phenanthrene	1-iodonaphthalene	Hexane	1100	$1{\cdot}3 \times 10^9$	(*b*)
Phenanthrene	1-iodonaphthalene	Ethanediol	1100	2×10^8	(*b*)
Phenanthrene	1-bromonaphthalene	Hexane	900	$1{\cdot}5 \times 10^8$	(*b*)
Phenanthrene	1-bromonaphthalene	Ethanediol	900	$1{\cdot}5 \times 10^7$	(*b*)
Naphthalene	1-iodonaphthalene	Ethanediol	800	$2{\cdot}8 \times 10^8$	(*b*)
Biacetyl	Coronene	Benzene	700	2×10^8	(*a*)
Biacetyl	1-Nitronaphthalene	Benzene	500	$1{\cdot}1 \times 10^8$	(*a*)
Phenanthrene	Naphthalene	Hexane	300	$2{\cdot}9 \times 10^6$	(*b*)
Phenanthrene	Naphthalene	Ethanediol	300	$2{\cdot}3 \times 10^6$	(*b*)
1-bromo-naphthalene	1-iodonaphthalene	Ethanediol	200	8×10^7	(*b*)
Biacetyl	2,2′-Binaphthyl	Benzene	150	$9{\cdot}7 \times 10^6$	(*a*)

(*a*) SANDROS and BACKSTRÖM *Acta chem. scand.* **12**, 828 (1958).
(*b*) PORTER and WILKINSON *Proc. R. Soc.* A264, 1 (1961).

quantitative information concerning the transfer process has been published. Triplet-triplet energy transfer has been suggested to explain measurements of sensitized biacetyl phosphorescence in the vapour phase which are brought about by various ketones, and benzene-sensitized isomerization of *cis*- and *trans*-2-butene.

A most thorough study of this type of transfer in the vapour phase has been the work of Ishikawa and Noyes on the system benzene-biacetyl. Benzene and cyclohexane both increase the emission from biacetyl when excitation is with 313 and 365 nm irradiation with similar efficiencies. Neither of these compounds absorbs at these wavelengths and their action is due to collisional stabilization of the excited biacetyl, i.e., removal of vibrational energy, and consequent reduction in the photochemical decomposition of the molecule prior to emission. With excitation at 254 nm cyclohexane continues to act in this way. However, benzene, which absorbs at this wavelength, produces a much larger effect and sensitizes biacetyl phosphorescence preferentially. As can be seen from Fig. 8.12, both singlet-singlet and triplet-triplet energy transfer from benzene to biacetyl are possible. From detailed studies under a variety of conditions, these authors showed that singlet-singlet energy transfer leads to sensitized decomposition of biacetyl, and deduced the extent of triplet-triplet transfer.

The yield of sensitized biacetyl phosphorescence was found to be 0·12

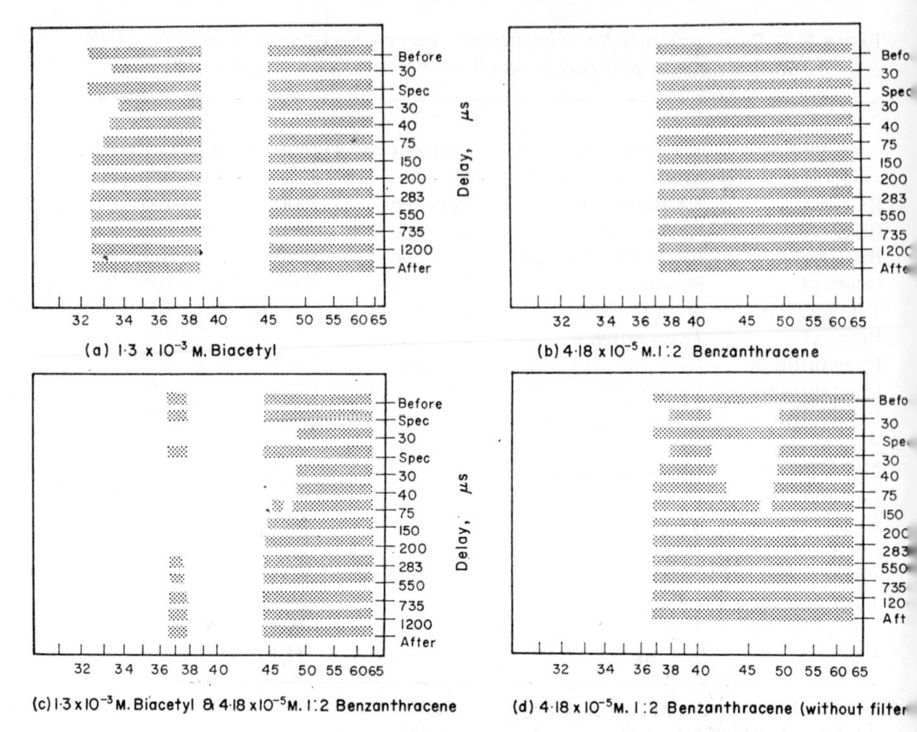

FIG. 8.11 Reproductions of flash photolysis showing triplet-triplet energy transfer from biacetyl to 1,2-benzanthracene in a filtered reaction vessel. Solvent: benzene at 25°C.

under certain conditions. Under these same conditions the quantum yield of fluorescence of biacetyl at 436 nm is 0·15 and the quantum yield of fluorescence of benzene is 0·22. If the quantum yields of triplet state production of biacetyl and benzene are unity and 0·78 respectively, the expected yield of sensitized biacetyl phosphorescence is $0{\cdot}15 \times 0{\cdot}78 = 0{\cdot}12$, which agrees surprisingly well with the measured value. These and other experiments indicate that triplet-triplet energy transfer occurs, in the vapour phase, with unit collisional efficiency.

(*d*) *Triplet-triplet Annihilation*

This process which can lead to delayed fluorescence has been discussed in Chapter 7. Although the process usually involves like species, mixed triplet-triplet annihilation has also been reported. This may be represented as follows:

$$^3D^* + ^3A^* \rightarrow ^1(DA^*) \rightarrow ^1D + ^1A^*$$

This process is allowed by an exchange mechanism. Triplet-triplet annihilation can also take place over large distances by coulombic interaction.

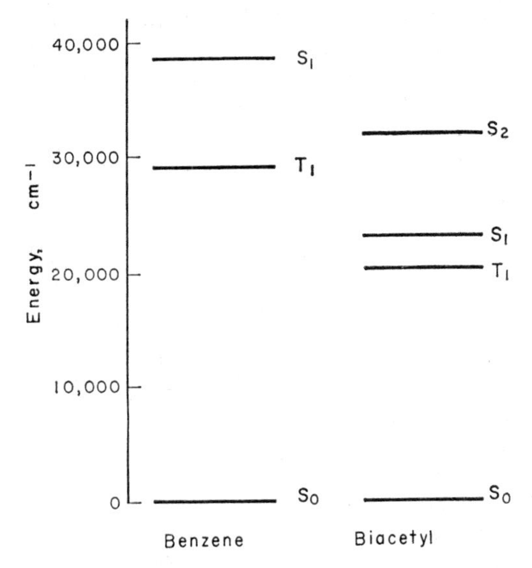

FIG. 8.12 Energy level diagram for the donor-acceptor pair benzene-biacetyl.

Sensitized anti-Stokes delayed fluorescence from naphthalene has been observed in dilute solution with phenanthrene as donor and naphthalene as acceptor. The most energetic quanta emitted are up to 5 kcal/mole greater than the energy of the exciting light.

The mechanism responsible for this is given below

$^1P \rightarrow {}^1P^*$	Absorption
$^1P^* \rightarrow {}^3P^*$	Intersystem crossing
$^1P^* \rightarrow {}^1P + h\nu$	Fluorescence
$^3P^* + {}^1N \rightarrow {}^1P + {}^3N^*$	Triplet-triplet energy transfer
$^3N^* \rightarrow N$	Intersystem crossing
$^3N^* + {}^3N^* \rightarrow {}^1N^* + N$	Triplet-triplet annihilation
$^1N^* \rightarrow {}^1N + h\nu'$	Fluorescence (delayed)

where P and N represent phenanthrene and naphthalene, and $h\nu' > h\nu$.

Triplet-triplet annihilation can also lead to delayed fluorescence in organic crystals. The singlet-triplet absorption of anthracene crystals can be excited by using ruby lasers. This leads to delayed fluorescence depending on the square of the light intensity. It is also possible to excite anthracene directly by biphotonic absorption to its singlet state. The concentration of singlet excitons excited in this way depends on the square of the light intensity. Intersystem crossing from the singlet state yields triplet excitons, and triplet-triplet annihilation produces delayed fluorescence with an expected fourth power intensity dependence as found experimentally.

The measured values for triplet-triplet annihilation rate constants in anthracene crystals range from 1×10^{-11} to 6×10^{-11} cm^3 s^{-1}, compared with a theoretical value calculated by Jortner of 2×10^{-11} cm^3 s^{-1}. From measurements of the lifetime of the delayed fluorescence in anthracene crystals, a diffusion coefficient for triplet excitation of 6×10^{-6} cm^2 s^{-1}, and a diffusion length of $3{\cdot}5 \times 10^{-4}$ cm have been obtained.

REFERENCES

Review type articles (see also references in these articles):

[1] FÖRSTER, TH. *Comparative Effects of Radiation* (edited by Burton, Kirby-Smith and Magee), p. 300, Wiley, New York (1960).
[2] FÖRSTER, TH. *Discuss. Faraday Soc.* **27**, 7 (1959).
[3] KASHA, M. *Radiat. Res.* **20**, 66 (1963).
[4] ERMOLAEV, V. L. *Usp. Fiz. Nauk.* **80**, 333 (1963).
[5] DAVYDOV, A. S. *Theory of Molecular Excitons*, translated by Kasha and Oppenheimer, McGraw-Hill, New York (1962).
[6] WILKINSON, F. *Quart. Rev.* **20**, 403 (1966) and *Adv. in Photochemistry*, **3**, 241 (1964).
[7] WINDSOR, M. *Physics and Chemistry of the Organic Solid State*, Vol. II, p. 345, Interscience, New York (1965).
[8] MCCLURE, D. S. *Solid St. Phys.* **8**, 1 (1959).
[9] DEXTER, D. L. *J. chem. Phys.* **21**, 836 (1953).
[10] VAN LOBEN SELS, J. W. and DUBOIS, J. T. *J. chem. Phys.* **45**, 1522 (1966).

Chapter 9

CHEMILUMINESCENCE OF SOLUTIONS

E. J. BOWEN

MOLECULES newly formed in chemical reactions have to get rid of their excess energy. This they usually do by sharing out the energy by colliding with solvent molecules, vessel walls, etc. Occasionally, however, excess energy is lost by radiation. Energetic reactions in gases, or shock waves generated mechanically, can often give rise to such high temperatures of the products that *thermal* radiation occurs (flames and explosions), but sometimes true chemiluminescence is observed (Chapter 4), characterized by a much stronger emission at a selective region of the spectrum than an ideally perfect thermal radiator would emit at the same temperature. In fact, here the concept of *temperature* becomes somewhat blurred because molecules of very high energy are being produced which radiate their characteristic spectra before they have properly equilibrated themselves to the bulk temperature energy of the system. Ordinary reactions in solution never get hot enough to emit thermally; any visible light observed must be chemiluminescence. The actual emission process must follow basic quantum principles, that is, molecules must be formed in an energetic quantum state of definite 'mean-life' (fixed by the degree to which transitions between levels are 'allowed'), and must resist de-activation and equilibration for a time long enough to allow radiation to take place. For emission in the visible region, individual molecules must be formed with energies of 40 (red) to 70 (blue) kcal/mole. The nature of the excitation must be *electronic*, and if the radiational transition to the ground state is 'allowed', the radiational lifetime is of the order of 10^{-8} s. Less-allowed transitions have longer lifetimes, up to a thousand million times or so longer for so-called 'forbidden' transitions. The length of time between collisions in liquids is about 10^{-12} s, so that radiation is possible only if the electronically excited molecules are able to ignore many thousands of collisions. The kind of molecule with this property is just that found for luminescence excited by light absorption. Allowed transitions, as singlet→singlet, are the ones usually to be expected in mobile solutions; the radiational lifetimes of forbidden transitions, as triplet→singlet, are so long that energy robbing (quenching) by traces of dissolved oxygen or other molecules present competes very effectively with light emission. Consequently in seeking the emitter in a chemiluminescent reaction, one must identify the nature of some *fluorescent* molecule responsible. This is commonly beset with difficulties since fluorescence emission bands, having little structure, are poor criteria for molecular identification. It is still more difficult to discover a fully acceptable process by which electronically excited molecules are formed, since most chemiluminescent reactions have complex mechanisms which often give many products, and light yields are usually very low.

The best-known chemiluminescent reaction, since the substance involved is commercially available, is the oxidation of 3-amino-phthalhydrazide, often called 'luminol'. Its solution in 0·03N NaOH, containing some hydrogen peroxide, gives a bright light emission when ferricyanide or hypochlorite solution is added. For many years confusion was caused by the close similarity between the emission (coming from an alkaline solution) and the fluorescence of luminol itself (although this is only shown by an acid solution). No other recognizable molecules from the products of oxidation in aqueous solution could be postulated as emitters. It was later found that even more intense luminescence was produced when strongly alkaline solutions in dimethyl sulphoxide were shaken with oxygen, and that the product formed was the ion of amino-phthalic acid. The emission may be green or blue depending on the degree of ionization and solvation of the product ion. Measurements of the light-yield (light quanta emitted/molecules oxidized) combined with the figure for the quantum yield of fluorescence of the product indicate that a large fraction of the product molecules are formed in their electronically excited state. The overall reaction may be represented as:

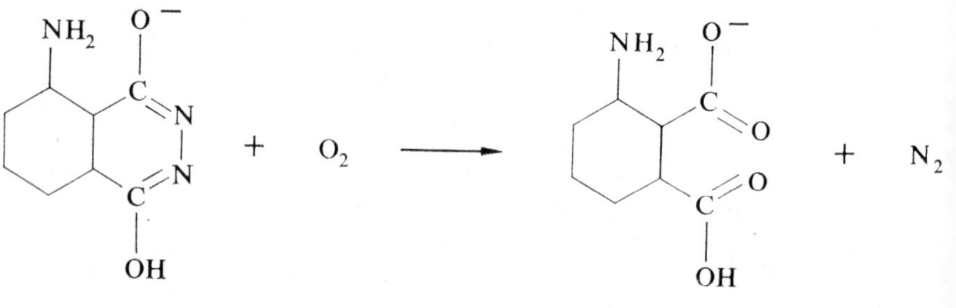

NH_2 O^- C N N C OH $+ O_2 \longrightarrow$ NH_2 O^- C O O C OH $+ N_2$

Evidence from electrochemical oxidation and from the usual methods for detecting radicals shows that, in aqueous solution, the first stage of reaction is a one-electron oxidation to give a semi-quinone-like radical. The later steps are more obscure, but they seem to involve a second one-electron oxidation, replacement of N_2 by O_2 (possibly to give the anhydride of 3-amino-peroxyphthalic acid), and a final two-electron reduction by hydrogen peroxide with oxygen liberation [1], [16].

The difficulties of elucidating the mechanism of chemiluminescence are clearly brought out in observations on the oxidation of indoles. In aqueous solution only alkaline peroxydisulphate gives rise to luminescence; none is observed with hydrogen peroxide or hypochlorite. In dimethyl sulphoxide solution, air and solid KOH are required. The following complex facts are found. There is no simple correspondence between the emission intensities observed for the same indole in aqueous or sulphoxide solution. Intensities vary by many powers of ten between indoles with different substituents. The kinetics indicate a reactant concentration-dependent first stage followed by a first-order intermediate decomposition to give the light-producing product. The nature of this substance has eluded detection, since the luminescence spectrum cannot be identified with the fluorescence of any of the recognizable oxidation products. The emitter must therefore either be formed as a minor side-product, or be liable to further rapid oxidation. The enormous variations of luminescence intensity between different indoles are intelligible in terms of widely different rates of various reaction stages and routes. This general position is one in which many chemiluminescent reactions are at present to be found [2].

A highly efficient chemiluminescent substance is N,N′-dimethyl biacridylium nitrate (lucigenin). With alkaline hydrogen peroxide the emission may be green or blue. The green emission occurs under mild conditions, and corresponds to the fluorescence of lucigenin itself. Here the substance is acting as a reversible catalyst which is decomposing hydrogen peroxide and being regenerated by a redox cycle. The blue emission corresponds to the fluorescence of N-methyl acridone, and is observed in hot solutions or in the presence of added catalysts, the lucigenin being irreversibly oxidized. The overall reaction may be represented as:

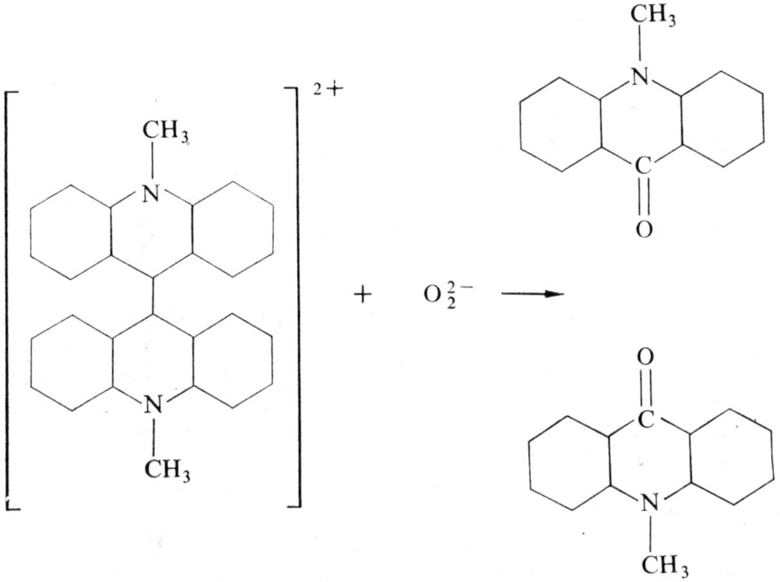

$$[\text{lucigenin } (CH_3 \text{–N, N–}CH_3)]^{2+} + O_2^{2-} \longrightarrow 2\ \text{N-methyl acridone } (CH_3\text{–N, C=O})$$

Lucigenin is electrochemically reducible to the uncharged diradical; its 2-electron reduction by HO_2^- to this diradical is the first chemical step with hydrogen peroxide [4]. Reoxidation of the diradical back to the parent substance explains its catalytic activity; the associated electronic excitation may be due to energy transfer from singlet O_2 molecules. Addition of oxygen to the diradical should give a cyclic peroxide with central structure

```
  \ /
   C—O
   |  |
   C—O
  / \
```

The acridone product may arise from a splitting across of this peroxide, or perhaps from interactions of two such radicals with O_2 evolution, which would be more exothermic.

Lucigenin can be used as a chemiluminescent indicator in titrations.

The thermal decomposition of organic peroxides is very commonly accompanied by chemiluminescence. The reactions are sufficiently energetic, but sometimes the products are not good emitters. The addition of fluorescent substances then may enhance emission. An interesting example is the decomposition of tetralin hydroperoxide in presence of the highly fluorescent substance zinc tetraphenylporphin. The porphin appears to supply an electron to the peroxide molecule to give OH^- and the radical

```
 \   /H
   C
 /   \O—
```

and subsequent electron return gives tetralone $>C{=}O$ and H^+. In the process the porphin is raised to its electronically excited state and emits red light. This double electron movement may be compared to that suggested above for semi-quinone luminescence. Whether the radicals involved really become free or whether the porphin molecule merely acts as an electron short-circuiting shunt is unknown; it definitely acts as a catalyst to the peroxide breakdown.

The uncatalysed thermal decomposition of hydroperoxides follows the course: R—O—O—H→R—O— + —O—H. In presence of oxygen, RO_2- radicals are formed. Similarly, in the thermal oxidation of hydrocarbons RH, initially-produced radicals R— react with oxygen to give RO_2–. The characteristic fate of these peroxy radicals is to react bimolecularly to give a ketone, an alcohol, and molecular oxygen, with an energy evolution of over 100 kcal/mole. If fluorescent substances such as 9:10 substituted anthracenes are simultaneously present, the reaction energy may be transferred to them so that they emit (chemi-) fluorescence. Since substitution by bromine enhances the emission one possibility is that ketone molecules in their triplet state (which is very rapidly formed from any excited singlet states) transfer energy on collision to the anthracene molecules, and bring them into excited singlet states because the heavy-atom effect of bromine allows of violation of the normal spin rules [9]. In the absence of added fluorescent substances, a weaker red luminescence is observed which is now recognized as coming from the molecular oxygen also liberated in an electronically excited state. Oxygen has the peculiarity of having a triplet structure, with two unpaired electrons, for its ground state, spectroscopically designated $^3\Sigma_g^-$. At 22·2 and 37·5 kcal/mole respectively above the ground state are two singlet states, $^1\Delta$ and $^1\Sigma_g^+$, and in certain reactions O_2 may be liberated in these states. This red luminescence, a narrow band at 634 nm, cannot be attributed to radiational transitions from either of these two states to ground because of their very low theoretical probability, and because they would emit at longer wavelengths. It now appears that long-lived $^1\Delta$ states collide in pairs and drop to two ground state molecules by an *allowed* transition (singlet : singlet→triplet + triplet), emitting *one* light quantum of energy $2 \times 22 \cdot 2$ kcal/mole, and of the observed wavelength [5]. This red emission is found not only in organic peroxy radical interactions, but also in certain inorganic reactions such as that between hypochlorite and hydrogen peroxide. It does not seem to occur in ordinary catalysed hydrogen peroxide decompositions, however violent. Excited oxygen double molecules may in fact be the intermediates in transferring energy to anthracene molecules in hydroperoxide decomposition.

When oxygen is strictly excluded from organic peroxide decompositions, so that no RO_2^- radicals are formed, chemiluminescence is either very weak, or absent. An interesting case is that of benzoyl peroxide in deoxygenated benzene solutions at 80°C. The major, homogenous, breakdown reaction gives no emission, but the vessel walls, especially if previously treated with alkali, give rise to a base-catalysed side-reaction which liberates oxygen in an excited state and showing the red luminescence [10].

It has long been known that luminescence is sometimes observable when oxidation of organic compounds with hydrogen peroxide, per-acids or ozone occurs in presence of fluorescent substances, even if the fluorescent substance itself is the reaction [12]. Evidence is here accumulating that the fluorescent molecules are excited by energy transfer from dimers of molecular oxygen [13]. $[O_2 : O_2]$ complexes in the states $\Delta_g : \Delta_g$, Δ_g : $^1\Sigma_g^+$ and $^1\Sigma_g^+ : {}^1\Sigma_g^+$ have energy contents of 44·5, 59·8 and 75·1 kcal/mole for each quantum of vibration, and these complexes are therefore potentially capable of inducing fluorescence in all parts of the visible spectrum. Many examples of chemiluminescent reactions where the reactant, a resultant, or some added fluorescent substance emits seem almost certainly explicable by such electronic energy transfer, and there remains a wide unexplored field relating to the yields produced by the various oxygen-dimer states for different reaction mixtures. The autoxidation of 9:10 dihydroxyanthracenes, the high temperature interaction of hydrogen peroxide with aromatic hydrocarbon solutions, and the thermal dissociation of trans-annular peroxides of anthracenes all seem to have oxygen-induced chemiluminescence effects. The last-named reaction is interesting because the photochemical *formation* of such peroxides requires singlet excited states of the oxygen molecule; the dissociation back into singlet-state products is therefore not surprising [14, 15]. The ion O_2^- undoubtedly plays a part in some examples of chemiluminescence; by restoring an electron to semi-quinones it may form hydroxy-substances in their electronically excited states. A reaction between a triplet organic molecule and singlet O_2 to give ground state O_2 and a fluorescing molecule is a possibility, but a difficult one to explore.

The autoxidation by air of tetrakis-(dimethylamino)-ethylene in solution can, under favourable solvent conditions, cause the emission of strong luminescence identical with the fluorescence of the parent compound. For demonstrations or practical applications this is probably the simplest chemiluminescent material; when the solution is smeared on to a surface exposed to air, a strong green glow persists for minutes. However, the chemical nature of the oxidation process is complex, with numerous final products. A mechanism has been proposed involving the formation of diradicals:

$$(CH_3)_2N{-}\overset{|}{\underset{|}{C}}{-}N(CH_3)_2$$

as well as other intermediates. The luminescence has been interpreted as the emission by excited molecules of the parent substance which are formed by the bimolecular association of some of these radicals [6, 7].

Oxalyl chloride and hydrogen peroxide form a chemiluminescent system if a suitable fluorescent additive, as 9:10-diphenylanthracene, is present. The effect arises from the breakdown of peroxyoxalic acid [8]. The reaction is believed to be of a free radical chain nature:

$$HO{-}\underset{\underset{O}{\|}}{C}{-}\underset{\underset{O}{\|}}{C}{-}OOH + R{-} \rightarrow 2CO_2 + RH + HO{-}$$

Simple decomposition to $2CO_2 + H_2O$ by a direct unimolecular process does not seem to occur. Three bonds are simultaneously cleaved in the molecular breakdown, and this is sufficient, and necessary, to accumulate energy for radiation in the blue region. A mechanism recently put forward for a similar type of reaction is shown opposite [18].

When aromatic hydrocarbons are electrolysed in dimethyl sulphoxide solution luminescence is due to anodic oxidation of anions: $A + e \rightarrow A^-$; $A^- + ox \rightarrow A^* \rightarrow h\nu$ [17]. A more complex system is dibenzanthrone in chloroform, treated with chlorine, to give the cation, and then with hydrogen peroxide and more chlorine; this emits the characteristic spectrum of dibenzanthrone itself. Here again, oxidation combined with reduction underlies the effect. A more understandable reaction is that between the anion of 9:10-diphenyl anthracene $(A)^-$ and 9:10 dihydrodichloro-diphenyl anthracene (ACl_2):

$$A^- + ACl_2 \rightarrow A + ACl{-} + Cl^-$$
$$A^- + ACl{-} \rightarrow 2A + Cl^-$$

$$>C(H)-C(=O)OR + \text{Base} \longrightarrow >C^{-}-C(=O)OR$$

$$\xrightarrow{+O_2}$$

$$>C(O-O^{-})-C(=O)OR \longrightarrow >C(-O-O-)C=O + RO^{-}$$

$$\longrightarrow CO_2 + >C=O^{*} \longrightarrow h\nu$$

One of the hydrocarbon molecules formed in the second stage is electronically excited and luminesces; alternatively the emission may be from the excimer AA* [11].

Some very weak luminescent effects of a very different nature have been observed when molten organic substances are allowed to crystallize; these are simply due to thermal decomposition processes giving radicals which oxidize in the air to produce the glow. The very high sensitivity of modern photo-multipliers has greatly increased the field of observation of weak emission effects, so that it is often difficult to achieve a satisfactorily high standard of purity of materials and cleanliness of apparatus to avoid spurious results. For example, when heated to 80°C in air, a glass surface contaminated by invisible fingerprints emits detectable chemiluminescence from an autoxidation reaction. Measurements at this level tend to outrun both preparative techniques and reliable interpretation of results.

REFERENCES

[1] Epstein, B. and Kuwana, J. *Photochem. and Photobiol.* **4**, 1157 (1965).

[2] Philbrook, G. E., Ayers, J. B., Garst, J. F. and Trotter, J. R. *Photochem. and Photobiol.* **4**, 869 (1965).

[3] McCapra, F., Richardson, D. G. and Chang, Y. C. *Photochem. and Photobiol.* **4**, 1111 (1965).

[4] Totter, J. R. and Philbrook, G. E. *Photochem. and Photobiol.* **5**, 177 (1966).

[5] ARNOLD, J. S., BROWNE, R. J. and OGRYZLO, E. A. *Photochem. and Photobiol.* **4**, 963 (1965).
[6] PARIS, J. P. *Photochem. and Photobiol.* **4**, 1059 (1965).
[7] URRY, W. H. and SHEETO, J. *Photochem. and Photobiol.* **4**, 1067 (1965).
[8] RAUHUT, M. M., SHEENAN, D., CLARKE, R. A. and SEMSEL, A. M. *Photochem. and Photobiol.* **4**, 1097 (1965).
[9] VASSIL'EV, R. F. *Progress in Reaction Kinetics*, Vol. 4, p. 307, Pergamon Press, Oxford (1967).
[10] BOWEN, E. J. and LLOYD, R. A. *Proc. R. Soc. Lond.* **A 275**, 465 (1963).
[11] CHANDROSS, E. A. and SONNTAG, F. I. *J. Am. chem. Soc.* **88**, 1089 (1966).
[12] MALLET, L. *Compt. Rend. Acad. Sci. Paris*, **185**, 352 (1927).
[13] KHAN, A. V. and KASHA, M. *J. Am. chem. Soc.* **88**, 1574 (1966).
[14] FOOTE, C. S. and WEXLER, S. *J. Am. chem. Soc.* **86**, 3879 (1964).
[15] BOWEN, E. J. *Chem. in Britain*, **2**, 249 (1966).
[16] WHITE, E. H. *J. Am. chem. Soc.* **86**, 940 (1964).
[17] HERCULES, D. M., LANSBURY, R. C. and ROE, D. K. *J. Am. chem. Soc.* **88**, 4578 (1966).
[18] MCCAPRA, F. *Q. Rev.* **20**, 485 (1966); *Chem. Comm.* 522 (1966), 1011 (1967).

Chapter 10

LUMINESCENCE IN BIOCHEMISTRY

G. K. RADDA and G. H. DODD

Introduction

THE basic light emission processes in biochemistry, and their mechanisms, are essentially similar to those of simpler chemical systems, that is, emission can occur from singlet or triplet excited states populated by introducing the energy in the form of light or chemical reactions. Therefore, in many cases, our discussions of phenomena may be regarded simply as extensions of previous chapters. This is certainly true when fluorescence and phosphorescence are used as analytical tools in estimating the concentration of biochemically important molecules or in following their rates of production or their disappearance. We shall not concern ourselves with these applications. Overriding in our treatment is the idea that the most fundamental difference between biological and chemical phenomena is the unique organization of chemical and physical processes in the cell. For instance, we know that the primary processes in photosynthesis require the gradually-evolved organized structure of chloroplast. Conversion of the energy liberated in the oxidation of foodstuff into the storable and transportable form of hydrolytic energy of adenosine triphosphate (ATP) is only effective within the highly specific organization of mitochondria. Similarly light energy in visual processes is converted into a nerve-impulse through physical and chemical reactions occurring within the intricate structure of the retina. Even the efficient catalysis of chemical reactions by isolated proteins (enzymes), or the duplication of nucleic acids, involves the specific three-dimensional arrangement of these macromolecules.

Most biochemical systems, then, may be regarded as periodic or aperiodic 'solids' or, in any event, as heterogeneous phases. We shall therefore concentrate on those apsects of luminescence which are characteristic of these. We shall have to discuss some simpler systems for comparison. It will be seen that the complexities of the biochemical phenomena can direct our chemical studies away from the chemistry of dilute solutions towards systems where intermolecular interactions, energy transfer and heterogeneous reactions occur. It is not surprising, for instance, that the interest in organic semiconductors was much stimulated by the possibility of such conduction in the primary light conversion act

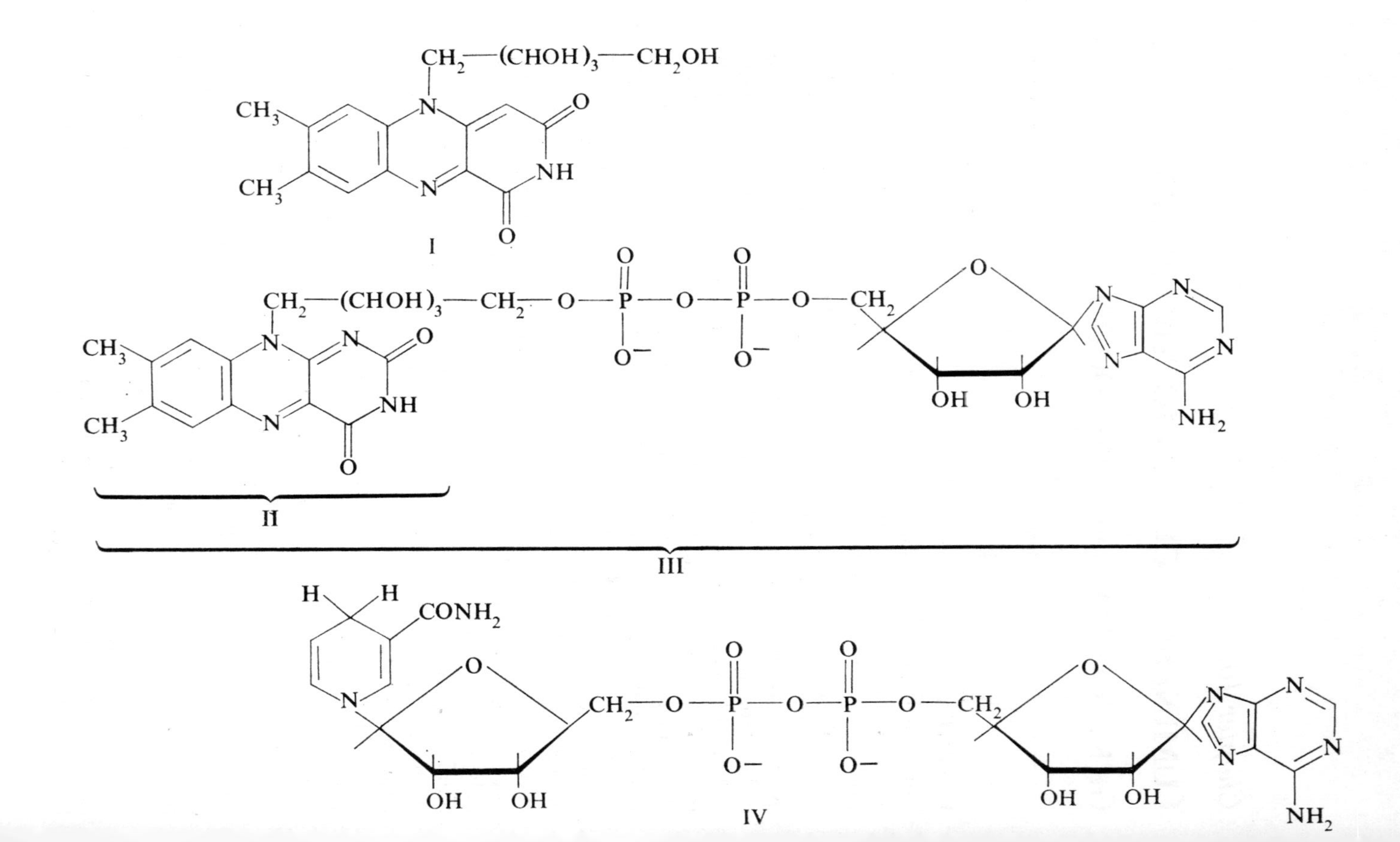
I
II
III
IV

in photosynthetic organisms, or that the development of X-ray crystallographic methods was so intricately linked with the realization that an understanding of the molecular structure of biopolymers can contribute immensely to the understanding of life in molecular terms.

Fluorescence and Phosphorescence of Small Molecules

Most biochemically important 'small' molecules are fairly complex organic compounds such as have been described in Chapter 3. We shall therefore only discuss a few examples which are relevant to our later sections, or which exhibit properties that are of particular interest to the biochemist. The compounds we have chosen are the redox coenzymes derived from flavins, nicotinamide, the photopigments chlorophyll and related compounds and vitamin A.

REDOX COENZYMES

Riboflavin (I), flavin mononucleotide (FMN, II), flavinadenine dinucleotide (FAD, III) and reduced nicotinamide-adenine dinucleotide (NADH, IV) are involved in many biochemical redox reactions as coenzymes. Flavins also participate in many photobiological processes such as photosynthesis, bacterial bioluminescence and, possibly, vision and phototropism.

Flavins in the oxidized form (I–III) show an intense fluorescence with $\lambda_{max} = 530$ nm. Riboflavin has a quantum yield of fluorescence of 0·25 in water, 0·52 in 90% dioxane/water and 0·71 in dioxane. The fluorescence of flavin nucleotides is weaker than that of riboflavin, FAD being considerably weaker. Only the uncharged forms of flavins are fluorescent; their anions and cations are not. Since their fluorescence is almost always quenched when they are bound to proteins in flavoenzymes, quenching by a variety of substances has been studied. Two types of quenching mechanism are recognized: collisional quenching, which causes a decrease in the lifetime of the excited state, and static quenching, brought about by the formation of a non-fluorescent complex in the ground state, when there is no change in the lifetime. Electrolytes (e.g., KI, $NaHSO_3$), metal ions (e.g., Fe^{3+}, Ni^{2+}, Cu^{2+}) and thiols (e.g., dihydrolipoic acid, thioglycolic acid) act as collisional quenchers. Many aromatic and heterocyclic compounds (e.g., adenine, naphthalene–2–sulphonic acid, tryptophan, tyrosine) quench by the static mechanism. This mechanism also explains the low fluorescence yield in FAD. In this case free rotations in the connecting sugar chain allow a conformation where the chromophoric groups come into close contact. This suggestion is supported by the observation of inductive resonance transfer between them. Similar observations have been made for NADH in which the photon energy absorbed by adenine is transferred with high efficiency to the nicotinamide part and is emitted as its fluorescence (V). These intramolecular complexes are formed only in aqueous solvents. This is clearly illustrated by comparing the fluorescence of FMN or riboflavin with that of FAD in mixed solvents (Fig. 10.1), and

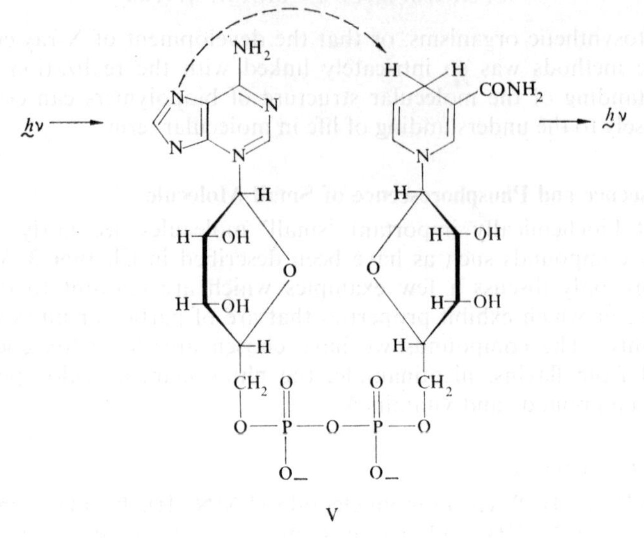

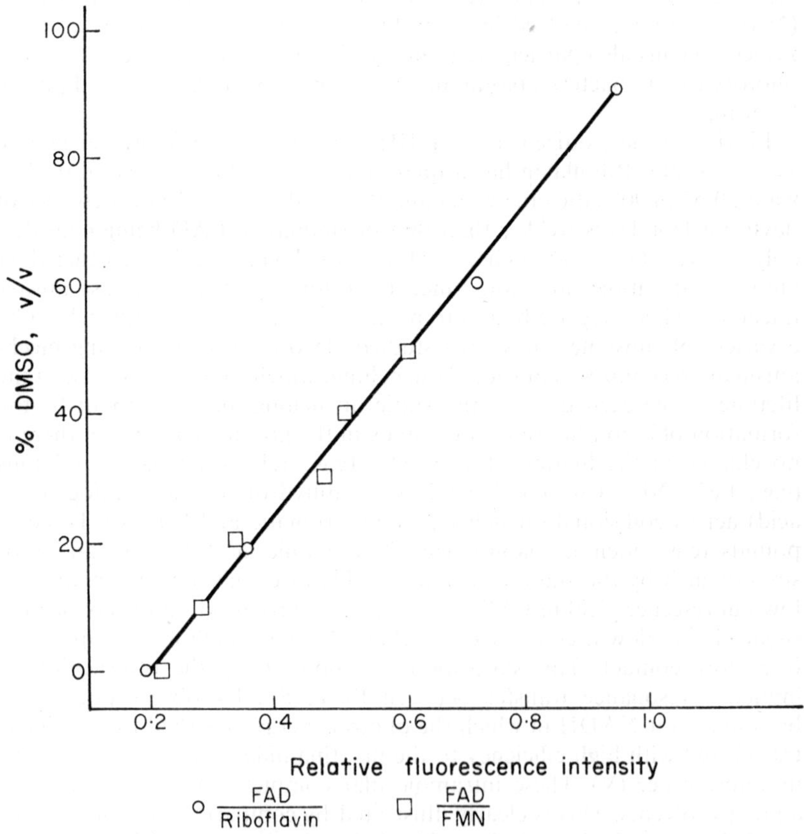

FIG. 10.1. Influence of DMSO concentration on FAD fluorescence.

indicates a gradual unfolding of the molecule as the water content of the solvent decreases.

A variety of other organic solvents have similar effects on the fluorescence of both FAD and NADH. The conformation of these coenzymes is probably highly relevant to their mode of action in stereospecific enzymic reactions.

Singlet-singlet energy transfer between different flavin derivatives has also been demonstrated. Since substituents in the flavin ring can affect the position of absorption and emission maxima (Table 10.1) the condition of favourable overlap can be satisfied.

TABLE 10.1. *Effect of substituents on the spectral properties of flavin derivatives*

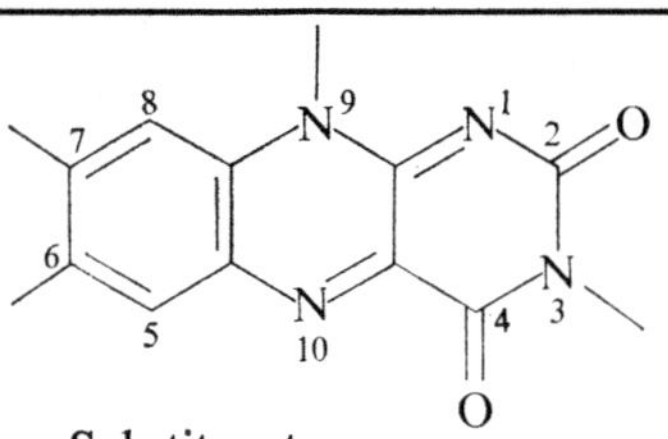

Compound No.	Substituents 1	2	3	4	5	6	7	9	λ_{max} absorption (nm)	λ_{max} emission (nm)
I	—	=O	H	=O	H	H	H	Me	348,433	520
II	—	=O	H	=O	H	Me	H	$(CH_2)_2NEt_2$	356,447	530
III	—	=O	H	=O	H	Me	Me	ribityl	375,447	530
IV	—	=O	H	=O	H	Cl	H	$(CH_2)_3NEt_2$	338,446	525
V	—	=O	H	=O	H	H	Cl	$(CH_2)_3NEt_2$	361,431	510
VI	—	=O	H	=O	H	Cl	OMe	$(CH_2)_2$—N(pyrrolidinyl ring)	445	505
VII	—	=O	—	—N(morpholine ring)O	H	Me	Me	Me	470	540
VIII	—	=S	H	=O	H	Me	Me	Me	364,490	545

For instance, the fluorescence emission of flavin VI is completely quenched by VII (Fig. 10.2) at a ratio of VII:VI = 20:80 when these two flavins are included in a methyl cellulose matrix in a transparent film. The calculated critical transfer distance (by Förster's formula) in this case is 2·1 nm. It is likely that in these films the flavins are randomly oriented in capillary channels of the methyl cellulose film, but are essentially in an aqueous environment, thus retaining their rotational freedom. This last point is derived from polarization of fluorescence measurements which shows that the microscopic viscosity of the chromophore environment is considerably less than the bulk viscosity of the film. This is the kind of model one might have to use to understand the arrangement of pigments in natural systems.

Flavins undergo many photochemical reactions and can act as sensitizers via their triplet excited states. Many of them exhibit pronounced phosphorescence at low temperatures and these triplets are directly observable by ESR in the forbidden $\Delta m = 2$ transition. (Table 10.2.)

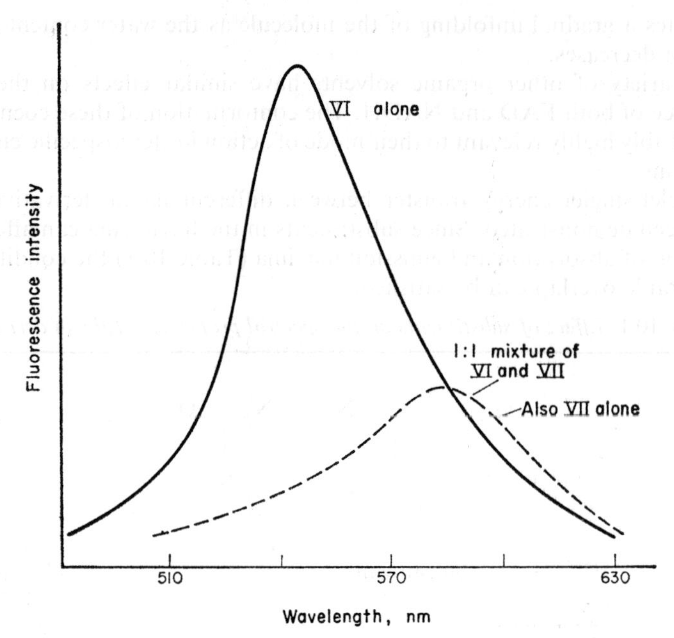

FIG. 10.2 Fluorescence of flavins in methyl cellulose films.

TABLE 10.2. *Phosphorescence and ESR of flavin triplets*

Compound	Solvent	H_0 (gauss)	D^*/hc (cm^{-1})	λ_{max} (nm)
Riboflavin	EtOH	1602	0·062	605
Tetra-acetylriboflavin	EtOH	1603	0·061	610
0–(2)–Methyl-lumi-flavin	EtOH	1596	0·066	598
2-Thiolumiflavin	EtOH	1596	0·066	585
7–Morpholino-7-nor-lumiflavin	EtOH—DMF	1585	0·072	620
FMN neutral	Propylene glycol—H_2O	1602	0·062	608
anionic	Do. with NaOH	1602	0·062	608
cationic	Do. with HCl	1582	0·074	555

D* zero-field splitting parameter.

A detailed analysis of the ESR spectrum gives the electron distribution in the triplet. (See reference in the reading list.)

FLUORESCENCE OF CHLOROPHYLL AND DERIVATIVES

Because of the unique function of chlorophyll in photosynthesis a great deal of detailed information on its fluorescence both *in vitro* and *in vivo* is available.

The overlap between the absorption and emission bands of chlorophyll is greater than it is for many other pigments. (λ_{max} absorption = 660 nm, λ_{max} emission = 664 nm). The fluorescence quantum yields and lifetimes for chlorophyll a and chlorophyll b in several solvents are summarized in Table 10.3.

TABLE 10.3. *Fluorescence characteristics of chlorophylls a and b*

Compound	Solvent	Fluorescent quantum yield	lifetime ($\tau \times 10^9$ s)
Chlorophyll a	Benzene (wet)	0·33	7·8
,,	Ethyl ether	0·32	5·1
,,	Ethanol	0·23	5·0
,,	Cyclohexanol	0·30	6·5
Chlorophyll b	Benzene	0·11	6·3
,,	Ethyl ether	0·12	3·9
,,	Ethanol	0·10	3·4

One of the most striking properties of chlorophyll is that it is strongly fluorescent when solvated, and is non-fluorescent in the unsolvated state. In carefully dried hydrocarbons the quantum yield of chlorophyll drops below 0·01. These solvent effects are also reflected in the absorption spectra of all of the metal-complexed chlorins and porphyrins, but chlorophyll is the only one to exhibit these unique fluorescence properties. The solvent effect on its fluorescence can be rationalized on the basis of the relative energies of the lowest nπ^* and $\pi\pi^*$ states (Fig. 10.3). When the $\pi\pi^*$ level is lower than the $n\pi^*$, direct radiative return from the $\pi\pi^*$ state to ground state competes efficiently with other paths (e.g., conversion to $\pi\pi^*$ triplet) fluorescence is observed. The radiationless de-excitation via the $n\pi^*$ triplet state on the other hand is more favoured when the $n\pi^*$ state is lowest.

The fluorescence of chlorophyll in wet benzene involves a specific association between chlorophyll and water molecules, and this complexing occurs only in the presence of the central magnesium atom. More generally chlorophyll forms addition compounds with bases (which include water and ethanol), to give a fluorescent molecule.

Fluorescence of Chlorophyll Aggregates and Oriented Chlorophyll Molecules

Studies of chlorophyll aggregates (e.g., colloids, crystals) and oriented monolayers are of particular interest since such aggregates are certainly involved in the primary quantum conversion in photosynthesis.

In moderately concentrated solutions of chlorophyll monosolvates, intermolecular energy transfer occurs efficiently, as shown by self-quenching and concentration depolarization, and by the sensitization of the fluorescence of chlorophyll a by chlorophyll b. Self-quenching and

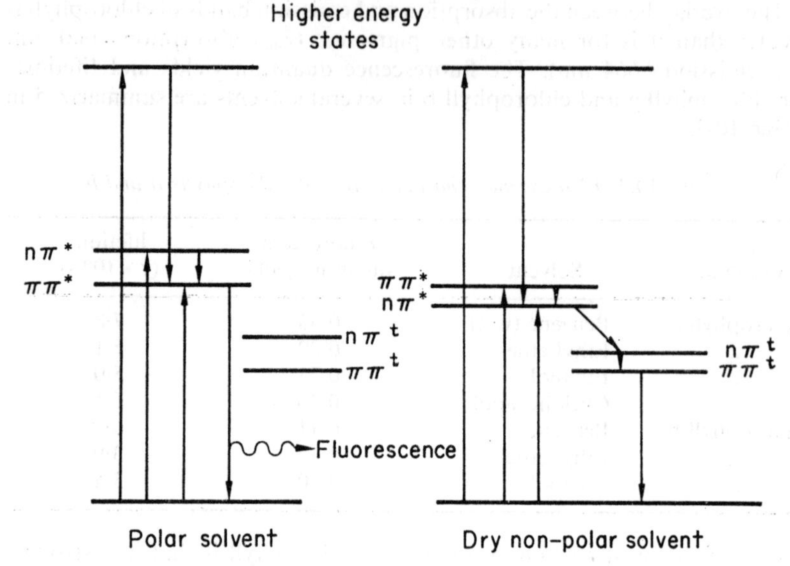

FIG. 10.3 Solvent effect on the energy levels of chlorophyll.

concentration depolarization occur in solutions with concentrations greater than 2×10^{-4} M. The efficiency of energy transfer from chlorophyll a to chlorophyll b in an equimolar mixture is maximal at 10^{-3} M concentrations (about 1/2).

The fluoresence of chlorophyll monolayers spread on water (when decyl alcohol is used to disperse chlorophyll) has been studied in great detail. These monolayers are two-dimensional liquids with no preferred direction in the plane of the films. From the angular distribution of the fluorescence it was found that the transition moment responsible for the red absorption band makes an angle of $<20°$ with the water surface, while that responsible for the blue absorption band describes an angle of $\simeq 20°$. Since both these angles are measured in the porphyrin plane the chlorophyll molecule must be oriented, at the interphase, with its aromatic plane describing an angle with the water surface and presumably with its hydrophobic phytyl side-chain in the organic layer. The energy from chlorophyll a in these monolayers is transferred to added copper pheophytin a, as seen from measurements of quantum yields (Fig. 10.4). In dilute films a single-transfer calculation on the Förster model describes the quenching satisfactorily, the range of chlorophyll-quencher interaction being 4 nm. For undiluted monolayers, where chlorophyll-chlorophyll interaction is more important, the single transfer model is not applicable.

From studies of absorption spectra, electric dichroism and ORD it appears that chlorophyll (or bacteriochlorophyll) exists *in vivo* in a condensed state. The aggregate may be two- or three-dimensional. The major portion of the total chlorophyll is probably a random aggregate, but a

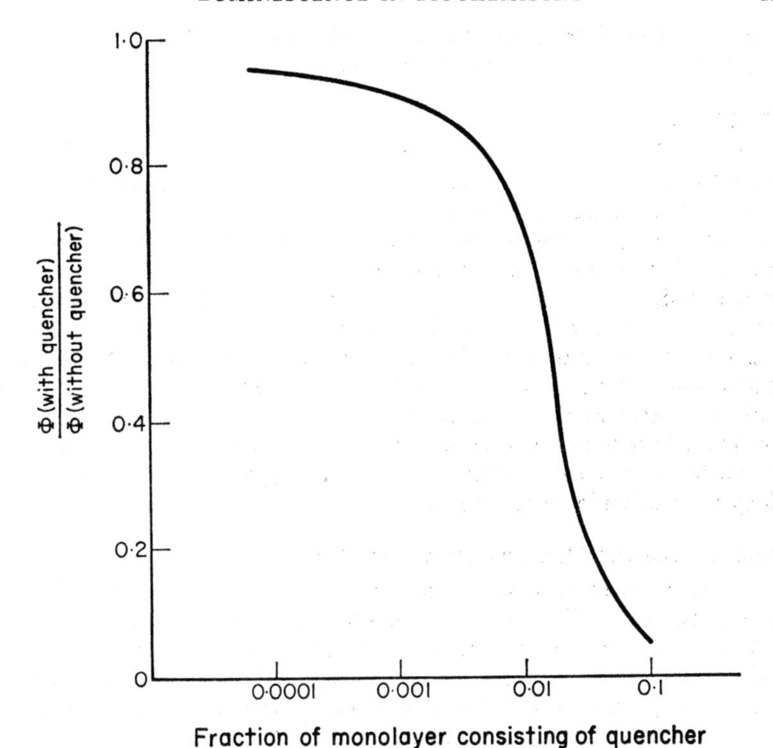

FIG. 10.4 Quenching of chlorophyll a fluorescence by Cu-pheophytin a in monolayers.

minor component absorbing at longer wavelength than the bulk of the chlorophyll is in an ordered arrangement. Fluorescence studies greatly contributed to these conclusions as, for instance, in Euglena chloroplast (oriented in their natural state within the cells) the fluorescence of the main chlorophyll fraction is unpolarized (~680–700 nm) but the fluorescence associated with the 720 nm pigment exhibits strong polarization.

During photosynthesis, fluorescence generally rises when the absorbed light energy cannot all be used for photosynthesis, or when photosynthesis is inhibited. Roughly similar effects are found in the delayed fluorescence observed under many conditions. Delayed fluorescence of chlorophyll in ethanol is mainly of the P-type (triplet-triplet annihilation) while in leaves the interpretation is more complex, and emission may be of either the P-type or the E-type. The quantum yield of fluorescence in living algae is 2% when extrapolated to infinitely dim exciting light, while in somewhat brighter light (but still below the saturating intensity for photosynthesis) the yield rises to 3%. The observed values of fluorescence lifetimes in chloroplast and algae range from $0{\cdot}6 \times 10^{-9}$ to $1{\cdot}6 \times 10^{-9}$ s. An excellent summary of the intricacies of light emission in photosynthetic systems is found in Clayton's book. (See reference list.)

Vitamin A

The highly specific organization of the visual pigment rhodopsin (a Schiff base composed of the protein opsin and vitamin A aldehyde, retinal) in the retina stimulated interest in the possibility of electronic energy transfer in visual processes. Rhodopsin fluorescence has not been observed, but retinal in solution has a radiative lifetime of 0·73 mμ s and a critical transfer radius of 1·7 nm in a medium of refractive index 1·3. If molecules are randomly distributed, their fluorescence should become 50% depolarized at 0·1 M concentration. This is apparently an optimistic estimate since appreciable polarization (at $-100°C$) is observed even at 0·5 M concentrations. Although the spectral properties of rhodopsin are different (a considerable red-shift is observed on binding retinal to the protein, opsin) the inefficiency of energy transfer for retinal suggests that it is not a very likely mechanism in visual systems.

Luminescence and Macromolecular Structure

Although different kinds of polymeric materials form part of living organisms, we shall almost exclusively limit ourselves to fluorescence and phosphorescence properties which are associated with polymers of amino acids and nucleotides. The chemical composition of these polymers and the sequence of the building blocks within them are collectively referred to as the primary structure, and on the whole will not directly concern us here. The detailed spatial organization (which is believed to be sharply defined by the primary structure) is generally discussed in terms of regular stereochemical repeating units within the molecule (secondary structure), and apparently irregular but specific residual structure largely stabilized, for instance, by the interaction of amino acid side-chains in proteins (tertiary structure). The most common regular structure involves helical forms such as the α-helix in proteins and the doubly stranded helix of deoxyribonucleic acid (DNA) held together by hydrogen bonding of complementary base-pairs (see Fig. 10.5). When we speak of conformational changes we refer to changes of either secondary or tertiary structure or both. By denaturation we understand the complete disruption of these, to give a completely random coil.

Fluorescence of Amino Acids

Of the twenty or so amino acids found in proteins, only three have chromophores that absorb and emit light in a convenient spectral range. The spectral and fluorescent properties of tryptophan, tyrosine and phenylalanine are summarized in Table 10.4. Attachment of the amino acid side-chain does not greatly alter the spectral characteristics of the parent compounds: indole, phenol and benzene.

The quantum yields of tyrosine and tryptophan fluorescence in proteins and polypeptides are considerably lower than those of the free amino

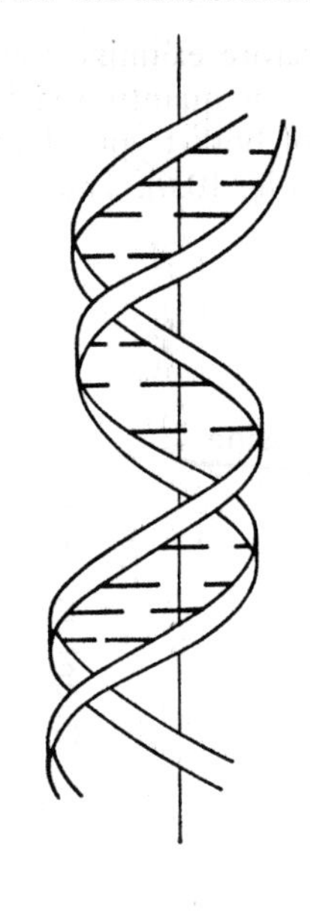

Protein α-helix

DNA double helix
(the lines perpendicular to the helix axis represent the nucleotide bases; the ribbon represents the phosphate + sugar chains)

FIG. 10.5 Schematic diagrams for protein and DNA helices.

TABLE 10.4. *Absorption and emission spectra of aromatic amino acids*

	Absorption		Emission	
Amino acid	λ_{max} (nm)	ε (litre/mole^{-1} cm^{-1})	λ_{max} (nm)	Quantum yield
	257	200		
	206	9,000	282	0·035
	187	58,000		
(Phenylalanine)				
	275	1,200		
	222	8,000	303	0·21
	192	47,000		
(Tyrosine)				
	280	5,500		
	220	32,000	350	0·20
	196	21,000		
(Tryptophan)				

acids. We shall therefore examine the various quenching processes that may account for the low quantum yields.

The fluorescence of both tryptophan and tyrosine is decreased in acid and alkaline media (Fig. 10.6). The —COOH and —NH_3^+ groups of the

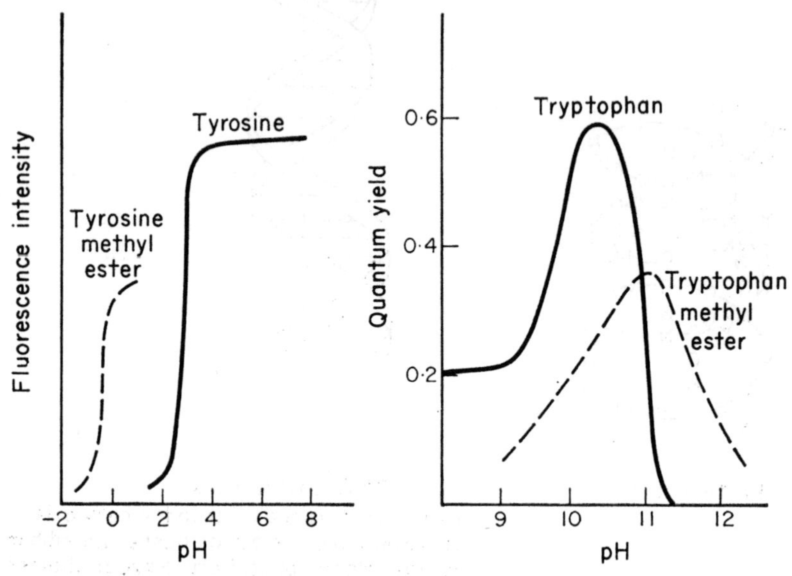

FIG. 10.6 pH dependence of fluorescence of tryptophan and tyrosine.

amino acids are also a potential source of quencher protons. This is shown by the fact that both tyrosine and tryptophan are half quenched at pH 2·3, the pK_a of the carboxyl group while their esters are quenched only at pH −0·55 and pH 0·5 in the two derivatives respectively. These quenching mechanisms probably involve a proton transfer to the excited ring. The sensitivity of tyrosine fluorescence to carboxylic acids in both the anionic and un-ionized forms should be compared to similar effects with phenol.

VI

(See Chapter 3.) Similar quenching processes by the —COOH and $—NH_3^+$ groups in glycyl peptides of both tyrosine and tryptophan may account for their low quantum yields (~0·04) which are increased to almost the normal value of 0·2 in a medium of high viscosity (such as propylene glycol at low temperatures), and this indicates that the short-range interaction between the rings and the relatively distant —COOH and $—NH_3^+$ groups requires a rotational diffusion of the side-chain to contact the aromatic ring.

The fluorescence of tyrosine, in contrast to that of tryptophan, is relatively insensitive to the polarity of the solvent. We can illustrate solvent effects on tryptophan by using acetyl tryptophan amide (VI), since this is a more realistic model for a tryptophan residue which is incorporated in a polypeptide chain. The fluorescence emission spectrum of this molecule is similar to that of tryptophan, and its intensity of fluorescence is susceptible to relatively small changes in the solvent (Fig. 10.7). Additives which

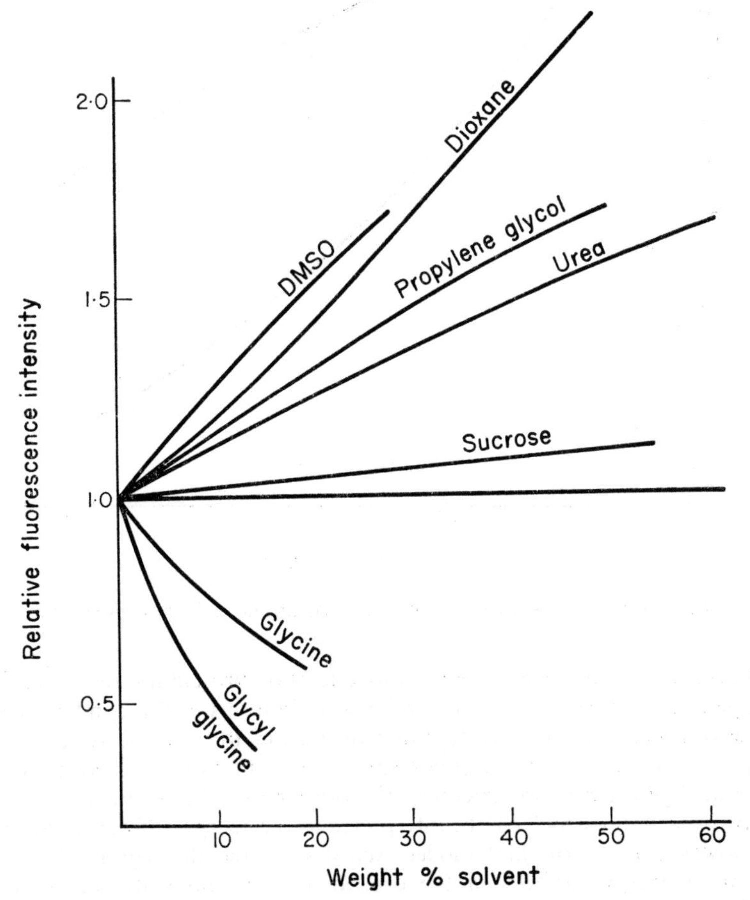

FIG. 10.7 Solvent effect on the fluorescence intensity of acetyl tryptophan amide.

decrease the dielectric constant of water result in enhancement of fluorescence, while increase in dielectric constant has the opposite effect. Urea, a common protein denaturing agent, enhances fluorescence despite the fact that its dielectric increment is positive. Some direct interaction may be involved here.

The intensity of the amino acid fluorescence is also significantly temperature-dependent (Fig. 10.8). This is not an unusual observation, but is

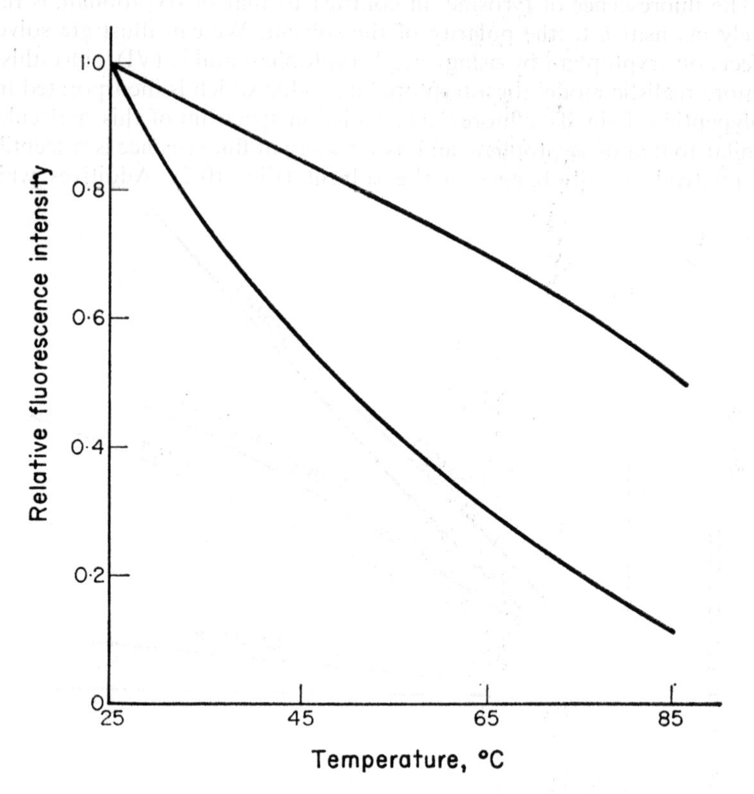

FIG. 10.8 Temperature dependence of amino acid fluorescence.

important for interpreting temperature effects on protein fluorescence. The temperature dependence can be understood in terms of three independent processes which contribute to the return from the excited state to the ground state. Two of these (fluorescence and one non-radiative process) have negligible activation energies. The other non-radiative process has an activation energy which can be interpreted as the energy difference between the lowest level of the first singlet excited state and the region where the potential energy surfaces of the excited and ground states cross. This activation energy, then, can be calculated, and is 8·1 kcal/mole for tryptophan and 7·1 kcal/mole for tyrosine.

FLUORESCENCE IN POLYPEPTIDES

Before discussing the modification of tyrosine and tryptophan fluorescence by incorporation into proteins, we shall briefly examine their fluorescence properties in simpler homo- and copolymeric peptides. They have often been used as models for interpreting other optical properties (e.g., absorption spectra, ORD) of proteins.

The fluorescence properties of poly-L-tyrosine and poly-L-tryptophan are significantly different from those of the monomers, not only in terms of quantum yields, but also in the shape of the emission bands. Thus poly-L-tyrosine in dimethylformamide exhibits a new emission band centred around 420 nm. This band is less pronounced in dimethylsulphoxide (Fig. 10.9). Similarly, poly-L-tryptophan exhibits a new emission band at

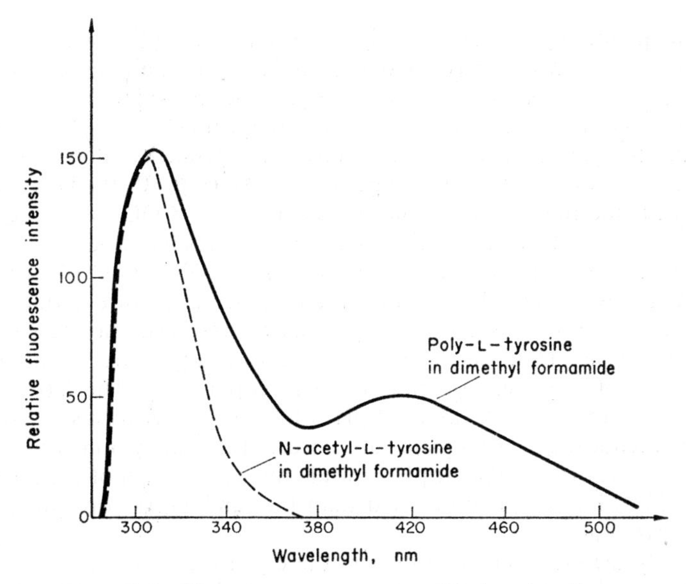

FIG. 10.9 Fluorescence spectrum of poly-L-tyrosine.

470 nm in dimethylformamide. In dimethylformamide both polymers are largely in the form of an α-helix (this can be shown by ORD), while their helical content is much reduced in dimethylsulphoxide. It is therefore likely that in the helical polymers the excited state is delocalized over more than one aromatic residue in the same way as in crystals of aromatic compounds, and that the new fluorescence emission occurs from this excimer state. Clearly fluorescence in this case is a sensitive method of following the conversion of the helix into the random coil form.

In copolymers of L-tyrosine with L-tryptophan tyrosine fluorescence is quenched by increasing amounts of tryptophan, while tryptophan fluorescence is sensitized by tyrosine. This is a good indication that energy transfer from tyrosine to tryptophan can occur under favourable conditions, and we shall examine the implications of this for proteins.

The quenching mechanisms already discussed are invaluable for interpreting the effect of pH on the fluorescence of copolymers of tyrosine with glutamic acid (VII) and with lysine (VIII), and of O-methylated tyrosine with the same amino acids.

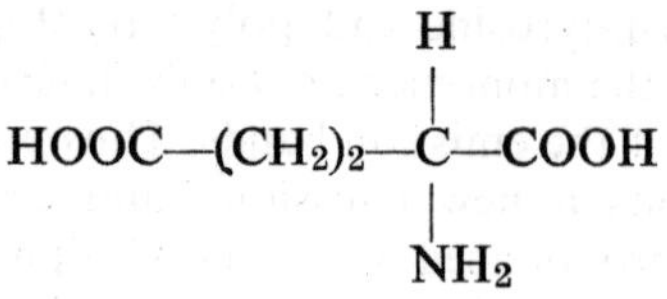

$$HOOC—(CH_2)_2—\underset{NH_2}{\overset{H}{C}}—COOH \qquad H_2N—(CH_2)_4—\underset{NH_2}{\overset{H}{C}}—COOH$$

(VII) 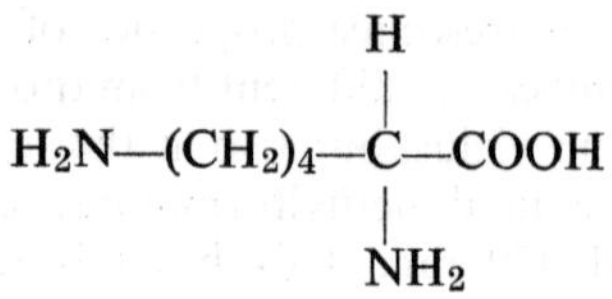 (VIII)

The quantum yield for a copolymer of L-tyrosine (4%) with L-glutamic acid (96%) is constant between pH 6·0 and 9·0 (where all the acid groups of glutamic acid are dissociated) but it approaches the value for free tyrosine at low pH, which indicates that the polypeptide chain *per se* has no significant quenching effect. In contrast the O-methyl derivative has a higher quantum yield in the range pH 6·0–9·0 (Fig. 10.10). This is consistent with the interpretation that quenching by $—COO^-$ ions requires a proton transfer from the excited state of the phenolic groups.

The sudden increase in fluorescence in both polymers at pH 4 must be associated with the helix-coil transition. A clearer picture about the role of polypeptide conformation in tyrosine fluorescence may be gained by comparing the pH dependence of copolymers of L-tyrosine-L-glutamic acid with that of L-tyrosine-DL-glutamic acid copolymers. The latter is in a random coil form at all values of pH, while the former forms an α-helix at pH 3. Identical quantum yields are observed for the two polymers in the random form ($\Phi = 0{\cdot}038$ at pH 7) while the helical form of the L-polymer has a 25% higher quantum yield than the random DL-copolymer at the same pH.

In L-lysine L-tyrosine copolymers there is a rapid increase in quantum yield around the pK_a of the amino group (which is absent in the O-methyl tyrosine copolymer) so that proton transfer to $—NH_2$ is again responsible for quenching of the phenolic fluorescence.

Fluorescence of Proteins

The quantum yields of fluorescence of proteins are very low, in the range 0·008 to 0·07, and this may be due to quenching processes of the type described for polypeptides. Simple proteins can be divided into two groups, on the basis of their fluorescence properties.

Proteins which only contain phenylalanine and tyrosine (of the aromatic amino acids) exhibit fluorescence spectra indistinguishable from that of free tyrosine, with $\lambda_{max} = 303$ nm. This reflects the insensitivity of tyrosine fluorescence to the polarity of the environment—insensitivity which is due

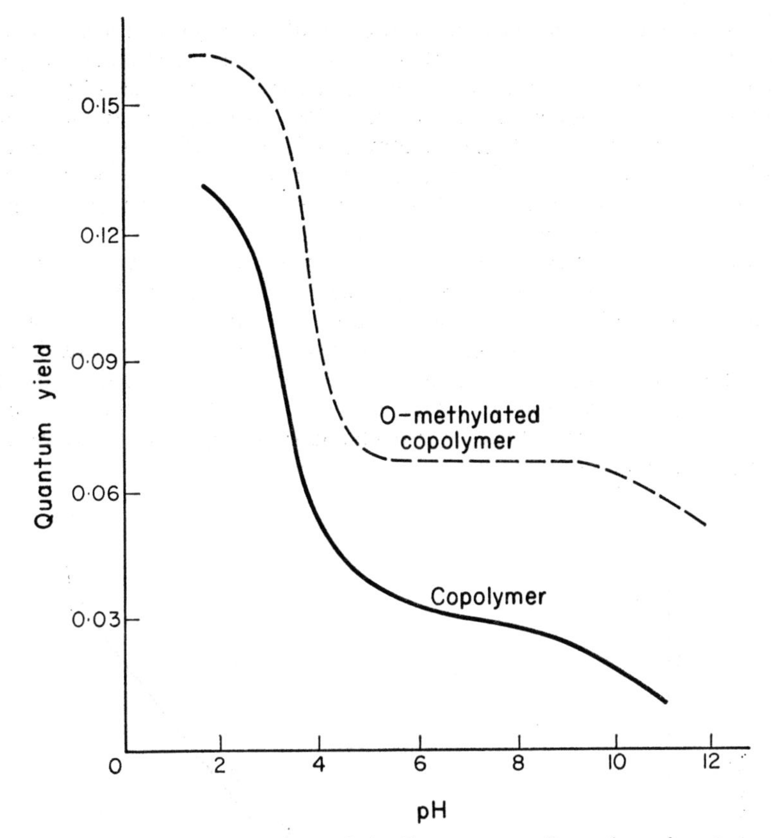

FIG. 10.10 pH dependence of the fluorescence of copoly-L-glu-L-tyr.

to the low polarity of its excited state. Proteins in this class include insulin, tropomyosin, zein and ribonuclease.

Proteins containing both tyrosine and tryptophan show a fluorescence spectrum which is characteristic of tryptophan and tryptophan alone. (Most globular proteins belong to this class.) The maxima of emission depend on the protein. For instance λ_{max} for chymotrypsin is 332 nm, and for pepsin 346 nm, and it is likely that, in pepsin, tryptophan residues are in contact with water while in chymotrypsin they are largely buried in the hydrophobic interior of the protein. The lack of tyrosine fluorescence in these proteins is independent of the protein conformation, as tyrosine emission is not observed in both native and denatured forms. The reasons for this are still unexplained.

The Influence of Conformation and Environment on the Fluorescence of Proteins

A knowledge of the detailed stereochemistry of proteins in solution is not only of intrinsic chemical interest, but is also a major factor in understanding the mechanism of enzyme catalysis. No single technique has yet been devised which enables us to unravel all the complexities of protein conformation in solution, as X-ray diffraction has been used to elucidate the stereochemistry of crystalline proteins. It is therefore important to realize that the value of fluorescence spectroscopy in studying protein structure is greatly enhanced when combined with other physical (e.g., spectroscopy, ORD) and chemical studies.

Two approaches have been used in evaluating the correlation between the fluorescence properties of proteins and their structure. In the first, generally referred to as 'solvent perturbation', the action on protein fluorescence is observed of relatively inert solvents which are unlikely to alter the conformation of proteins. Alternatively by bringing about structural transformations in the protein in a predictable direction (e.g., denaturation by urea, heat or pH-changes), changes in fluorescence emission are recorded.

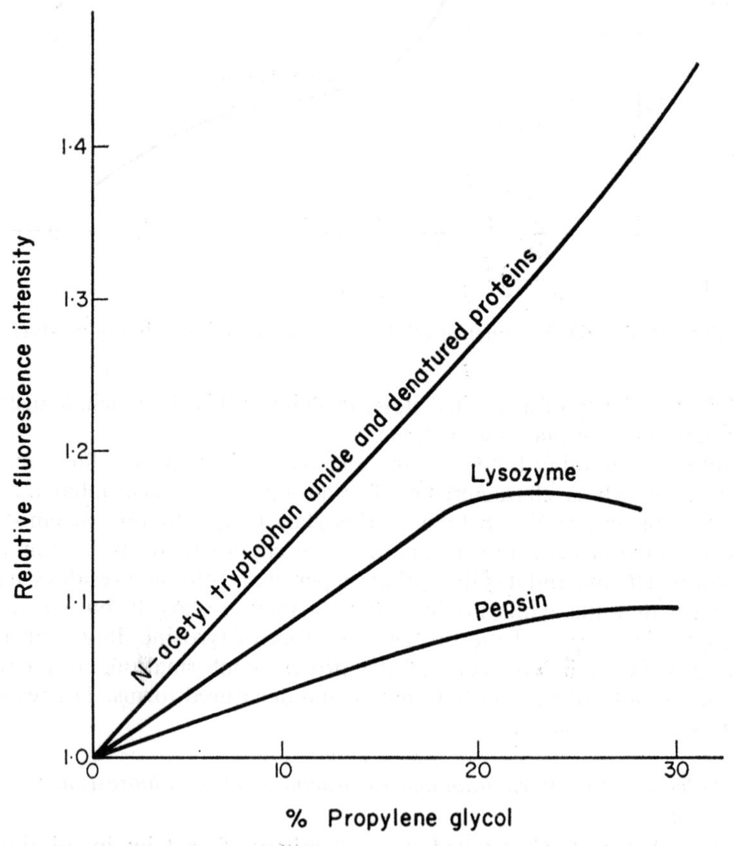

FIG. 10.11 Effect of propylene glycol on protein fluorescence.

The idea behind solvent perturbation is basically simple. The native forms of globular proteins are relatively compact and impermeable to solvent. It is therefore expected that the electronic states of aromatic residues which are embedded in the interior of the protein should be insensitive to solvent composition. In contrast, residues on the protein surface will be influenced by the solvent. Clearly there must be intermediate stages where the aromatic residues lie in cavities that are accessible only to solvent molecules of certain dimensions. Most organic solvents which influence the fluorescence of tryptophan (cf. Fig. 10.7) can be used to explore the environment of this chromophore in native proteins. The effect of propylene glycol for instance, on several native and denatured proteins is compared in Fig. 10.11 with that on acetyl tryptophan amide. In no case is the effect of the solvent as large on the fluorescence intensity of the native protein as on acetyl tryptophan amide or on the denatured protein. The initial slopes of the lines may be regarded as a measure of that fraction of the volume surrounding each chromophore which is, on average, occupied by solvent. If this model is correct, the implication is, for instance, that the tryptophans of pepsin are less available to solvent than are those of lysozyme. Perturbation studies on absorption spectra show similar trends.

In contrast, the effect of urea on tryptophan containing proteins is highly variable. We have seen that urea enhances the fluorescence of acetyl tryptophan amide, while with proteins, in some cases a large decrease, in others an increase, in the fluorescence emission is observed (Fig. 10.12). In cases where pronounced quenching is observed (e.g., bovine serum albumin, lysozyme) the intrinsic exalting effect of urea is countered by the effects arising from the loss of organized structure in denaturation.

Thermal transformations in proteins may also lead to denaturation. This is illustrated by observing the differences in the heating curves for amino acids (Fig. 10.8) and those for proteins. Again, for the latter the effects are highly variable, but, generally, pronounced changes are observed in the temperature range where denaturation occurs. Ribonuclease, for instance, which contains only tyrosine as its fluorescent chromophore, shows a sharp increase in fluorescence intensity within the same narrow temperature range as it undergoes an increase in laevorotation and a decrease in absorbency at 280 nm.

Many of the details of these observations are not yet clearly understood, but the examples serve to illustrate the potential usefulness of fluorescence methods in structural studies.

Phosphorescence in Proteins

Protein phosphorescence contains emission from both tyrosine and tryptophan. The data available so far are limited, but by assuming that the emission spectrum for the tyrosine residues in all proteins is the same as is that from ribonuclease (which contains only tyrosine and no tryptophan), it is possible to analyse the contribution from tyrosine emission in many proteins (Table 10.5).

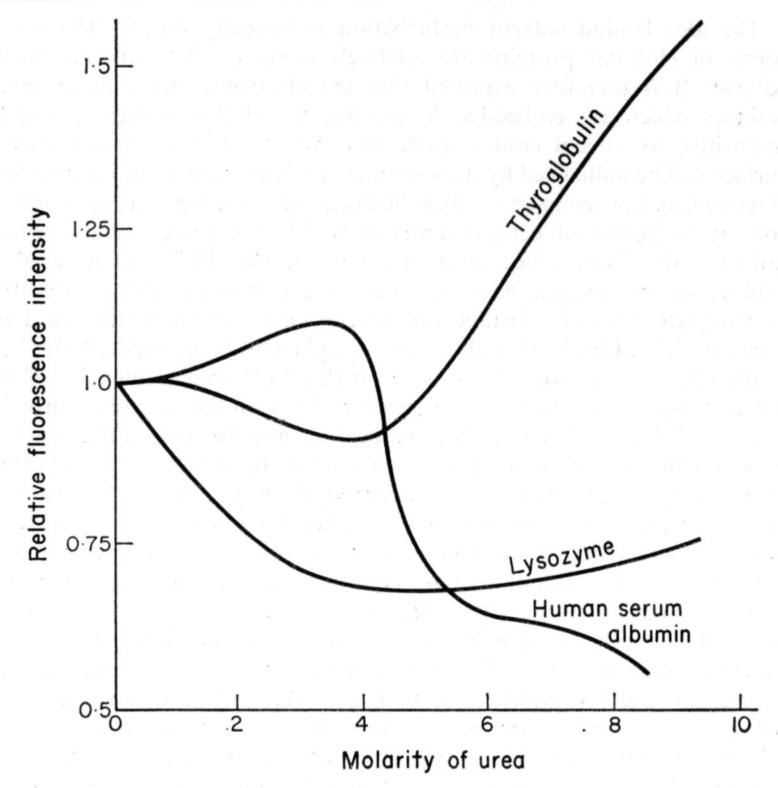

FIG. 10.12 Effect of urea on protein fluorescence.

TABLE 10.5. *Percentage of phosphorescence from tyrosine in proteins*

Protein	% Tyrosine phosphorescence excitation λ (nm)	
	210	280
Alcohol dehydrogenase	90	60
Chymotrypsin	0	0
Lysozyme	58	14
Trypsin	13	8
Carboxypeptidase	48	46

In most of the proteins studied, a significant fraction of the energy in excited tyrosine residues is not transferred to tryptophan. Phosphorescence, therefore, is a promising method for studying tyrosine emission, although it is not so for fluorescence.

THE INTERACTION OF PROTEINS WITH SMALL MOLECULES

Non-covalent Specific Interaction

Proteins can form complexes with a variety of small molecules. Some of these occur in specific regions of the protein structure and are related to the catalytic functions of enzymes and the mode of antigen-antibody interactions. Less specific interactions are also observed between organic dyes and proteins, and can be utilized among other things in dyeing wool and in microscopic staining techniques. The intermolecular forces responsible for these interactions can be ionic, van der Waals, hydrophobic or charge-transfer, and often fluorescence studies can aid in choosing between these alternatives. An attempt will be made rather to illustrate the type of information that studies of these non-covalent interactions can give us, than to review this large field comprehensively.

Changes in Protein Fluorescence on Binding

Present theories of enzyme catalysis rely on the assumption that proteins can exist in several possible conformations, and that interconversion between these conformations can be brought about by the action of the reactants (substrates, coenzymes, cofactors) or catalyst inhibitors.

In general, from fluorescence we can learn about the nature of the site at which binding occurs, and/or structural alterations in the protein, which derives its energy from the enzyme-substrate interaction.

For instance, the tryptophan fluorescence of lysozyme decreases on binding the substrate tri-N-acetyl-D-glucosamine. The substrate does not quench the fluorescence of free tryptophan in solution. It is thus likely that several of the tryptophans in lysozyme are placed at, or close to, the substrate binding site. This conclusion is confirmed by X-ray crystallography, which shows that three of the six tryptophan residues in lysozyme are involved in substrate binding.

In contrast there is an increase in the fluorescence emission of phosphoglucomutase on interaction with its substrate (glucose-6-phosphate), or of yeast enolase with Mg^{2+} or Mg^{2+} and substrate (2-phosphoglyceric acid) (Fig. 10.13). This indicates an alteration in the protein conformation which is reflected in a change of polarity in the tryptophan environment. Both transitions appear to transfer some tryptophans from an aqueous environment into the less polar interior of the protein. This theory is supported by spectral and chemical studies for phosphoglucomutase and by the observation of a change in the protein rotational relaxation times for yeast enolase.

Enzyme-Coenzyme Interactions

The fluorescence properties of flavin and reduced nicotinamide nucleotides are strongly modified when these are bound to enzymes for which they act as coenzymes in redox reactions. While the fluorescence of flavins is almost always completely quenched, that of NADH is modified in several ways on binding. In almost all cases it is enhanced (glyceraldehyde-3-phosphate dehydrogenase being an exception) and the position of its

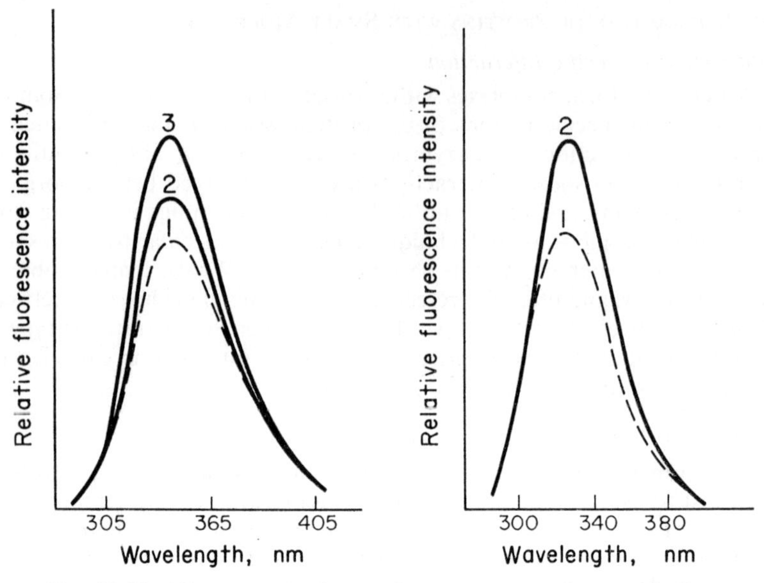

FIG. 10.13 Fluorescence changes in enzymes on substrate binding.

emission maximum is displaced. The increase in quantum yield is probably due to the loss of vibrational relaxation processes, and the shifts in emission reflect the polarity of the binding sites. The binding is also accompanied by an increase in fluorescence polarization since the period of rotation for the complex is considerably increased. All these changes in fluorescence properties can be utilized in measuring equilibrium constants for the complexes and estimating the number of molecules of coenzyme bound to each protein molecule. A typical binding curve using fluorescence polarization is shown in Fig. 10.14. The limiting value of polarization represents the property of the bound coenzyme only. It will depend not only on the size and shape of the protein molecule, but also on the rotational freedom of the coenzyme within the complex. In most instances this is probably an insignificant factor, although the low polarization value for glyceraldehyde-3-phosphate dehydrogenase (Table 10.6) may be due to such an effect. This also may be the case for glutamate dehydrogenase (Fig. 10.13) for which the two limiting values are thought to represent different conformational states of the protein, the lower value being associated with the enzymically active form. The conformation of the bound coenzyme can be evaluated by studying the efficiency of energy transfer between its adenine and nicotinamide rings. Some of the relevant information for a number of dehydrogenases is collected in Table 10.6.

For flavin coenzymes, other intricate details of the interaction can be evaluated from the degree of quenching equilibrium constants and from the rate of quenching on recombination with the protein. For instance, when FAD is combined with its apoenzyme in D-amino acid oxidase, the

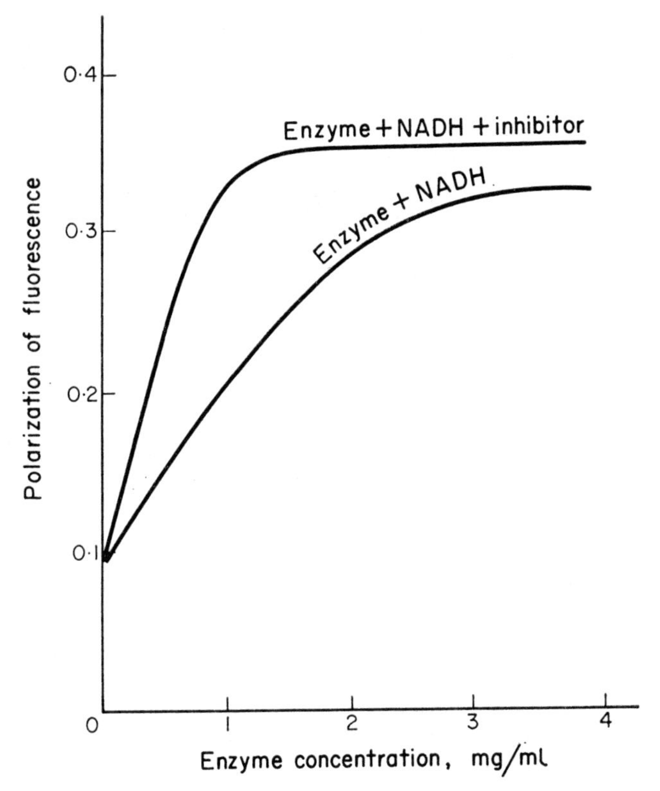

FIG. 10.14 Binding of NADH to glutamate dehydrogenase.

TABLE 10.6. *Luminescence of bound NADH*

Enzyme	Dissociation constant (μM)	Absorption (nm) λ_{max}	Emission of bound NADH λ_{max} (nm)	Emission of bound NADH Enhancement	Polarization of fluorescence	Conformation
Lactate Dehydrogenase	0.3	335	440	+	0·47	Open
Alcohol Dehydrogenase	0.4	325	450	+	0·42	Open
Glutamate Dehydrogenase	3.0	340	445	+	0·44	Open
Glyceralde-3-phosphate Dehydrogenase	0.2	340	465	–	0·28	Closed
Free NADH in water		340	465		0·1	Closed

kinetics of quenching are biphasic. The fast second-order process (combination of coenzyme with protein) is followed by a slow process that depends only on protein concentration (probably representing an isomerization of the coenzyme-protein complex).

Fluorescent Probes for Protein Structure and Conformation

The fact that the fluorescence of small chromophores is modified on interaction with polymers is utilized in a somewhat different approach to explore the secondary or tertiary structures of native proteins. The quantum yields of fluorescence of 1-anilino-8-naphthalene sulphonate (ANS) and 2-*p*-toluidinyl-6-naphthalene sulphonate (TNS) are very low in water, but are considerably enhanced in non-polar solvents (Table 10.7). Similar large enhancements are observed when these dyes are bound to proteins. The increase in quantum yield is a measure of the polarity of the binding sites, binding being strongest at non-polar (so called hydrophobic) regions on the protein surface. An example will illustrate the usefulness of the method.

Table 10.7. *Fluorescence quantum yields for ANS and TNS in various solvents*

TNS		ANS	
Solvent	Quantum yield	Solvent	Quantum yield
H_2O	0·0008	EtOH	0·37
MeOH	0·34	90% EtOH—H_2O	0·24
EtOH	0·52	80% EtOH—H_2O	0·13
CH_3COOH	0·18	50% EtOH—H_2O	0·039
Ethylene glycol	0·14	20% EtOH—H_2O	0·007
Pyridine	0·002	H_2O	0·004

Five moles of ANS are bound to bovine serum albumin to give a ~150-fold increase in fluorescence quantum yield. The polarization of fluorescence of the bound chromophore decreases as the average number of bound chromophores/mole of protein is increased. This is due to energy transfer among the ligand molecules, and the relation of the hydrophobic binding sites may be evaluated by means of two simple and justifiable assumptions: (*a*) that random distribution of ligand molecules prevails among the protein binding sites; (*b*) that a single transfer of the excited state is responsible for depolarization. On this basis the average distance between a pair of binding sites is 2·1 nm, and the average angle between the two emission oscillators is 33°.

Since the size and nature of the fluorescent chromophore may be varied at will, this technique provides a powerful tool for exploring many details of protein conformation in solution.

Covalent Fluorescent Conjugates of Biopolymers

Fluorescent dyes, when covalently linked to proteins and nucleic acids in a random way, may be used to characterize the polymer. The fluorescence polarizations of these conjugates depend on the size and shape of the polymer. If the polymer is treated as a rigid ellipsoid the fluorescence polarization depends on the three characteristic relaxation times of the particle according to Perrin's formula:

$$\frac{1/p+1/3}{1/p_0+1/3}=1+\tau\left(\frac{1}{\rho_1}+\frac{1}{\rho_2}+\frac{1}{\rho_3}\right)$$

when p = polarization of fluorescent light at a given temperature and viscosity, unpolarized exciting light being used. p_0 = limiting value of p at high values of relaxation. τ = excited lifetime of fluorescent group. ρ_1, ρ_2, ρ_3 are the relaxation times. It is therefore possible to obtain information about the overall size and shape of the polymer.

If there are internal degrees of rotational freedom in the polymer the value of the mean harmonic relaxation time ρ_h

$$\left(\frac{1}{\rho_h}=\frac{1}{\rho_1}+\frac{1}{\rho_2}+\frac{1}{\rho_3}\right)$$

will be reduced over the value expected from a rigid particle of the same size. If ρ_h is known from independent measurements (e.g., hydrodynamic data), a measure is available of the departure of the particle from complete rigidity, and hence of the extent of disruption of the organized secondary and tertiary structure of the polymer.

The most commonly used reagents for preparing fluorescent conjugates of proteins are 1-dimethylamino-naphthalene-5-sulphonyl chloride and fluorescein isothiocyanate. The fluorescent conjugate of bovine-γ-globulin with the former reagent in neutral solutions yields rotational relaxation times which are much lower than expected from the translational dynamic properties of the molecule. The value of the harmonic mean relaxation time obtained from fluorescence polarization is in the range $4–13 \times 10^{-8}$ s, while it is 45×10^{-8} s when computed from dielectric dispersion data. This suggests that there is considerable general flexibility within the molecule of bovine-γ-globulin.

It follows that this method can also be used to detect structural changes in polymers and to study specific regions in the polymer if the reagent is made specific with respect to its reactivity towards certain groups of the polymer. One such reagent is pyridoxal-5′-phosphate which forms Schiff bases with free amino groups of proteins. These, on reduction with sodium borohydride, give a strongly fluorescent conjugate. Fluorescent labelled proteins can also be used in a variety of other fields including immunochemistry, in histochemistry in localizing certain polymers in the cell, and have many medical applications.

Fluorescence and Phosphorescence of Nucleotides and Nucleic Acids

The fluorescence of nucleotides and their polymers can be observed only at low temperatures in glasses. The ultra-violet absorption spectra of dinucleotides under these conditions are essentially identical with those of an equimolar mixture of their constituents, and that of DNA resembles that of a mixture of its constituent nucleotides, apart from a decrease in the extinction coefficient at 260 nm in DNA (hypochromism). The fluorescence of many dinucleotides and DNA, on the other hand, differ qualitatively from their constituent nucleotides. The dinucleotides UpA, ApC, CpC and TpT at pH 7 are stacked (cf. nucleotide coenzymes) while in CpC at pH 2 and TpT at pH 12 there is very little stacking (as shown by ORD). (The abbreviations are as follows: A: Adenosine; C: Cytidine; T: Thymidine; U: Uridine; P: Phosphate.) The fluorescence spectra of the dinucleotides in the stacked state, of polynucleotides and of DNA are considerably broader and are red-shifted from those of their controls (i.e., constituent nucleotides) (Fig. 10.15). These observations have been interpreted as

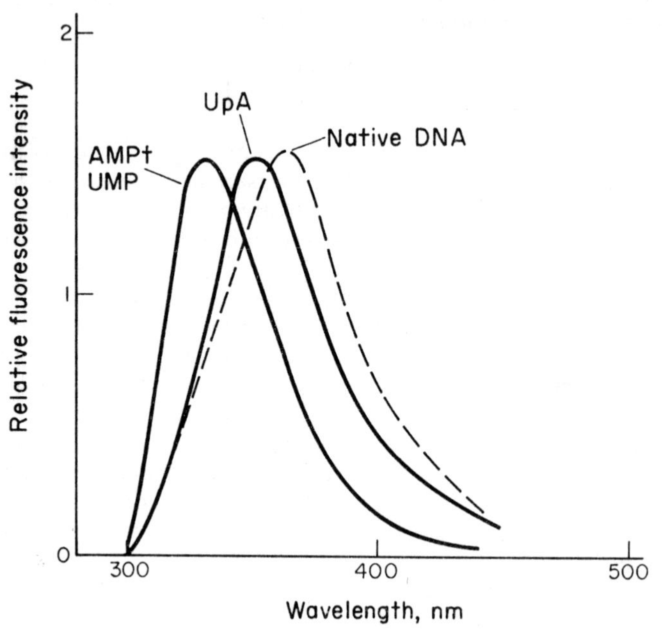

FIG. 10.15 Fluorescence of nucleotides and DNA.

being due to excimer fluorescence brought about by stacking interactions.

The phosphorescence of nucleotides is very weak, and no emission can be observed from deoxycytidilic acid, uracil, cytosine and thymine at low concentrations. Guanine derivatives phosphoresce with a characteristic decay time of 1·2 s (λ_{max} = 410 nm) and adenine derivatives have a decay time of 2·3 s. In dinucleotides of adenine with cytosine and cytidine, kinetic measurements of the decay indicate two emissions each characterized by a different lifetime. Since cytosine alone shows no phosphorescence, it was suggested that in the dimer it is excited by triplet-triplet energy transfer from the adenine chromophore. The phosphorescence of DNA is essentially a sum of a guanine and adenine type emission and it is quenched efficiently by Mn^{2+} and Fe^{3+}. The ratio of purines quenched to cation is greater than 10, and on this basis again a delocalized triplet excited state is favoured in DNA.

Fluorescence and Phosphorescence of DNA-Dye Complexes

DNA forms strong complexes with dyes such as proflavine and acridine orange. While in principle these dye-polymer complexes might be thought to be similar to those already discussed, they will be described here, as they differ in many respects. These differences arise from the fact that DNA provides a more regular structure than most proteins, and therefore DNA-dye complexes are good models for studying the effects of coupling between chromophores. As suggested earlier, the regular ordering of chromophores on a polymeric substrate might be the key to the mechanism of energy transfer both in the primary steps in photosynthesis and in the process of vision. In addition, these complexes are thought to be responsible for the mutagenic action of proflavin.

The fluorescence properties of acridine dyes bound to DNA can be best interpreted on the basis of the so-called intercalation model. In this, the dye molecules are assumed to be positioned between the stacked bases of the DNA helix, the plane of the aromatic dye being perpendicular to the helix axis. Good evidence for this model is obtained by comparing the fluorescence polarization of DNA-dye complexes when the DNA molecule is oriented (by flow through a narrow capillary) and when its orientation is random. If a particular transition is rendered parallel to the relevant prism orientation, the probability of observing transition will increase from the average value for random orientation. Fluorescence intensities in both transitions of the bound acridine dye quinacrine decrease during flow when the prisms are parallel to the flow, and both transitions increase when the prisms are perpendicular (Table 10.8). Since both transitions are contained in the plane of the dye molecule they must be oriented perpendicular to the direction of flow, and hence of the helix axis.

Some other interesting features of the DNA-dye interaction can be briefly summarized as follows.

(*a*) The relative quantum yield of emission at low dye : nucleotide base ratio (D : B-ratio) is different from one. (*b*) The relative quantum yield at high D : B-ratio is zero. (*c*) The variation of the relative quantum yield as a function of log D : B-ratio is sigmoidal in shape. (*d*) The polarization of fluorescence of the dye also decreases with increasing D : B-ratio, and closely follows variation in quantum yield. (*e*) The quantum yield is a function of the wavelength of excitation, and is higher when the excitation takes place in the region of absorption by the DNA itself. These observations can be explained by dye-to-dye energy transfer in visible excitation (the observations are similar to concentration quenching and depolarization) and nucleotide base-to-dye energy transfer in ultra-violet excitation.

At low D : B-ratios at 77°K in 50% glycerol-H_2O, phosphorescence from the bound dye is observable. Delayed fluorescence of the dye is also seen, but only when excited by light absorbed by the DNA itself. This delayed fluorescence closely parallels the behaviour of the DNA phosphorescence

TABLE 10.8. *Fractional change due to flow in the fluorescence of quinacrine bound to DNA*

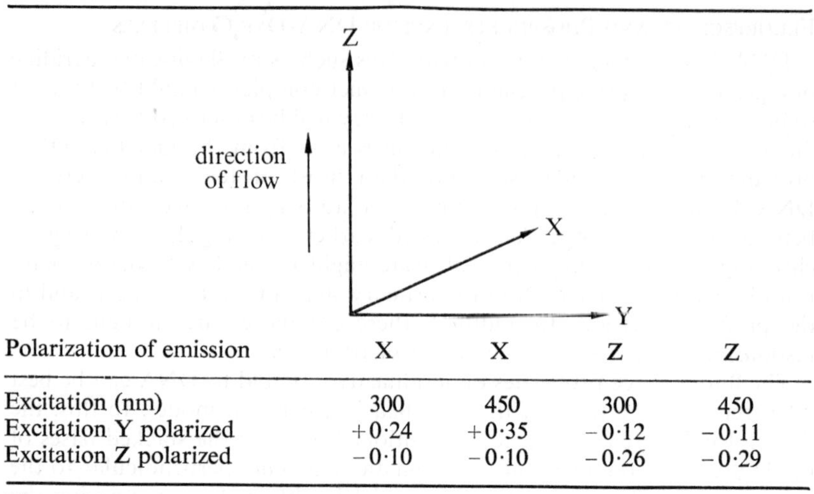

Polarization of emission	X	X	Z	Z
Excitation (nm)	300	450	300	450
Excitation Y polarized	+0·24	+0·35	−0·12	−0·11
Excitation Z polarized	−0·10	−0·10	−0·26	−0·29

The transitions excited at 450 nm and 300 nm are parallel and perpendicular, respectively, to the transition of emission.

without the dye. This gives the interesting possibility of a triplet-singlet energy transfer from nucleotide to dye according to the following scheme:

$$\text{purine} \xrightarrow{h\nu} \text{(singlet)* delocalized} \rightarrow \text{(triplet)* delocalized}$$
$$\text{purine (triplet)*} + \text{dye} \rightarrow \text{dye (singlet)*}$$

Bioluminescence

The emission of light by certain enzyme catalysed reactions occurs in many different living organisms. The light emission is simply due to the fact that fluorescent molecules become electronically excited by receiving reaction energy in a cell-oxidation process. Because of either insufficient concentration of energy, or the lack of a suitable fluorescent acceptor, most cell reactions do not emit visible light; however, bioluminescence is found very irregularly distributed over a wide range of organisms, e.g., bacteria, fungi, crustacea, insects, fish. In the more organized creatures, natural selection has put the accidental possession of luminescence to good uses (e.g., sex attraction, warning, advertising occupancy of territory). Luminescent organs of great variety and complexity have been developed, often with associated mechanisms for physiological control.

This topic will be discussed only in outline, not because of the lack of work or interest in it, but partly because an excellent summary is available (see reading list), and partly because we are here concerned with it only in relation to the chemistry of the processes.

The common features of most (or possibly all) bioluminescent reactions are that they require an enzyme (luciferase), a small organic molecule (as luciferin) whose oxidation product is of a fluorescent type, and oxygen. Therefore they are clearly analogous to chemiluminescence except that the oxidations are enzyme catalysed. The process most clearly understood is that of firefly luminescence. The oxidation product of firefly luciferin (IX) is probably dehydroluciferin (X). Besides luciferin and luciferase the

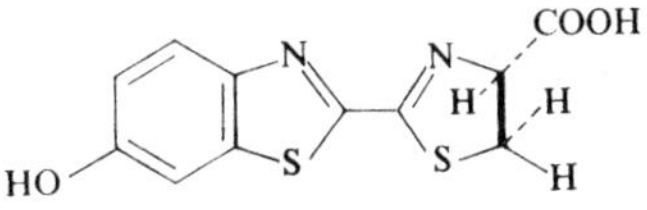

D (−) luciferin

IX

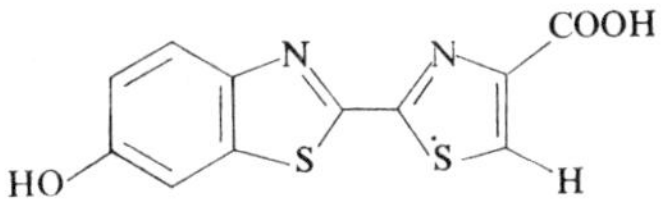

Dehydroluciferin

X

reaction requires adenosine triphosphate (ATP) and Mg^{2+} for light emission. Only the D(−) form of luciferin can give rise to luminescence. The mechanism of the reaction is briefly summarized in the scheme below.

[ATP][Enz][Luc.]⇌[AMP][Enz][Luc:] + inorganic phosphate.

↳ + O_2→[AMP][Enz][Luc:peroxide]

↳[AMP][Enz][Dehydroluc:]* + H_2O

↳$h\nu$

The salient features of the reaction are the formation of an enzyme-luciferin-adenosine monophosphate (AMP) complex with the liberation of inorganic pyrophosphate, and the formation of a peroxy intermediate which then decomposes with emission of light. The light-emitting species is an enzyme luciferin complex. This is based on the observation that the same luciferin molecule, with luciferase obtained from different species, can give rise to light of different spectral distribution. The light-yield approaches unity. How the reaction energy so efficiently produces the electronically excited state of the dehydroluciferin molecule is not clear. (See Chapter 9.)

In contrast, bacteria utilize a very different mechanism for light emission. The reaction requires two molecules of reduced flavin mononucleotide ($FMNH_2$), a long-chain aldehyde, oxygen and luciferase. Neither the role of aldehyde nor that of $FMNH_2$ is clearly understood, but since the emission is again species-dependent, it is likely that it occurs from an enzyme-bound small molecule (FMN or a derivative of it).

Conclusions

As our understanding of biological phenomena advances more and more towards the molecular level, the extension of standard chemical techniques to more complicated structural and dynamic problems becomes essential.

One possible direction in which fluorescence might be developed is the application of quantitative fluorescent microscopic techniques. This might provide an essential link between the study of isolated biochemical reactions and the processes that occur within an intact living cell.

BIBLIOGRAPHY

Only leading references are given when possible, from which the original papers can be followed up.

Redox Coenzymes

VELICK, S. F. in *Light and Life* (edited by W. D. McElroy and B. Glass), Johns Hopkins Press, Baltimore (1961).

PENZER, G. R. and RADDA, G. K. *Quart. Rev.* (London), 21, No. 1 (1967).

Fluorescence of Chlorophyll and Derivatives

LIVINGSTON, R. *Quart. Rev.* (*London*), **14**, 174 (1960).

TWEET, A. G., BELLAMY, W. D. and GAINES, G. L. *J. chem. Phys.* **41**, 2068 (1964).

PARKER, C. A. and JOYCE, T. A. *Nature*, **210**, 701 (1966).

CLAYTON, R. K. *Molecular Physics in Photosynthesis*, Blaisdell, New York (1965).

Vitamin A

'Energy transfer with special reference to biological systems', *Discuss. Faraday Soc.* No. 27 (1959).

Fluorescence in Amino Acids, Polypeptides and Proteins

WEBER, G. and TEALE, F. W. J. in *The Proteins* (edited by H. Neurath), Vol. 3, Academic Press, New York (1965).

Biopolymers Symposia, No. 1 (1963).

YEARGERS, E., BISHAI, F. R. and AUGENSTEIN, L. *Biochem. Biophys. Res. Comm.* **23**, 570 (1966).

COWGILL, R. W. *Biochim. Biophys. Acta*, **112**, 550 (1966).

KONEV, S. V., *Fluorescence and Phosphorescence of Proteins and Nucleic Acids*, Plenum Press, New York (1967).

Interaction of Proteins with Small Molecules

LEHRER, S. S. and FASMAN, G. D. *Biochem. Biophys. Res. Comm.* **23**, 133 (1966).

BREWER, J. M. and WEBER, G. *J. biol. Chem.* **241**, 2550 (1966).

YANKEELOV, J. A. and KOSHLAND, D. E. *J. biol. Chem.* **240**, 1593 (1965).

MASSEY, V. and CURTI, B. *J. biol. Chem.* **241**, 3417 (1966).

Fluorescent Probes for Protein Structure and Conformation

DANIEL, E. and WEBER, G. *J. biol. Chem.* **241**, 1893 (1966).

STRYER, L. *J. molec. Biol.* **13**, 482 (1965).
MCCLURE, W. O. and EDELMAN, G. M. *Biochemistry*, **5**, 1908 (1966).

Covalent Fluorescent Conjugates of Biopolymers

WEBER, G. and TEALE, F. W. J. in *The Proteins* (edited by H. Neurath), Vol. 3, Academic Press, New York (1965).
CHURCHICH, J. E. *Biochim. Biophys. Acta*, **102**, 280 (1965).

Fluorescence and Phosphorescence of Nucleotides and Nucleic Acids

EISINGER, J., GUERON, M., SHULMAN, R. G. and YAMANE, T. *Proc. natl Acad. Sci. U.S.A.* **55**, 1015 (1966).
BERSOHN, R. and ISENBERG, I. *Biochem. Biophys. Res. Comm.* **13**, 205 (1963).

Fluorescence and Phosphorescence of DNA-Dye Complexes

LERMAN, L. S. *Proc. natl Acad. Sci. U.S.A.* **49**, 94 (1963).
WEILL, G. and CALVIN, M. *Biopolymers*, **1**, 401 (1963).
ISENBERG, I., LESLIE, R. B., BAIRD, S. L., ROSENBLUTH, R. and BERSOHN, R. *Proc. natl Acad. Sci. U.S.A.* **52**, 379 (1964).

Bioluminescence

SELIGER, H. H. and MCELROY, W. D. *Light: Physical and Biological Action*, Academic Press, New York (1965).

Chapter 11

FLUORESCENCE MICROSCOPY AND HISTOCHEMISTRY

P. J. STOWARD

IN the past, chemists have not been interested in histochemistry, that is, in the identification of macromolecular substances as they actually occur in cells and tissues. Recently, however, chemists and biochemists on the one hand, and biologists and pathologists on the other, have come to realize that many problems of biology and pathology cannot be solved until it is possible to determine the chemical structures of biological macromolecules *in situ* instead of in the test-tube. To do this, sensitive and specific methods must first be developed for detecting the principal chemical groups of such macromolecules, and here fluorescence histochemical techniques are of especial value.

The more promising of these techniques are usually based either on the affinity of the negatively-charged, anionic groups of tissue macromolecules for positively-charged, basic fluorescent dyes (or the converse, on the binding of acid fluorescent dyes on to the cationic groups of tissue macromolecules), or on reactions involving the condensation of amine groups (usually of aromatic hydrazines) with carbonyl groups (particularly of aldehydes). In this chapter the use of the basic dye coriphosphine for detecting the sulphate groups of acid mucosubstances will be described in detail as an example of the first type of technique. The second type will be illustrated by the reactions of salicylhydrazide with the aldehyde and aliphatic ketone groups which by various means can be specifically engendered from, or introduced into, most of the various types of chemical groups present in tissue marcromolecules. Fluorescence histochemical reactions utilizing other reagents will also be briefly described.

The manipulations of histochemistry, which differ from those of biochemistry and organic chemistry, are outlined first. There follows a description of the special features of the fluorescence microscope.

Histological Techniques

PREPARATION OF TISSUES

The first step in a histochemical investigation is to obtain fresh tissue from, say, a recently-killed animal or, for human material, from a surgical operation or from an autopsy. This material is placed, as soon as possible, in a 'fixative' solution that 'fixes' or coagulates, or otherwise renders indiffusible, the various soluble compounds (including complex polymers) in the tissue. Chemists sometimes criticize biologists for fixing biological materials, arguing that the fixation process alters the chemical and physical nature of biological macromolecules, and that therefore the histochemical properties observed subsequently for such fixed macromolecules bear no relationship to the true chemical properties in the living tissue. In practice, however, fixation does not always alter the chemical properties of macromolecules *in situ* as much as might be expected. Even where such alterations do occur, it is possible, by using several different types of fixative, to infer the true chemical properties of the unfixed macromolecule.

There are roughly two types of fixative in common use: those which precipitate or coagulate macromolecules directly, and those which combine chemically with the macromolecule to form an insoluble complex. In routine work, solutions containing formalin are used. Formalin fixes by combining mostly with the terminal amino groups of tissue proteins to form cross-linked methylene bridge compounds which, presumably, enmesh other tissue compounds. This is undesirable for fluorescence histochemical work. It is better to precipitate tissue macromolecules with, for example, ethyl alcohol or one of the higher ethers.

After fixation, the tissue is cut, on a microtome, into thin sections, from 1 to 10μ thick. However, most tissues are fragile, and disintegrate on cutting; therefore they are usually first infiltrated with, and embedded in, a relatively hard solid, such as paraffin wax. This requires dehydration by the use of several changes of absolute ethyl alcohol, removal of the alcohol with several changes of the aromatic hydrocarbon xylene containing increasing amounts of paraffin wax, and finally immersion in molten paraffin wax (m.p. about 60°C) itself. In the final step, the paraffin wax containing the tissues is cooled to room temperature, at which it forms a hard, solid block. Thin sections can be cut from these blocks and affixed to glass slides with albumin solution. These sections stick to the glass quite firmly if they are dried in an oven at 37°C for 2–3 days.

Before a histochemical reaction is performed on tissue sections, the paraffin wax in the section is removed. This is achieved by dissolving the wax out with several changes of xylene, removing the xylene by immersing the sections in two or three changes of absolute ethyl alcohol, and finally placing them in decreasing strengths of alcohol down, usually, to water. The hydrated sections are last placed in solutions containing the appropriate histochemical reagent (e.g., a fluorescent dye). Afterwards the section is examined in a fluorescence microscope. To attain the maximum resolution of the microscope (i.e., to observe the maximum amount of structural detail in the section), the section has to be 'mounted' in, or infiltrated with, an inert, non-fluorescent liquid (capable ultimately of hardening to a translucent solid) of refractive index higher than about 1·5. Practically all such mounting media are immiscible with water. Therefore after sections have been in contact with aqueous reagent solutions, they are dehydrated, usually with several changes of absolute ethyl alcohol or isopropyl alcohol, and are next immersed in several changes of xylene; finally a drop of xylene-soluble mounting medium (e.g., D.P.X.) is placed on the tissue section and covered with a very thin cover-glass.

Fluorescence Microscopy

THE THEORETICAL BASIS OF FLUORESCENCE MICROSCOPY

When a suitable specimen is illuminated or 'excited' with light of short wavelength (in the ultra-violet or blue regions), the image field normally consists of visible fluorescent light emanating from the object itself, and excess transmitted illumination. However, when the latter is absorbed by a suitable filter transmitting only the fluorescent light (as it is in the fluorescence microscope), the final image is that of a self-luminous object.

The advantages of this system over conventional light microscopy are improved contrast and sensitivity. The enhanced contrast can be illustrated by a simple example. Suppose that a specimen absorbs 1% of the light falling on it: the comparative light intensities of the object and its background will be 99:100, which means that the contrast will be low. But if the excess incident light passing through the object and its background is removed before reaching the observer, and if the light absorbed by the object is re-emitted as fluorescence, the relative intensities will be 1:0 and theoretically, the contrast will be infinite. The *relative* contrast in the second system appears much greater than in the first; the eye can distinguish large differences at low light levels much more easily than small differences at high light levels.

THE FLUORESCENCE MICROSCOPE

The three main features of a fluorescence microscope are the light source, the isolating and barrier filters, and the lenses. Light from the source (usually a mercury lamp) passes first through a heat filter and then through a filter which transmits only the ultra-violet or blue components emitted by the source. From there, this component, the exciting illumination, is reflected from the substage mirror of the microscope and collected by a substage condenser whence it is concentrated as an intense spot falling on to the specimen mounted on a glass slide. The fluorescence light emitted by the specimen and the excess exciting illumination enters the objective, behind which is placed a complementary or barrier filter which absorbs the excess exciting illumination but not, ideally at least, visible fluorescence emanating from the specimen. The observer sees an image of a self-luminous object against a black 'background'.

LIGHT SOURCES

The prime requirement for achieving maximum fluorescence emission is to illuminate the specimen with the brightest possible light at wavelengths near the specimen's absorption maxima—usually in the ultra-violet or blue regions. However, the ultra-violet (u.v.) and blue content of the tungsten lamps used in conventional microscopes is insufficient to induce much fluorescence from most objects. Even under optimum conditions, fluorescence is often weak.

The best results are obtained by using the intense spectral lines emitted by high-pressure mercury discharge lamps. The choice of a particular lamp is still a matter of compromise between stability, intrinsic brilliance, and physical convenience. The Osram HBO 200-watt lamp is the one used in most commercial fluorescence microscopes. Since about 1964, some manufacturers have used the cheaper and more convenient quartz-iodine lamps. These are very useful for studying microscopic objects that fluoresce appreciably on excitation with blue light as, for example, in routine diagnoses of suspected cancer smears stained with acridine orange. However, such lamps emit comparatively little ultra-violet light, and consequently are of limited value in the broader applications of fluorescence microscopy.

Very little attention is usually paid to the collection of the maximum light available from a given lamp. The apertures of most commercial collector lens systems are too small, so that only a fraction of the available light ultimately enters the microscope substage condenser. The potential efficiency of lamp units incorporated into commercial fluorescence microscopes is usually sacrificed for convenience and compactness of design of the complete apparatus. The collector lenses are made of heat-resisting glass or quartz, rather than ordinary glass which may crack in the intense heat of the lamp.

FILTERS

It is preferable to absorb the heat emitted by a mercury lamp before isolating the intense 366 nm group of mercury lines—the one usually employed for u.v. fluorescence microscopy. Suitable heat filters are the Schott K 1 or K 2. Another suitable but perhaps less convenient filter is a solution of copper sulphate, which has the added advantage of partly isolating the ultra-violet and the blue mercury lines. After absorbing most of the heat, the group of lines around 366 nm is isolated. The Schott UG 4 or UG 1, or the Chance OX 1, does this very efficiently.

Microscopic objects appearing yellow by transmitted white light are sometimes best excited with blue light. The Schott BG 12 filter is often used for isolating the mercury 435 nm line for this purpose.

The complementary filter for absorbing the excess exciting illumination is best placed in the body of the microscope, above the objective and below the ocular. The ideal complementary filter for u.v. excitation would completely absorb the 366 nm line, but transmit all the visible region above, say, 400 nm. The light yellow Wratten 2B filter meets this requirement fairly well. When blue light is used for excitation, a deeper yellow or an orange complementary filter (e.g., a Schott OG 1) is necessary.

Lenses

Special lenses are not necessary for fluorescence microscopes, although statements are sometimes made to the contrary. As it is desirable to excite a specimen with the maximum amount of illumination, a condenser with the largest possible numerical aperture (NA) is called for. Simple Abbé glass condensers are adequate for this and will transmit wavelengths down to 320 nm. The lens cement of more complex condensers, on the other hand, may absorb much ultra-violet light and fluoresce appreciably.

Several workers have advocated the use of dark ground condensers, although much of the light entering the condenser never reaches the specimen. However, the system possesses some compensating advantages. It directs the light away from the objective so that none of the exciting illumination, apart from scattered light, reaches the image. Thus a wider spectral band can be used to stimulate fluorescence, while the complementary filter can be relatively thin, often colourless, and still produce a black background.

Substage condensers can be avoided by illuminating the object from above. This system of direct illumination is used in fluorochemical methods for examining metallurgical, geological and similar specimens. In some commercial fluorescence microscopes it can be combined with illumination from below via the normal optical system.

The most important criterion underlying the choice of objective is the light-gathering power; this is related to the quantity

$$(\mathrm{NA}/\mathrm{magnification})^2,$$

so that the highest possible NA compatible with the lowest convenient magnification should be employed.

The only important practical point about eyepieces is that a monocular head is preferable. There is little value in using a binocular eyepiece, since this reduces, by more than half, the intensity of what is already a weak image.

Commercial Fluorescence Microscopes

Conventional microscopes can, with little alteration, be used for fluorescence work. All that is needed is a free-standing mercury lamp fitted with the appropriate collector lenses and heat and isolating filters, a front-surface aluminized mirror to clip on to the normal substage mirror (silver has a low reflectivity in the ultra-violet region), and complementary filters placed in the body of the microscope.

Many people cannot, or do not, exploit the several advantages of fluorescence microscopy as a routine technique, simply because they prefer ready-made fluorescence microscopes which are at present very expensive. Others are put off by the mystique which some practitioners in the field have hinted, without any justification, to be necessary for successful fluorescence microscopy. Fluorescence microscopy will not be used as widely as it should until the microscope manufacturers can provide a comparatively cheap and simple fluorescence microscope; that is, one without the 'extras', without dual-illumination, but with provision for both ultra-violet and blue light excitation.

Fluorescence Histochemistry

The Scope of Histochemistry

Before discussing fluorescence histochemical reactions for detecting specific chemical groups, it is useful to consider what groups may be present in biological tissues. Biological materials are made up from roughly four classes of organic compounds: proteins, nucleic acids, carbohydrates and lipids—usually in some form of combination with each other. The elementary precursors and breakdown products (metabolites) of these four major classes are also present in most biological tissues, especially in fresh, unfixed materials.

Proteins constitute the major proportion of the organic matrix of biological materials. They are built up from various amino-acid units condensed in unique sequences through peptide linkages. Some of the chemical groups most likely to occur in common proteins are illustrated in the hypothetical peptide shown in Fig. 11.1. They are amines, thiols, disulphides, phenols, and side-chain and C-terminal carboxyl groups.

There are two types of nucleic acid in biological tissues: ribonucleic acid (RNA) and deoxyribonucleic acid (DNA). Both have three chemically-reactive centres: negatively-charged phosphate groups, amine groups and heterocyclic bases. DNA differs from RNA in at least two respects. First, the base thymine is present only in the former. Secondly, when DNA is hydrolysed with a dilute mineral acid, an aldehyde is revealed at the C_3 position of the sugar fraction. RNA, which has a hydroxyl group at the C_2 position of its ribose fraction, does not reveal an aldehyde when hydrolysed in this way.

Some saccharides are secreted by, and stored in, two types of special mucus-secreting tissue cells known as mucous and serous cells. Other saccharides constitute part of many connective tissues such as collagen fibres and the matrix of cartilage. Occasionally saccharides are stored as pure polysaccharides such as glycogen (in mammals) or starch (in plants). More usually saccharides are secreted as part of complex polymers called, by some, mucosubstances, and by others (particularly in the older literature), mucopolysaccharides. According to recent views, many mucosubstances consist essentially of a protein backbone or core to which are attached oligosaccharide side-chains linked to amino-acid residues such as serine and threonine. Some of the more common saccharide units which have been identified so far in acid mucosaccharides extracted from connective-tissues are illustrated in Fig. 11.2.

The most important chemical groups in such units are primary and secondary hydroxyls (particularly *vicinal* hydroxyls, two secondary hydroxyls on adjacent saccharide carbon atoms), uronic acids, the

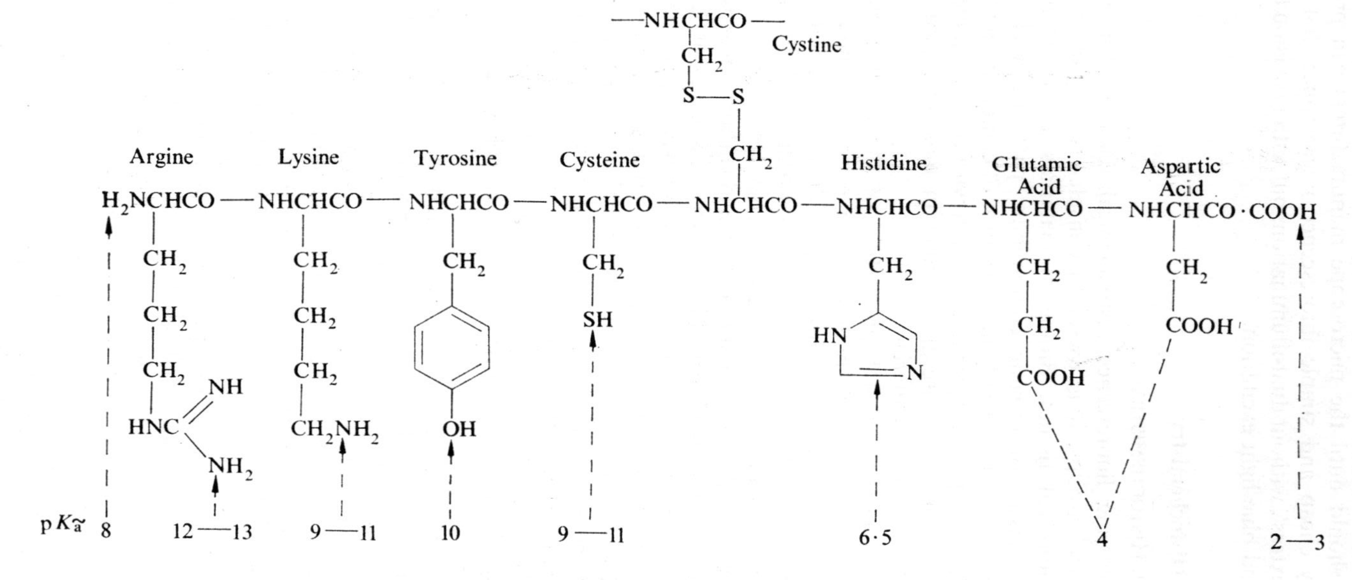

FIG. 11.1 A hypothetical peptide sequence.

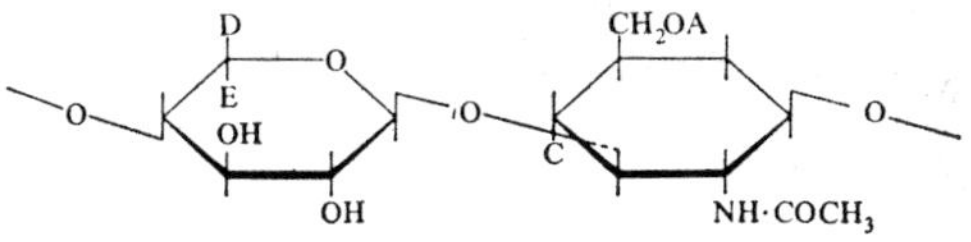

	A	B	C	D	E
Hyaluronic Acid	H	H	OH	COOH	H
Chondroitin Sulphate A	H	$O\cdot SO_3H$	H	COOH	H
" " B	H	$O\cdot SO_3H$	H	H	COOH
" " C	SO_3H	H	H	COOH	H

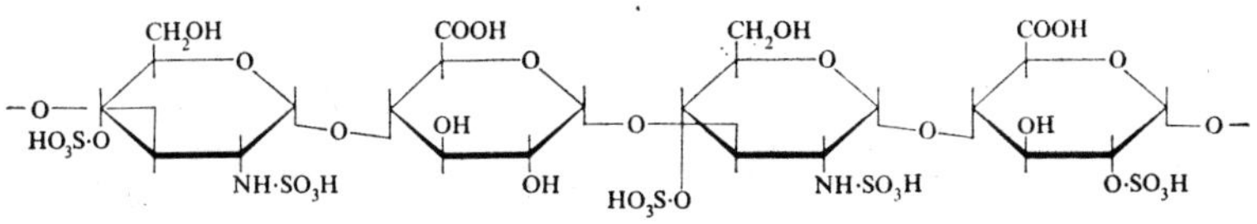

Heparin

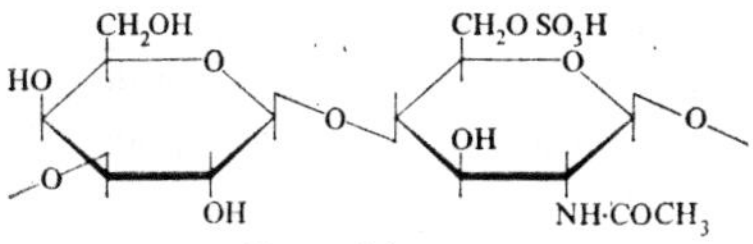

Keratosulphate

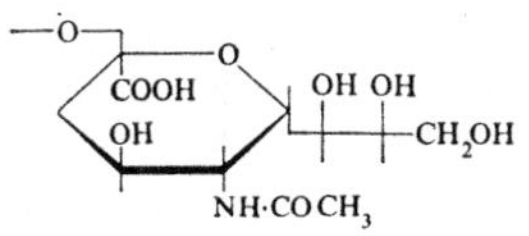

O-Sialic acid (N-acetyl neuraminic acid)

FIG. 11.2 Repeating structural units of some acid mucosaccharides.

carboxyls of sialic acid residues, sulphates (which strictly speaking are half-ester sulphates) and N-acetylamines.

Acid mucosubstances are compounds which have 'acid' properties, as characterized by the affinity of their negatively-charged anionic groups at pH 4 and above, for positively-charged basic dyes such as azure A or Alcian Blue. Those possessing sulphate groups are often referred to as sulphated mucosubstances or sulphomucins, and those which do not possess sulphate groups are usually—sometimes on insufficient evidence—called sialomucins (i.e., polymers of N-acetylneuraminic acid). It is generally assumed that sulphomucins also contain uronic acid groups. However, one sulphomucin which has been isolated and found to have no uronic acid groups is keratosulphate. In contrast, a sulphate-free acid mucosubstance that does not contain sialic acid residues is hyaluronic acid, which is secreted in such diverse tissues as cock's comb and human umbilical cord.

Mucosubstances having no affinity for basic dyes are called neutral mucosubstances. They usually have no sulphate or uronic acid groups or terminal sialic acid residues (Fig. 11.3). Their most important, and sometimes only identifiable, chemical groups are *vic*-glycols.

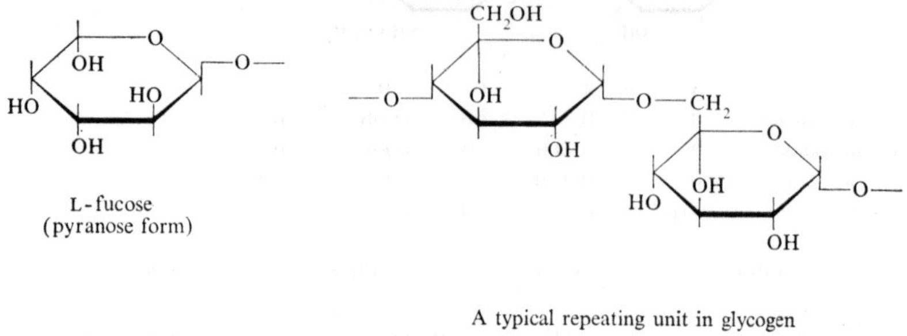

L-fucose
(pyranose form)

A typical repeating unit in glycogen

FIG. 11.3 Examples of residues present in neutral mucosubstances.

The protein fractions of mucosubstances *in situ* are virtually undetectable with the histochemical techniques available at present. Even chemists have concentrated most of their efforts on studying the carbohydrate fraction of mucosubstances. Consequently, histochemical investigations of mucosubstances reduce to a study of the properties of the *vic*-glycol, sulphate, uronic acid, and sialo-carboxyl groups of the carbohydrate fraction.

Lipids are not likely to be found in the fixed and paraffin-embedded tissue sections used in routine histological research. However, when they are present, they can sometimes be characterized by the affinity of their long hydrocarbon chains for protoporphyrin IX or 3, 4-benzpyrene or by the oxidation of their olefinic groups to aldehydes or, in the case of certain ketosteroids, by the reactions of their ketone groups with salicylhydrazide.

THE DEVELOPMENT OF HISTOCHEMICAL TECHNIQUES

At the moment, new fluorescence histochemical techniques are being developed along the following lines. First, reactions which appear to be specific for a particular chemical group *in vitro* are tried out on sections of fixed tissue. Second, if a fluorescence reaction appears to take place, the results are compared against those obtained with other, more conventional histochemical techniques whose chemical specificities most biologists would accept. Third, if a good correlation is obtained, the mechanism and specificity of the histochemical course of the chosen fluorescence reaction is investigated as rigorously as possible.

It should be emphasized that at the moment histochemical techniques are more limited than those used in organic chemistry, for the following reasons:

(1) Practically all histochemical reactions are essentially reactions between reagents in the solution phase and tissue substrates in the solid phase.
(2) The products obtained from such reactions should be insoluble and

indiffusible in either aqueous or alcoholic solvents—preferably in both.

(3) The pH range over which histochemical reactions can be carried out must be between 1 and 8 or, at the most, 9, as otherwise the morphology of the tissue becomes seriously distorted.

(4) The maximum temperature at which histochemical reactions are carried out should not exceed 60°C. Above that temperature, tissue components tend to be extracted from tissues.

(5) There are few histochemical techniques which can be confidently used for comparative purposes; the specificities of most of those now in use (reviewed comprehensively by Pearse) would not satisfy the criteria of organic chemistry. However, there are a few apparently specific reactions (some of which have been reviewed by Spicer) and these have been used to substantiate the dogmatic statements in this chapter that, say, such and such a mucosubstance contains sulphate groups. Nevertheless, when a histochemical reaction is applied to the detection of a particular chemical group in a particular tissue site, the possibilities that other groups in that site may also react should always be considered. This point will be illustrated later.

(6) It is very rarely possible to investigate the physico-chemical properties (e.g., light absorption) of reaction products in tissue sections, and therefore one can never be absolutely certain of the identities of the products obtained in histochemical reactions. This is largely due to technical difficulties. Such reaction products are present only in very small amounts (about 10^{-7}–10^{-9} g at most in a particular cell in 5μ sections) and even these are distributed heterogeneously. Suitable commercial instruments (e.g., absorption microspectrophotometers) are now being developed, and should no doubt be used, but unfortunately they are extremely expensive.

The Use of the Basic Dye Coriphosphine for the Detection of the Sulphate Group of Mucosubstances

Normally the sulphate, uronic acid and sialo-carboxyl groups of acid mucosubstances are negatively-charged (i.e., ionized) at pH 4 (2 and above for many sulphate groups) as are also the phosphate groups of nucleic acids, the carboxyl groups of proteins, and the sulphinic and sulphonic acid groups of sulphur-containing proteins (e.g., keratin). Therefore in tissue sections, acid mucosubstances should, and do, take up basic fluorescent dyes (which, in salt-free solutions, are almost always positively charged) in much the same way as they bind visibly-coloured dyes such as azure A or Alcian Blue. However, with the exception of coriphosphine, acridine orange and thioflavine T adsorbed on to sulphomucins, and with the possible exception of hyaluronic acid, fluorescent dyes adsorbed on to acid mucosubstances never fluoresce. Acriflavine, for example, stains sulphated mucosubstances visibly and distinctly, but the mucosubstance-acriflavine addition compound does not fluoresce. Even auramine O, which in solution or when combined with most substrates emits an extremely intense, greenish-yellow fluorescence, has its fluorescence completely quenched when adsorbed on to acid mucosubstances.

When coriphosphine (I) or acridine orange (II) is adsorbed on to sulphated mucosubstances in tissue sections it usually emits a characteristic red or yellowish brown fluorescence. Unfortunately the colour of this fluorescence is markedly affected by impurities in the dye and by the method used for fixing the tissue. Even when these two variables are eliminated or taken into account, the fluorescence of the sulphomucin-dye conjugate cannot always be distinguished from that emitted by proteins and nuclei, which under certain conditions is of a similar colour.

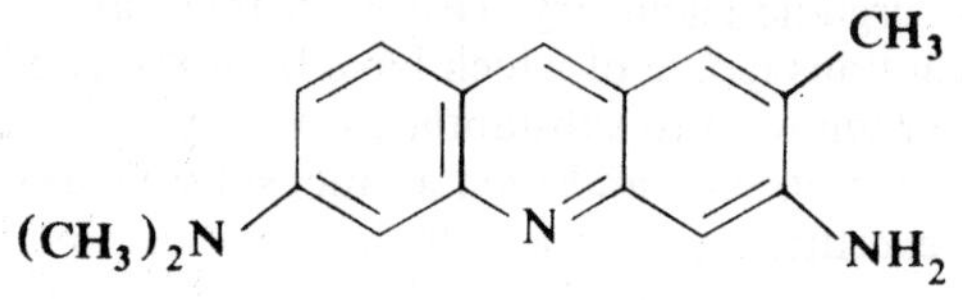

CH_3

$(CH_3)_2N$ N NH_2

I

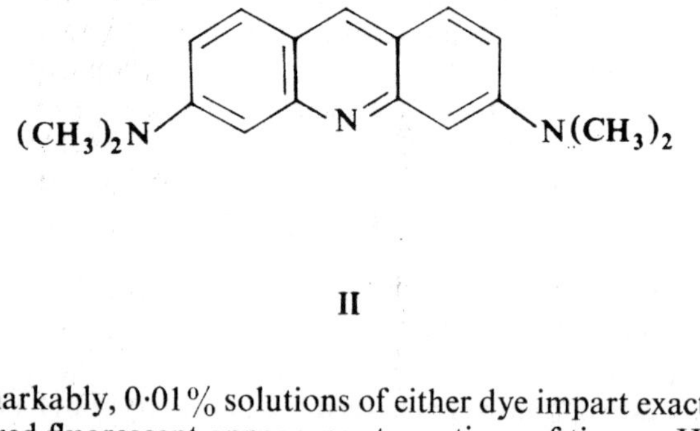

$(CH_3)_2N$ N $N(CH_3)_2$

II

Quite remarkably, 0·01% solutions of either dye impart exactly the same multi-coloured fluorescent appearance to sections of tissues. However, the fluorescence of sections stained with coriphosphine fades much more slowly on prolonged exposure to ultra-violet light than does the fluorescence of those stained with acridine orange.

When sections of mammalian tissue that have been fixed in a coagulant (such as Carnoy's fluid consisting of equal parts of absolute ethanol, glacial acetic acid and chloroform) are stained at pH 1·5–4·0 in a dilute (0·01%) solution of 'impure' coriphosphine (i.e., as supplied commercially), tissue cells known to contain high concentrations of sulphomucins either emit a yellowish fluorescence, or they do not fluoresce at all. The extent of the fluorescence never corresponds to the distribution of mucosubstance sulphate groups revealed by other histochemical methods. This effect is illustrated for the glossal glands of hamster tongue in Figs. 1 and 2 of the frontispiece. The sulphomucins secreted within these glands fluoresce hardly at all when stained with impure dye at pH 1·5–3, and not at all when stained at pH 4.

Some sulphomucins, for a reason as yet unexplained, do not take up basic dyes with molecular weights less than about 400 until the pH of the dye solution is about 3·5. Tissue sites known to contain such sulphomucins do not usually emit any fluorescence when stained at this pH with 0·01% solutions of 'impure' coriphosphine. If the dye adsorbed at this pH fluoresces at all, its fluorescence is weak, and dull brown in colour. The intensity of the fluorescence emitted by nuclei is also much less after staining at pH 3·5 and above than after staining at pH 1·5.

Commercial samples of coriphosphine usually contain appreciable amounts of zinc chloride, and consequently, when tissue sections are stained in dye solutions containing this impurity, zinc ions are probably co-adsorbed on to sulphomucins where they tend to quench the fluorescence of the adsorbed dye.

However, solutions of purified coriphosphine which are free from metals impart a distinctive, often intense, fluorescence to sulphomucins *in situ* that is quite different from that obtained with solutions of impure dyes.

Unfortunately, in tissues which, of necessity, have been stained with a solution of pure coriphosphine at pH 3·5 (because their sulphomucin content would not take up dye below that pH), the colour and intensity of the fluorescence emitted by sulphomucins matches that emitted by nuclei and sometimes cytoplasmic proteins. Thus, those who are not familiar with the histological appearance of the small intestine of the rat, for example, would not be able to distinguish the orange fluorescence emitted by the sulphated mucosubstances in the mucus cells (shown towards the top of Fig. 4 of the frontispiece) from the diffuse and similarly-coloured fluorescence emitted by nuclei. Therefore before coriphosphine or acridine orange can be used for demonstrating sulphomucins specifically, the uptake of these dyes by tissue polyanions other than sulphate group of sulphomucins must be 'blocked' (as biologists would say) or substantially reduced.

In sections of fixed tissue from which lipids have been extracted, the following polyanionic groups, in theory at least, need to be blocked: carboxyls of proteins, sulphinic and sulphonic acids of oxidized sulphur-containing proteins, phosphate groups of DNA and RNA, sialo-carboxyls and, perhaps, uronic acids of acid mucosubstances. With 0·01% (but not 0·1%) solutions of coriphosphine of pH below 3·5, the colour of the fluorescence of the conjugate formed with protein carboxyls is blue or, more usually, green. It is rarely red or yellow, and normally can be distinguished easily from the fluorescence emitted by coriphosphine taken up by sulphomucins. On the other hand, the conjugates formed with RNA, nuclear DNA, oxidized sulphoproteins and some sialomucins, often emit a red or yellowish-brown fluorescence. This, however, can be largely blocked by treating the tissues in such a way that the phosphate groups of DNA in nuclei no longer take up the dye. Three ways of doing this will be described. They involve the use of meta-phenylene diamines, anions for removing adsorbed dye, and reducing agents.

The Use of Meta-phenylene Diamines for Blocking the Affinity of DNA Phosphate Groups for Basic Fluorescent Dyes

Nuclei do not take up basic dyes such as coriphosphine and acridine orange if they are first submitted to a Feulgen hydrolysis (five minutes treatment with 5N hydrochloric acid at room temperature) and then treated overnight with a solution of N,N-dimethyl-*m*-phenylene diamine near pH 6. It is believed that the aldehydes exposed in the DNA macromolecule by the Feulgen hydrolysis condense with the unsubstituted amino-groups of the diamine, and that, in the product thus formed (III), the positive charge on each dimethylamino group 'neutralizes' the negative charge on the nearest phosphate group of the DNA polymer.

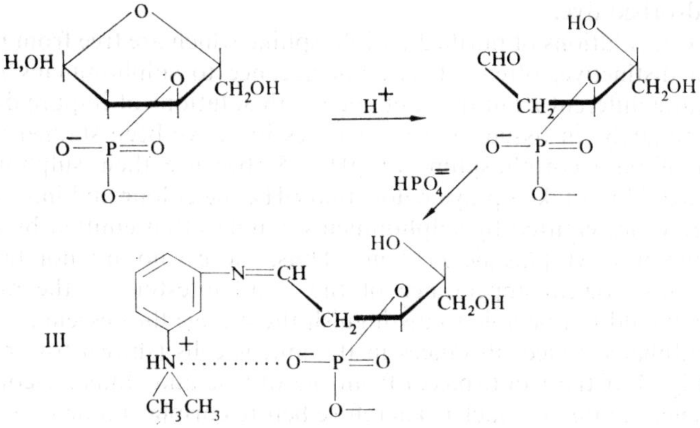

When such pretreated tissue sections are stained in a solution of coriphosphine, the reddish-brown fluorescence of most sulphomucins stands out clearly and distinctly against the green fluorescence of the connective tissue and the non-fluorescent nuclei (e.g., Fig. 6, frontispiece). Unfortunately with this procedure tissue cells known to contain only sialomucins also emit a reddish-brown or orange fluorescence. According to Spicer, the carboxyls of sialomucins adsorb meta-phenylene diamines substantially, and therefore it is puzzling that here such pretreated sialomucins still take up coriphosphine. A similar fluorescence is also emitted by sites known to contain only partly-oxidized sulphoproteins such as occur in hair shafts.

Although meta-phenylene diamines can be used successfully for blocking the affinity of nuclear DNA for basic dyes, they are of limited value in methods intended for the specific demonstration of sulphomucins with coriphosphine.

Reduction with Lithium Aluminium Hydride

The sulphate group of mucopolysaccharides can, in theory, but not always in practice, be demonstrated specifically with the following procedure:

(*a*) First, sections are methylated, to convert all tissue polyanions to their methyl esters.

(*b*) Secondly, they are reduced with lithium aluminium hydride dissolved in hot (60°) dioxane for 48 hours or more, to reduce or remove all tissue polyanionic groups and their methyl esters *except* the full methyl ester sulphate groups on sulphomucins.

(*c*) Next they are saponified with an alcoholic solution of potassium hydroxide, to remove the methyl groups from the esterified sulphate groups of sulphomucins.

(*d*) Finally they are stained in a solution of coriphosphine (pH above 5) to reveal the presence of mucosubstance sulphate groups, in theory at least, the only groups remaining in the tissue which are able to take up basic dyes. The other polyanionic groups should, according to *in vitro* precedents, have been reduced to, or replaced by a hydroxyl group which, of course, has no affinity for basic dyes.

The methylation step can be accomplished by treating tissue sections with a 2% solution of thionyl chloride in absolute methanol at room temperature. This is preferable to an ethereal solution of diazomethane, which is the reagent most chemists would use. Although diazomethane does rapidly esterify most polyanionic groups in tissue sections, such esterified groups, for some unknown reason, cannot be saponified (to regenerate the original anionic groups) with conventional saponifying reagents such as alcoholic solutions of potassium hydroxide. In contrast, however, tissue polyanions which have esterified with methanolic solutions of thionyl chloride can be saponified with such reagents.

For the reduction stage, purified LAH is preferable. If technical grade LAH is used (containing, probably, lithium and aluminium hydrides), the reduced sections disintegrate, for reasons still undetermined, as soon as they are placed in saponification reagents. In addition the dioxane solvent should not be too dry or 'pure'; really dry dioxane (e.g., freshly-opened, AnalaR grade) rapidly builds up moderate concentrations of peroxides which apparently interfere with the reduction process, perhaps by oxidizing LAH or, alternatively, by oxidizing the thiol groups of sulphur-containing proteins (existing in the same site as mucosubstances) to sulphinic or sulphonic acids.

When sections of tissue have been processed through the three-stage LAH reduction technique, stained in coriphosphine at pH 1·5–4, and illuminated with ultra-violet light, most, but by no means all, sulphomucins emit a brownish-yellow or dull brown fluorescence (Fig. 3, frontispiece), whereas nuclei and protoplasmic proteins do not, initially at least, fluoresce at all, or at the most only very weakly. On prolonged exposure to ultra-violet light, however (as is needed for photomicrography), the cytoplasmic proteins acquire a green fluorescence.

The theory of the LAH reduction procedure as outlined here is vitiated by the observation that sulphate-free sialomucins, such as are secreted in the sublingual gland of the female Syrian hamster, emit an intense orange-brown fluorescence when they have been subjected to the LAH reduction procedure, stained with coriphosphine at pH 4·0 and excited with ultraviolet light. If sialomucins are reducible by LAH, after reduction they should not take up coriphosphine and therefore, contrary to observations and to all expectations, they should not emit any fluorescence.

Although the results obtained so far are disappointing, the LAH-coriphosphine method is still worthy of further study since it may, under the right experimental conditions, prove to be an absolute histochemical technique for detecting the sulphate groups of sulphomucins. Its limitations at the moment are undoubtedly due to complications caused, among other things, by some desulphation in the initial methylation stage, by competing and perhaps predominant side-reactions during the LAH reduction, and by the fixation of the tissue.

The Coriphosphine-Thiazol Yellow Method for Detecting Tissue Sulphomucins

When sections of fixed tissue are stained with basic fluorescent dyes and then washed in a solution containing an anion chemically similar to one of the polyanions in the tissue, the dye adsorbed on to that tissue polyanionic site is often extracted first, sometimes preferentially and exclusively. This generalization holds for many simple inorganic anions, but organic anions sometimes behave differently. For example, at very low concentrations (0·001%) the sulphonated anionic dye thiazol yellow (IV) will, at pH 2 but

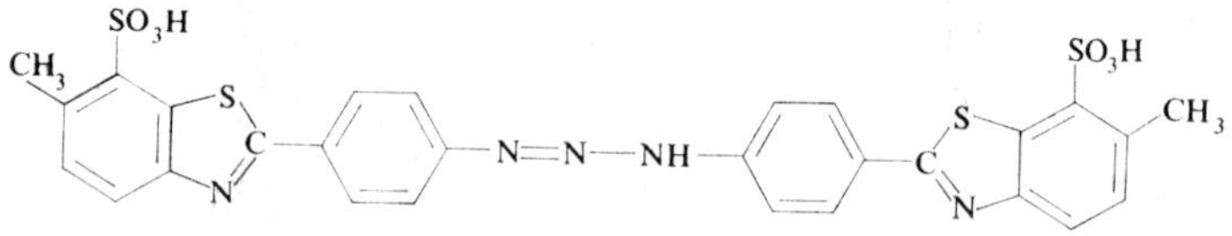

IV

not at higher pH levels, extract acridine orange or coriphosphine adsorbed on to tissue proteins, nucleic acids and oxidized sulphoproteins, but not that adsorbed by sulphomucins.

Thus sulphomucins can be demonstrated selectively with a coriphosphine-thiazol yellow sequence. One example is illustrated in Fig. 5 of the frontispiece (cf. Fig. 4) in which it can be seen that nuclei after losing their adsorbed coriphosphine, bind thiazol yellow on to their basic amine ($—NH_3^+$) groups and emit fluorescence which is of a moderately intense green (with u.v. excitation), or yellowish-green (with blue light excitation). Connective-tissue proteins bind thiazol yellow similarly and initially emit a bright blue fluorescence, but this fades quickly on exposure to ultraviolet or blue light, and can never be recorded on photomicrographs.

A few sulphomucins, such as those present in the surface epilthelial cells of the Syrian hamster vagina, hardly fluoresce after the coriphosphine-thiazol yellow sequence when excited with ultra-violet illumination. However, they fluoresce distinctively if they are excited with blue light. Oxidized sulphoproteins (keratins), on the other hand, never emit the yellow to red fluorescence characteristic of sulphomucins, as they do in sections stained only in coriphosphine, whether excited with ultra-violet or with blue light.

Of the techniques described in this chapter, the coriphosphine-thiazol yellow sequence, although essentially an empirical technique, is the easiest to use and appears to be the most specific. Moreover the pH of the coriphosphine solution, and the manner of excitation, can be varied in order to differentiate various types of sulphomucin.

It is, of course, necessary to prove that, in the various methods described, the fluorescence emanating from sulphomucin sites is actually due to a sulphate-coriphosphine conjugate and not to coriphosphine adsorbed on to other anionic groups such as uronic acids present in the same mucosubstance macromolecule. Unfortunately such an unequivocal proof is not yet available.

Sulphinic Acids

Another 'sulphate' group which can be detected with coriphosphine is the sulphinic acid (or perhaps sulphonic acid) group formed by the oxidation of the thiol and disulphide groups of proteins with peracetic or performic acids. These sulphinic acids differ from all other tissue anionic groups in being the only ones which cannot be esterified or hydrolysed off even under the most drastic conditions. Thus if tissue sections are first oxidized in peracetic acid and then methylated for 24 hours with a methanolic solution of thionyl chloride, the only groups which will take up basic fluorescent dyes are the sulphinic acids present in sulphoproteins. If stained in coriphosphine, for example, they emit a characteristic flame red fluorescence.

The Histochemical Demonstration of *Vic*-glycols and Aldehydes

When tissue sections are oxidized with a solution of periodic acid at a pH between 3 and 5, the *vic*-glycols (V) of mucosubstances are cleaved into 'dialdehydes' (VI) which in some instances may undergo instantaneous hydration and internal condensation into hemialdals (VII) or, if primary hydroxyl groups are present in the saccharide units (as in glucose residues), into hemiacetals (VIII).

There are two fluorescence histochemical methods for demonstrating the aldehydes thus engendered. One method utilizes basic fluorescent dyes, and the other salicylhydrazide.

Engendered aldehydes react with sulphurous acid at pH 2–3 to form, it is believed, alkyl sulphonic acids (IX).

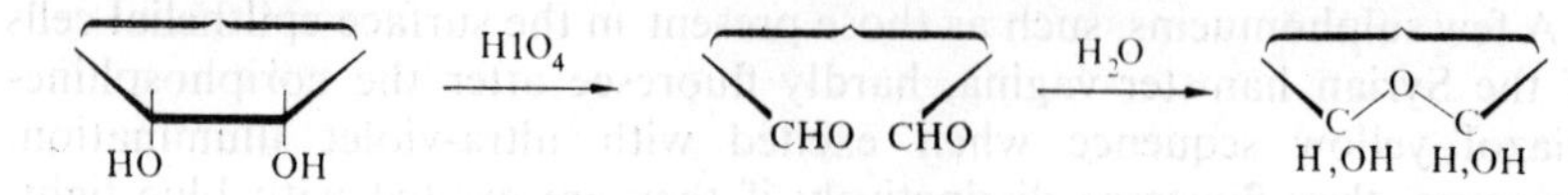

HO OH
V
HIO_4
CHO CHO
VI
H_2O
H,OH H,OH
VII

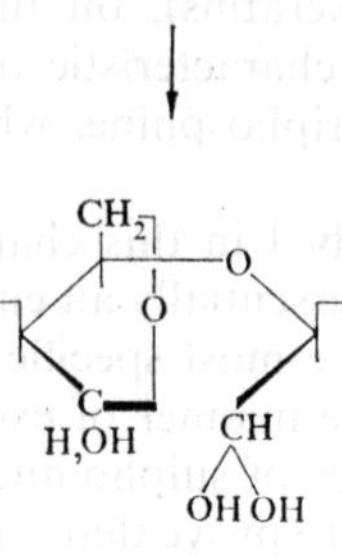

CH_2
H,OH CH
OH OH
VIII

$$R.CHO + H_2SO_3 \rightarrow R.CH(OH)(SO_3H)$$

IX

These products have a strong affinity for basic fluorescent dyes (B^+) at a pH near 2.

$$IX \xrightarrow[-H^+]{?} R.CH(OH)(SO_3^+) \xrightarrow{B^+} R.CH(OH)(SO_3^- \ldots B^+)$$

However, if tissue sections are oxidized in periodic acid, and next either treated with sulphurous acid and then stained in a solution of any basic fluorescent dye (e.g., benzoflavine X), or immersed in a solution (pH about 3) of any basic fluorescent dye saturated with sulphur dioxide, it is not always possible to differentiate the fluorescence induced in mucosubstance sites from the remainder of the tissue whose anionic groups have taken up the dye by simple electrostatic forces as described earlier for coriphosphine.

Fortunately this uptake can be prevented by treating tissue sections with

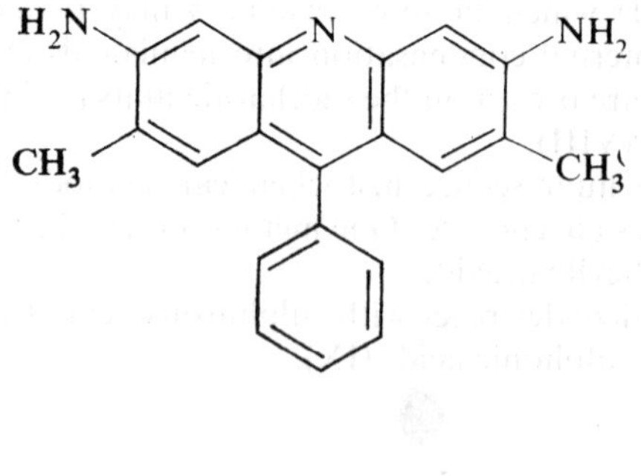

H_2N N NH_2
CH_3 CH_3

a methanolic solution of thionyl chloride (which converts all tissue anionic groups to their appropriate methyl esters) before carrying out the periodate oxidation (cf. Figs. 7 and 8, frontispiece). The methylating reagent apparently does not methylate saccharide hydroxyl groups.

One odd, so far inexplicable, property of the alkyl sulphonic acid intermediates (IX) formed from periodate-engendered aldehydes is that although (as shown by the increased intensity of their fluorescence emission) they will take up, in increasing amounts, basic fluorescent dyes such as benzoflavine, acridine yellow or acridine orange as the pH of the dye solution is raised from 1·5 to 3·0, they do not fluoresce when stained at pH 4 and above. If the periodate-oxidized mucosubstance-sulphurous acid intermediates do, as it is claimed here, have the general formula IX, then the alkylsulphonic groups of the intermediates should, if anything, be ionized more at pH 4 and consequently, contrary to observation, have a greater affinity for basic fluorescent dyes at this pH level than at pH 3 and below.

At first sight the failure to fluoresce might be thought to be due to a breakdown of the intermediates IX at pH 4, with a subsequent inability to bind fluorescent dyes. That this is not so can be proved by showing that most periodate-oxidized mucosubstances fluoresce quite strongly when treated first with a sulphurous acid at pH 4 and second with a solution of a basic fluorescent dye at pH 3.

Another explanation is that the alkyl sulphonic acids actually condense with the amine groups of basic fluorescent dyes, especially as the greatest uptake of dye occurs at pH 3 which is usually regarded as the optimum pH for Schiff condensations. This is improbable for at least two reasons. First, the amine groups of most fluorescent dyes are virtually unreactive chemically; for example, they will only take up hydrogen ions in concentrated mineral acids. Second, the periodate-oxidized mucosubstance-sulphurous acid intermediates take up acridine orange (II), a dye which does not possess any free amine groups (Fig. 9, frontispiece). This last fact also disproves a theory, current in the histochemical literature, which concerns the reaction between aldehydes and Schiff's solution itself.

Schiff's solution is essentially a colourless solution of the triphenylmethane dye pararosaniline that has been saturated with sulphur dioxide and reacts with aldehydes to form magenta-coloured derivatives. According to several opinions, the sulphur dioxide in the solution combines with amine groups of the dye ($R.NH_2$) to form colourless N-sulphinic acids (XI) that react with aldehydes (R.CHO) to form products of general formula XII. An older view, until recently rejected by many, is that the sulphur dioxide present in Schiff's solution reacts with aldehydes to form alkyl sulphonic acids which can condense with the anilinium-like ion of pararosaniline. The product thus formed is coloured because the pH of the reaction (usually 1·5–3·0) is simply not low or high enough to keep the product in a colourless form. This view was substantiated in 1960 with spectrophotometric evidence, as it is also by the fluorescence histochemical work quoted here.

$$R.NH_2 + SO_2 \rightarrow R.NH{-}SO_2H$$

XI

$$XI + CHO.R' \rightarrow R.NH.SO_2CH\begin{smallmatrix} R' \\ OH \end{smallmatrix}$$

XII

Solutions of fluorescent aminoacridine dyes saturated with sulphur dioxide (which are sometimes erroneously called Schiff-type solutions) will also react with periodate-engendered aldehydes of tissue mucosubstances. However, the intensity of fluorescence emitted by the products thus formed very rarely equals that emitted by the corresponding derivatives obtained with the sulphurous acid-basic dye sequence just described. Perhaps with Schiff-type solutions the periodate-engendered aldehydes react with the sulphurous acid present in solution to form alkyl sulphonic acids as before, but in addition the sulphurous acid may also become loosely attached to the electrophilic centres of the dye to form a sort of sulphonic acid such as that illustrated below for acridine yellow (XIII). It is highly probable that the intermediates (e.g., XV and IX) would, instead of combining, tend to repel each other.

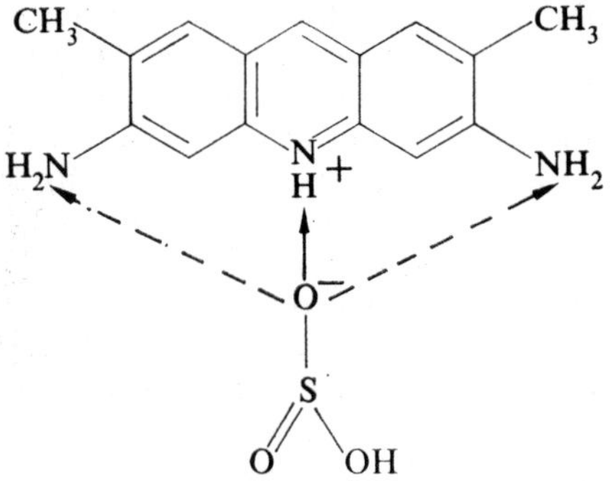
CH3
CH3
H2N
N
H
+
NH2
O
S
O
OH

XIII

XIV

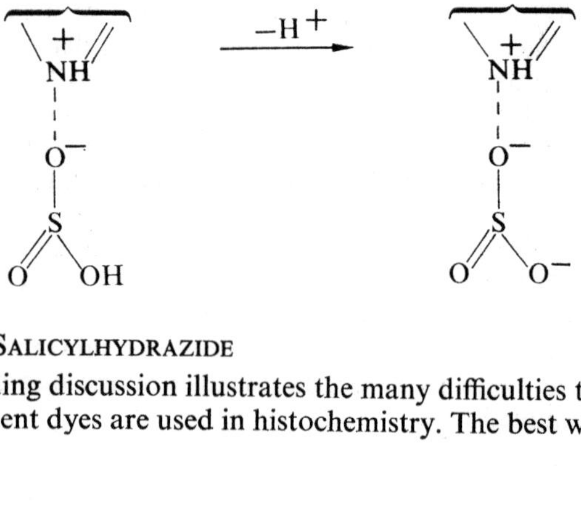
+
NH
O
S
O
OH
−H+
+
NH
O
S
O
O

XV

The Use of Salicylhydrazide

The preceding discussion illustrates the many difficulties that arise when basic fluorescent dyes are used in histochemistry. The best way of avoiding such difficulties is not to use dyes at all. Reagents which actually condense with tissue groupings are much more likely to be specific and also easier to control experimentally. Thus a simpler method of demonstrating aldehydes such as those engendered in tissue mucosubstances after a periodate oxidation is to condense them with salicylhydrazide (XVI). The hydrazones (XVII; $R_2 = H$) thus formed usually emit a characteristic

$$\text{(2-OH)}C_6H_4\text{-}CONHNH_2 + O{=}C\begin{matrix} R_1 \\ R_2 \end{matrix} \longrightarrow \text{(2-OH)}C_6H_4\text{-}CONHN{=}CH\begin{matrix} R_1 \\ R_2 \end{matrix}$$

VI XVII

light-blue fluorescence whose intensity is enhanced after complexing the hydrazones with aluminium ions, but if they are derived from periodate-engendered aldehydes in the hemialdal or hemiacetal forms (VII or VIII), the hydrazones emit a less intense, royal-blue fluorescence (Fig. 11, frontispiece) which, in contrast, tends to be quenched by solutions of aluminium salts.

The aldehyde salicylhydrazone—aluminium complexes derived from periodate-oxidized sulphomucins and sialomucins emit a very intense fluorescence (Fig. 10, frontispiece) but the complexes derived from either periodate-oxidized neutral mucosubstances, or from the aldehydes exposed in nuclear DNA after a brief hydrolysis in hot hydrochloric acid, emit a *comparatively* weak fluorescence. This illustrates one of the several unusual properties of salicylhydrazide. As a general rule, the fluorescence of compounds (e.g., fluorescent basic dyes) adsorbed on to or near indigenous sulphate or sialo-carboxyl groups of tissue mucosaccharides tends to decay or be quenched, whereas the intensity of the fluorescence emitted by derivatives of tissue compounds in which these groups are absent tends to be enhanced. With salicylhydrazide, the generalization is reversed.

Salicylhydrazide is perhaps the most useful reagent so far discovered for fluorescence histochemical work. It can be used in several ways for the practical demonstration of nearly all types of chemical group that are likely to occur in tissues. This usefulness is based on two unique properties of salicylhydrazide. First, although it reacts with all compounds containing a chemically reactive carbonyl group, only aldehyde derivatives emit a stable, intense fluorescence, in contradistinction to aliphatic ketone salicylhydrazones, which do not fluoresce until they have been complexed with zinc ions at a pH between 5 and 6. Minor exceptions to this generalization are the derivatives formed from ketosteroids. 17-ketosteroid salicylhydrazones fluoresce without zinc treatment, but this fluorescence is bluish-green, comparatively weak, and quenchable by alkali, and it quickly fades after about 10 minutes' exposure to ultra-violet light. Many

3-ketosteroid salicylhydrazones, on the other hand, fluoresce only when they have been treated with a strong alkali.

The second unique property of salicylhydrazide is that it appears to be the only acid hydrazide whose aldehyde hydrazones react with diazonium salts in the presence of pyridine to form highly-coloured formazans.

The essential problem in using salicylhydrazide for detecting a particular chemical group in tissue macromolecules is to convert or modify that group first, so that it contains a carbonyl group. The following are examples of groups where this is possible.

Amine Groups

Terminal amine-acid groups of proteins (XVIII) can be converted into aldehydes by brief oxidation with a solution of sodium hypochlorite at pH 7·0–7·5.

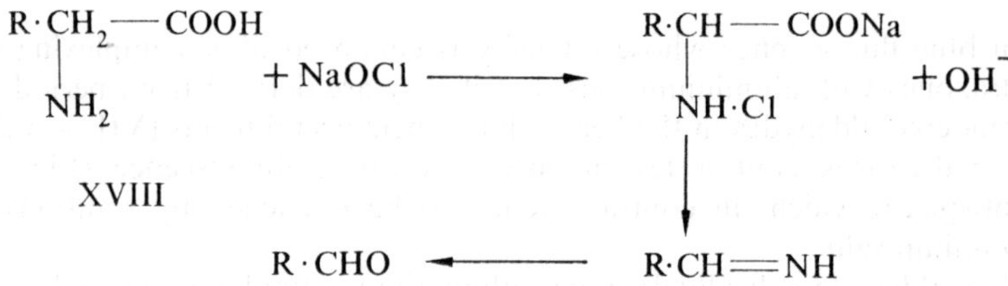

$$R\cdot CH(NH_2)-COOH \ (\text{XVIII}) + NaOCl \longrightarrow R\cdot CH(NH\cdot Cl)-COONa + OH^-$$

$$R\cdot CH(NH\cdot Cl)-COONa \longrightarrow R\cdot CH{=}NH \longrightarrow R\cdot CHO$$

One interesting feature of this oxidation is that when the oxidized proteins in tissue sections are condensed with salicylhydrazide, discrete granules of blue fluorescent hydrazones are formed: the whole connective-tissue does not fluoresce as would be expected if the distribution of terminal amino-acids were uniform (at the molecular level) throughout the protein, as indeed is indicated from other histochemical tests (e.g., those using acid fluorescent dyes, such as thiazol yellow, at neutral or acid pH levels).

Thiol Groups

Many protein thiol groups (R.SH) react with N-ethylmaleimide at pH 7·2–7·4 to form a product (XIX) which possesses an unconjugated

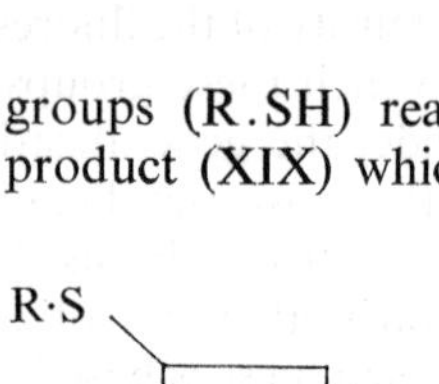

XIX: 3-(R·S)-N-ethylsuccinimide (five-membered ring with R·S on C-3, two C=O groups flanking N, N–Et)

ketone group. This product condenses with salicylhydrazide to form a non-fluorescent hydrazone which, however, emits a green fluorescence when treated with a solution of zinc acetate. Amino groups and imidazole and histidine residues of proteins may also react with N-ethylmaleimide, but they do not yield products with an unconjugated ketone group available for reaction with salicylhydrazide.

C-terminal Carboxyl Groups

A mixture of hot acetic anhydride and pyridine converts the C-terminal groups (XX) of proteins (originating mostly from aspartic and glutamic acid residues) into methyl ketone derivatives (XXIII) via, it is thought, azlactones (XXI) or Ψ-oxazolones (XXII) as intermediates which undergo acetylation, ring opening and decarboxylation.

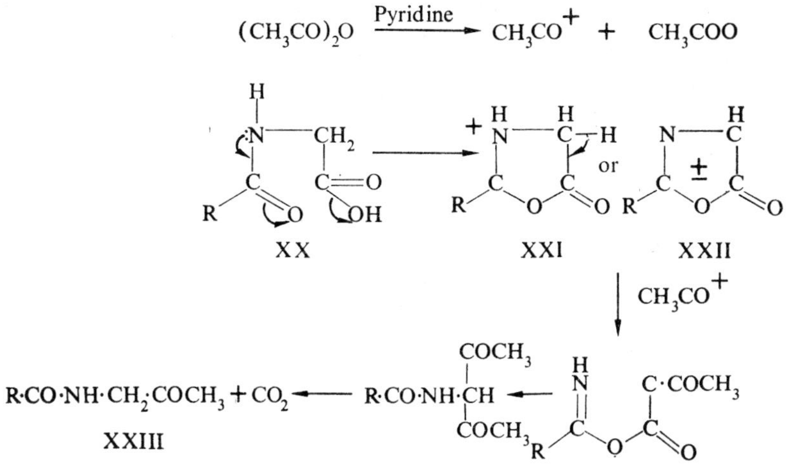

$$(CH_3CO)_2O \xrightarrow{\text{Pyridine}} CH_3CO^+ + CH_3COO$$

XX → XXI or XXII

XXII $\xrightarrow{CH_3CO^+}$ → $R{\cdot}CO{\cdot}NH{\cdot}CH(COCH_3)_2$ → $R{\cdot}CO{\cdot}NH{\cdot}CH_2{\cdot}COCH_3 + CO_2$

XXIII

Like other aliphatic ketones, the methyl ketones thus formed also react with salicylhydrazide to give a hydrazone which fluoresces only when it has been complexed with zinc ions.

According to biochemical analyses, C-terminal carboxyl groups occur sparsely in proteins, but this histochemical technique reveals high concentrations of such groups in, for example, the proteins constituting muscle fibres (Fig. 12, frontispiece).

The success of this technique, together with other evidence, vitiates an alternative theory proposed recently in the histochemical literature that hot acetic anhydride does *not* convert C-terminal carboxyl groups to methyl ketones, but on the contrary, reacts with side-chain carboxyl groups (XXIV) of proteins to give mixed acid anhydrides (XXV) which, if formed,

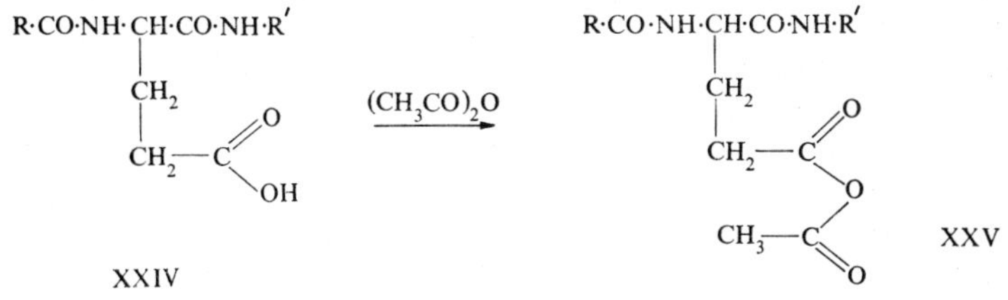

$R{\cdot}CO{\cdot}NH{\cdot}CH(CH_2CH_2COOH){\cdot}CO{\cdot}NH{\cdot}R'$ (XXIV) $\xrightarrow{(CH_3CO)_2O}$ $R{\cdot}CO{\cdot}NH{\cdot}CH(CH_2CH_2CO{\cdot}O{\cdot}COCH_3){\cdot}CO{\cdot}NH{\cdot}R'$ (XXV)

could acylate salicylhydrazide to form a mixture of dihydrazide derivatives. However, there are several lines of experimental evidence which indicate that if XXV are ever formed (which they may be, under certain conditions), they do not emit an intense blue fluorescence, nor do their salicyldihydrazide derivatives nor their zinc-salicyldihydrazide complexes.

Olefinic Groups of Lipids

Peracetic acid oxidizes unsaturated lipids (XXVI) in two stages. In the first stage, peroxides (XXVII) are formed, which, presumably, are then cleaved into hydroxyketo- or dihydroxy-compounds (XXVIII) that subsequently break down into an aldehyde (XXIX) and an ω-hydroxy compound.

The aldehydes (XXIX) thus formed in tissue sites after oxidation with per-acids can be condensed with salicylhydrazide (XVI) to yield blue fluorescent hydrazones.

$$\underset{\text{XXVI}}{R{\cdot}CH{=}CH{\cdot}R'} \longrightarrow \underset{\text{XXVII}}{R{\cdot}\overset{\displaystyle O}{\overset{|}{C}H}{-}\overset{\displaystyle O}{\overset{|}{C}H}{\cdot}R'} \quad (\text{O}{-}\text{O bridge})$$

$$\underset{\text{XXIX}}{R{\cdot}CHO + R'{\cdot}CH_2OH} \longleftarrow \underset{\text{XXVIII}}{R{\cdot}\overset{\displaystyle OH}{\overset{|}{C}}{-}\overset{\displaystyle OH}{\overset{|}{C}H}{\cdot}R'}$$

OTHER USEFUL FLUORESCENCE HISTOCHEMICAL REACTIONS

The following groups and compounds can be detected without recourse to coriphosphine or salicylhydrazide.

Protein Amine Groups

An *extremely* intense green fluorescence is emitted by tissue proteins treated with a methanolic solution of pyridine containing a few drops of an aqueous solution of cyanogen bromide, bromine, and potassium cyanide.

The chemistry of the reaction is complex. The simplest interpretation is that cyanogen bromide opens the pyridine ring and an aldehyde intermediate (XXX) is formed which condenses with protein amine groups to give a fluorescent polymethine anil.

Tissue proteins also react with salicylaldehyde (XXXI) to give a thermolabile, green fluorescent product, presumably an anil.

When tissues are treated with a complex formed by mixing a solution of certain Solochrome Black dyes with a solution of alum, proteins and

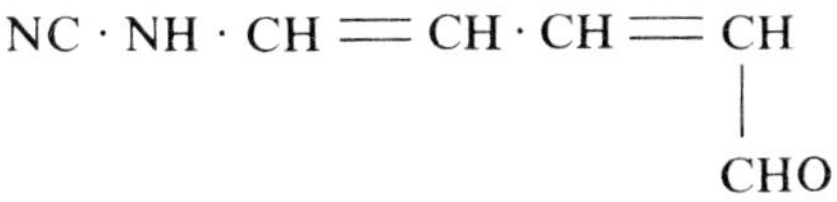

$NC \cdot NH \cdot CH{=}CH \cdot CH{=}CH(CHO)$

XXX

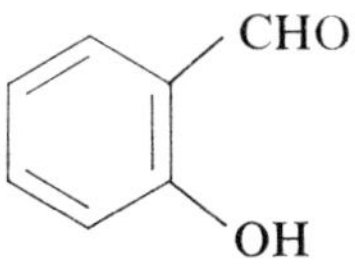

CHO
OH

XXXI

nuclei emit a distinctive red or purple fluorescence, the colour depending on the particular Solochrome Black used.

It is believed that in the mixed solutions, aluminium ions become chelated with the phenol and azo groups of the normally non-fluorescent Solochrome Black (e.g., 6BN, XXXII) and render the dye fluorescent. The

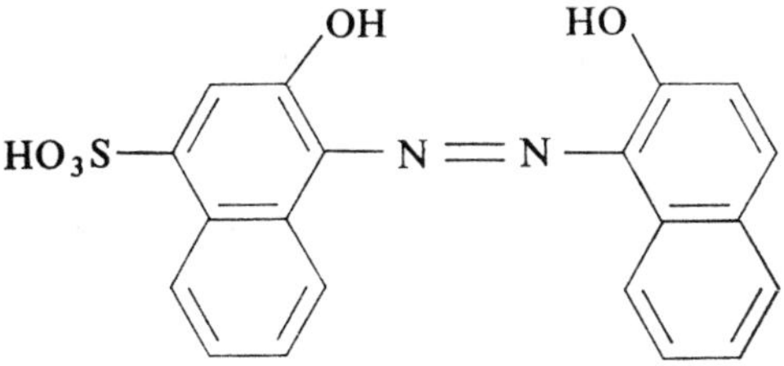

OH HO
HO_3S N=N

XXXII

dye-aluminium complex contains negatively-charged sulphonic acid groups which, at acid pH levels, presumably bind on to the positively-charged amine (—NH_3^+) groups of tissue proteins and nucleic acids.

Tryptophan Residues

Tryptophan residues in tissue proteins emit a characteristic purple fluorescence after being treated with a solution of dimethylaminobenzaldehyde in hydrochloric acid. In this complex reaction perhaps 3-indoyl derivatives are formed first, which subsequently undergo internal condensation into β-carbolines (XXXIII), known from *in vitro* experiments to be fluorescent.

Phenols

Protein tyrosine residues with free meta or ortho positions undergo ring substitution with 1-nitroso-2-naphthol to form unstable, green fluorescent products of indeterminate structure.

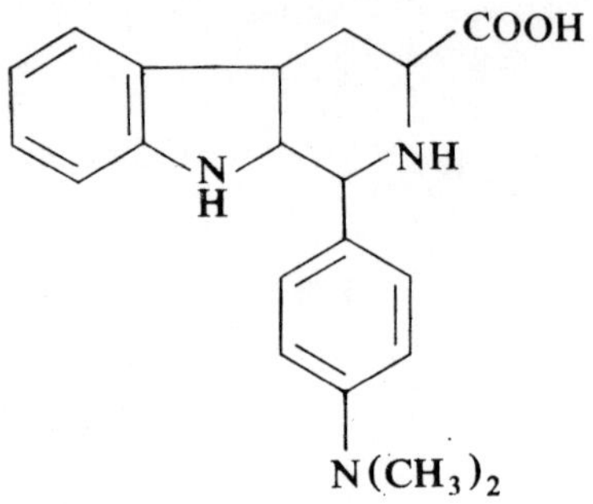

COOH NH N H $N(CH_3)_2$

XXXIII

Catecholamines

Sections of fresh-frozen tissucs, such as adrenal or brain, after being exposed to formaldehyde vapour, emit an intense yellow or green fluorescence in sites containing catecholamines. Such amines (e.g., dopamine XXXIV) condense with formaldehyde even under mild conditions to form an intermediate which almost immediately undergoes ring closure to give a fluorescent tetrahydroisoquinoline derivative (e.g., XXXV).

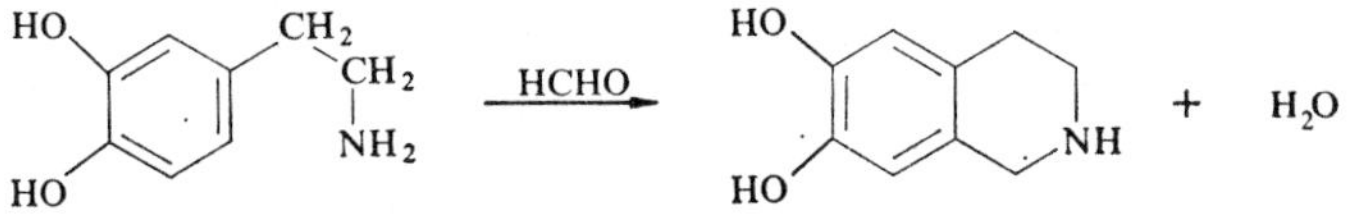

HO HO CH_2 CH_2 NH_2 HCHO HO HO NH + H_2O

XXXIV XXXV

Uronic Acids

The only commonly-occuring group which cannot be detected easily in tissue sections is the uronic acid group. A few, but by no means all, acid mucins thought to contain uronic acid groups emit an intense blue fluorescence after being treated carefully with concentrated sulphuric acid at 60–70°C. The chemical nature of the fluorescent product, where it is formed, is unknown. Hot concentrated sulphuric acid also renders certain steroids (e.g., cholesterol) fluorescent.

DIFFICULTIES ASSOCIATED WITH THE DEMONSTRATION OF NUCLEIC ACIDS

Distinctive and different fluorescence colours can be observed in sites known to contain nucleic acids that have been stained in dilute (0·01%) solutions of either acridine orange or coriphosphine at a pH near 4. Unfortunately the actual colour in any structural feature depends, as has been mentioned earlier, on the fixation of the tissue. It is rarely possible to assign a particular fluorescence colour unambiguously to a given nucleic acid.

In frozen sections, for example, nuclei (containing DNA) emit a green fluorescence after being stained with acridine orange, whereas cytoplasmic RNA emits a red fluorescence. But the fluorescence colours in fixed sections are often the reverse (see, for example, the orange fluorescent nuclei in the frontispiece, Fig. 2). Nevertheless DNA appears to be the only tissue component which emits a reddish-orange fluorescence in fixed sections which have been stained in Alcian Blue at pH 2·5, and then in coriphosphine at pH 3. Alcian Blue is a high-molecular-weight dye of the copper phthalocyanine type, and is taken up at pH 2·5 selectively by sulphomucins and sialomucins. The bound Alcian dye quenches the fluorescence of any coriphosphine adsorbed subsequently on to such mucins.

There are no unequivocal fluorescence methods for detecting the chemically-reactive centres of either DNA or RNA in tissue sections. Contrary to popular belief, it is practically impossible to detect RNA with any degree of certainty. Even the disappearance of certain types of staining in tissues pretreated with ribonuclease can be interpreted in several and controversial ways.

Some Medical and Industrial Applications of Fluorescence Microscopy

The applications so far described in this chapter are, in most cases, new ones, and are of more interest to chemists than to those engaged in, say, the medical field. However, since about 1940 fluorescence microscopy has been used extensively in medical research, for fluorescent immunoprotein tracing. When foreign proteins or antigens are injected into an animal, they induce the production of other proteins (globulins), known as antibodies, which counteract or neutralize the invading antigens. These antibodies can be extracted from the blood, purified and 'labelled' with a suitable fluorescent compound such as fluorescein isothiocyanate, which condenses with a small proportion of the terminal amine and thiol groups of the antibody protein molecule. If a solution of fluorescent labelled antibody is flooded on to a tissue section containing the original antigen, the antibody is precipitated on to that antigen, and thus the distribution of the antigen can be recorded by fluorescence microscopy. Nairn's book contains a detailed discussion of many applications of this technique.

Fluorescence microscopy is also widely used for the routine screening of certain types of cancer in which smears containing suspect cells are stained with acridine orange. It has been claimed that malignant cancer cells contain a comparatively large amount of an acidic substance which combines with acridine orange to form an unusually intense fluorescent complex. Normal cells form the same complex, but to a much smaller extent. The difference in total fluorescence output per cell is sufficiently large for semi-automated scanning methods to be used for the routine screening and approximate cell counts of stained smears.

Fluorescence microscopy has been used, too, for following the course and fate of fluorescent drugs such as aspirin or carcinogenic compounds injected into the body.

There have also been numerous empirical applications in industries processing such materials as metals, minerals, textiles, and paper. For example, fine etchings in an apparently polished surface have been shown up by smearing the surface with a fluorescent grease or oil and subsequently examining it by incident illumination. Another example is the detection of non-metallic inclusions in mineral ores as developed by a number of Russian workers. In their technique a drop of water is placed on a microscope slide and a small quantity of mineral (quartz, corundum, magnetite, etc.) placed in it. This is dried, a cover slip placed over it, and the mineral examined for fluorescence. A drop of fluorescein dissolved in alcohol is then introduced under the cover glass. Quartz emits a red fluorescence originally, but a bright golden-yellow fluorescence after treatment with fluorescein. Silicates exhibit a multi-coloured primary fluorescence. Corundum gives off a dark green fluorescence before, and a light green fluorescence after treatment.

Many other industrial applications are mentioned in the book by Radley and Grant.

Conclusion

The applications of fluorescence microscopy are legion. One application, largely neglected by chemists until now, is in histochemistry. This field would prosper much more rapidly than it does at present if chemists, rather than biologists, would investigate the mechanisms of the several fluorescence histochemical reactions described in this chapter. This would enable these reactions to be so refined that in due course they could be used with a fair degree of certainty for detecting the chemical functional groups of macromolecules occurring in biological tissues, and eventually for determining the chemical structures and identities of such macromolecules. Fortuitously and fortunately there are only two types of fluorescence histochemical reaction that, for the time being, need to be investigated in detail. One is the binding of basic fluorescent dyes on to the polyanions of tissue macromolecules, and the other is the condensation of carbonyl groups with aromatic hydrazines. These investigations can mostly be accomplished with the dye coriphosphine or with salicylhydrazide respectively. With these two reagents most of the chemical functional groups present in tissues can somehow be detected.

BIBLIOGRAPHY

Review of fluorescence microscopy

PRICE, G. R. and SCHWARTZ, S. *Physical Techniques in Biological Research*, Vol. 3, p. 91 (edited by G. Oster and A. W. Pollister), Academic Press, New York (1956).

Reviews of histochemical techniques in general use

PEARSE, A. G. E. *Histochemistry*, 2nd edn. Churchill, London (1960).

SPICER, S. S. 'Histochemical differentiation of mammalian mucopolysaccharides', *Ann. N.Y. Acad. Sci.* **106**, Art. 2, 379–383 (1963).

Theory and experimental details of fluorescence histochemical techniques

Several papers by P. J. STOWARD in *J. Roy. micr. Soc.* and *Histochemie.* Also D.Phil. thesis, Oxford (1963).

Review of potential industrial applications of fluorescence microscopy

RADLEY, J. A. and GRANT, J. *Fluorescence Analysis in Ultra-Violet Light*, 4th edn, Chapman and Hall, London (1954).

Review of immuno-fluorescent techniques

NAIRN, R. C. (ed.) *Fluorescent Protein Tracing*, 2nd edn, Livingstone, Edinburgh (1964).

INDEX

INDEX